The Home Satellite TV Book

THE HOME SATELLITE TV BOOK

How to Put the World in Your Backyard

ANTHONY T. EASTON

Manufactured in the United States of America.

First edition.
Wideview Books/A Division of PEI Books, Inc.

Library of Congress Cataloging in Publication Data
Easton, Anthony T.
The home satellite TV book.
1. Earth stations (Satellite telecommunication)—Amateurs' manuals. 2. Home video systems. I. Title.
II. Title: Home satellite TV book.
[TK5104.E18 1982b] 621.388'35 81–50336
ISBN 0–87223–730–3 (pbk.) AACR2

Contents

APPENDICES

Introduction

Welcome to the world of satellite television! From its beginnings less than three decades ago to the landing of the first man on the moon, the U.S. space effort has been a fabulous success. Thousands of satellites have been launched to map the earth's surface, locate oil and mineral deposits, maintain navigational aids for ships at sea, and monitor the politically troubled areas of the world. But perhaps the most exciting application of satellites has been in the area of telecommunications. Through the use of our domestic and international communications satellites, a television viewer in New York City, for example, can be instantly in touch with world events as they occur in London or in Los Angeles.

With the rapid development of cable television systems throughout North America, the home viewer can now tune in some 15 to 30 television programs, and a whole new industry has sprung up overnight to provide the cable television companies with original and creative programming. Such organizations as Home Box Office, Showtime, The Movie Channel, Cable News Network, Entertainment Sports Programming Network, C-SPAN, Superstation WTBS, and Christian Broadcasting Network are feeding their new television network signals to cable companies throughout the nation, and these network feeds are being beamed to the world by satellite. Currently, more than 50 television channels are transmitted from space, and others are turning on every month. By installing a satellite television receiving station next to your house or on the roof of your apartment building, you can bring the whole world into your living room.

This book is directed to the home satellite television pioneer, yet it does not require a sophisticated background in electrical engineering. After reading it, you will know how to pick up the world's television just by making a few simple phone calls. Dozens of dealers now provide turnkey installations anywhere in North America, or you can spend a few weeks (and save some money) by putting together your own system from component parts.

This book was also written for the future mini-CATV operator. As part of a condominium association, an apartment building owner, or a trailer park operator, you can legally create your own mini-CATV system by installing a satellite television receiving antenna and charging a monthly service fee to the viewers of the programs. Restaurant and tavern owners, hotels, churches, hospitals, and other community organizations can also own their own satellite TV systems. There is a chapter in this book to cover each of these possibilities.

Finally, you may become so excited about this dynamic new field that you will want to go into business for yourself as a satellite television dealer. There is a chapter for you too.

At the end of the book are a complete set of appendices listing useful information; names and addresses of equipment manufacturers, program suppliers, and other organizations; a bibliography; and a glossary.

As always in writing a book of this scope, the author is indebted to many people who have provided support, encouragement, and constructive feedback. I wish first to thank my research assistant, Cheryl Orvis, for her help in organizing so much of the confusion. My secretary, Joanne Tupper, spent many Saturdays working on the book with me. James Lentz contributed many illustrations. The good people at Downlink, Inc., and especially Wes Thomas, have provided much information. The president of SPACE (Society of Private and Commercial Earth Station Users), Professor Taylor Howard, shared his knowledge and a delightful videotape of his own backyard experiences. Likewise, SPACE's general counsel, Richard Brown, was most helpful in clarifying some of the legal and regulatory questions concerning this field. I am most pleased to be a member of that association.

Then, of course, there is Bob Cooper, Jr. What can one say about this man? Almost single-handedly he has started the home satellite TV revolution. Through the efforts of his excellent magazine, *Coop's Satellite Digest,* and his outstanding seminars and technical manuals, he has encouraged the birth of a vital and dynamic industry and increased the choices that we as television consumers now have available to us. We owe him much.

Throughout the development of this book, the people at Wordpower and LineStream in San Francisco have provided yeoman service in working on often illegible manuscripts late into the evening. And my friends John Kinick and Jim Kennedy have helped at various stages by providing technical expertise and feedback.

Finally, I wish to thank my wife, Susan, for her patience and support during the madness of writing this book.

Good luck, and happy satellite television viewing!

Anthony T. Easton
San Francisco, California
January 1982

Mr. Easton can be reached at The Easton Corporation, 559 Pacific Avenue, Suite 32, San Francisco, CA 94133.

The Home Satellite TV Book

Chapter 1

Introduction to the World of Satellites: History of Satellite Technology

By Isaac Newton's day (1642–1727), the basic space mechanics of celestial objects were understood, but the ability to place an artificial satellite into earth orbit was to be limited until sufficiently powerful launching vehicles (rockets) could be built. It was the Russian physicist K. E. Tsiolkovsky, whose first papers appeared in 1903, who proposed the use of high-energy liquid-fueled rockets. But it was not until 1926 that Robert H. Goddard, an American scientist, launched the first liquid-propellant rocket as a scientific experiment.

By the 1950s both the U.S. and Russian governments were hard at work using the German V-2 project technology and scientists captured at the end of World War II to construct new and powerful missiles. The Russians were first in space in 1957 with the successful launch of SPUTNIK I, a satellite about the size of a football that carried a tiny radio transmitter aloft. And within months the United States had launched its first satellite, EXPLORER I, whose on-board instruments detected the presence of the hitherto unknown Van Allen radiation belt, which surrounds and protects the earth.

By the early 1960s the cold war space race was on in earnest. Although the attention of the media became rapidly focused on manned space efforts such as the MERCURY, GEMINI, and APOLLO programs in the United States, and the VOSTOK, SALYUT, and SOYUZ programs in the U.S.S.R., the unmanned satellite programs were pressing quietly ahead.

Early on, military strategists saw the advantage of using satellites as surveillance devices. And cartographers hoped to accurately map the surface of the earth through electronic "eyes in the sky," because scientists were still unsure of the exact shape of the earth, or even of the distance between continents! In fact, some world maps were off by as much as a mile or more. Satellites would help to change all this.

Geologists were interested in using satellite detectors and cameras to locate oil and mineral deposits. The U.S. Navy expressed interest in providing navigational assistance to ships at sea by utilizing satellites to accurately pinpoint the ships' exact locations. Weather organizations and the world meteorological agencies saw the satellite as a way of looking down at massive storm formations from hundreds of miles in space. Even agriculture was interested in analyzing plant growth and predicting harvest yields by using infrared photographic pictures relayed back from circling satellites.

Private and governmental organizations also foresaw the use of satellites to provide telephone, telex, and even television communications, both within the borders of a country and internationally.

By the mid-1960s hundreds of satellites had been launched by the United States and the U.S.S.R. The French, Chinese, Indians, Italians, and others were soon to follow with their own rocket-launching facilities. The cultivation of space had become more than an element of national pride—it had become an economic necessity.

The Mechanics of Launching a Satellite

It is fascinating to consider that a 1000-pound metal box of components and batteries can be lifted from the earth and placed into an orbit circling the planet. How does it all work? Why doesn't the satellite simply fly off into space, or even fall back to earth?

The answer lies in the laws of gravity and orbital mechanics, first described accurately by Sir Isaac Newton in 1687. Gravity is simply the physical attraction of two bodies toward each other. The larger the body's size, the greater the attraction. The force of gravity is commonly measured in units of weight such as pounds or kilograms. To successfully lift a 1000-pound satellite into earth orbit, the rocket motors must have sufficient thrust to move this box several thousand miles straight up, directly opposing the force of the earth's gravity to pull the object back.

To most readily accomplish this, the rocket accelerates the satellite

Various uses of communications satellites (top to bottom): satellite television, communication with aircraft, sending mail electronically. (Courtesy of TeleSat Canada)

to a speed of several thousand miles per hour. The laws of gravity indicate that to break free from the pull of the earth, the rocket must be traveling at a speed, or velocity, of over 7 miles per second, or 25,000 miles per hour!

At speeds slower than that the satellite will be drawn back toward the earth, falling into a high-speed orbit, and circling the planet closer and closer until it ultimately reenters the earth's atmosphere and is consumed by the heat caused by buffeting friction of the thin air molecules found at high altitudes. If the satellite has a speed greater than the 7-mile-per-second escape velocity, it will fly off into space, never to return to the earth.

Satellites launched into low-altitude orbits circle the earth faster than

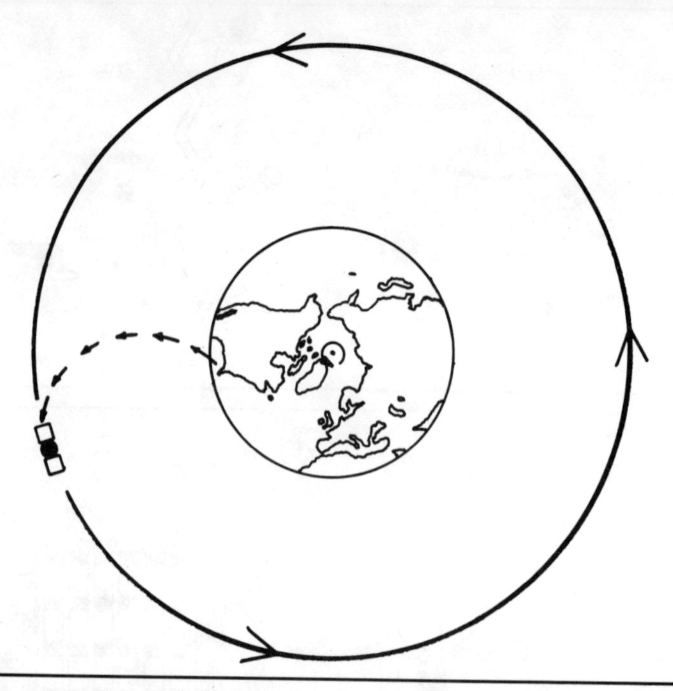

Gravity and escape velocity. If velocity is less than 25,000 MPH, earth orbit results; if greater than 25,000 MPH, the spacecraft leaves the orbit for deep space. (Courtesy of James Lentz)

the earth turns; spy satellites flying at 100-mile altitudes may only take 90 minutes to complete one revolution. The satellites that are launched into high-altitude orbits (above 23,000 miles) circle the earth very slowly, taking over a day to make a complete revolution. At one specific orbit altitude, roughly 22,300 miles above the equator, a satellite will rotate around the earth precisely once every 24 hours; this orbit is called the geosynchronous orbit, or Clarke orbit in honor of the science-fiction writer Arthur C. Clarke.

The Clarke Orbit

In 1945 Clarke wrote an article in the British publication *Wireless World,* suggesting that an international communications system could be created using three geosynchronous satellites carefully located at three strategic points 22,300 miles above the equator. Clarke argued that to an observer or a tracking station on the earth's surface, these satellites would appear to stand still in the heavens, as the satellites would complete one revolution around the earth every 24 hours and the earth itself would be turning underneath at precisely the same rate.

The Clarke article presented a revolutionary idea, and it was treated as science fiction at the time. In those days telephone communications between North America and Europe were relayed over notoriously

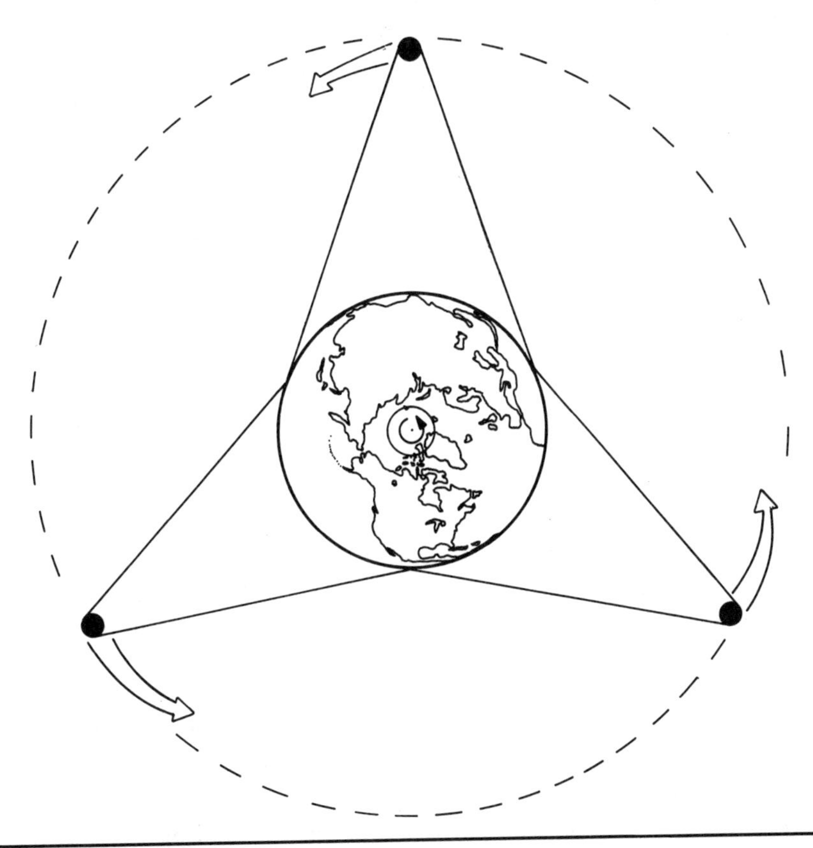

The Clarke orbit—looking down at the north pole. Three satellites spaced equidistantly 22,300 miles above the equator can provide communications coverage to the entire earth. (Courtesy of James Lentz)

inferior high-frequency radio signals. The first transatlantic telephone cable would not be laid until the following decade, in 1956.

On July 10, 1962, the American Telephone and Telegraph Company (AT&T) launched TELSTAR I from Cape Canaveral. The several-dozen-pound satellite was carried into space by a NASA Thor-Delta rocket. Although not put into a geosynchronous orbit, TELSTAR did bring the first live television pictures from Great Britain to the United States. Circling the earth at a very low altitude, TELSTAR could only transmit for a few minutes at a time, however, and required massive earth station antennas to track its motions as it whizzed across the sky.

A few weeks after this spectacular success, the U.S. Congress adopted the Communications Satellite Act, which created a new private corporation, COMSAT, charged with the goal of building a global communications satellite network. Two years later, in 1964, an international consortium of countries called INTELSAT was formed to own the satellites that COMSAT would build and operate. Unlike TELSTAR, the INTELSAT satellites would be placed into geosynchronous orbits positioned strategically over the Atlantic, Pacific, and Indian oceans. Today more than a dozen INTELSAT satellites are operational, providing telephone, telex, and television communications to all points of the world.

Radio, Television, and Satellites (How Radio Frequencies Work)

To receive signals from the earth successfully and relay them back again, satellites use very-high-frequency radio waves operating in the microwave radio-frequency band.

Everything in life oscillates according to some periodical cycle, whether it is the electrons spinning around in the atoms of your bloodstream or the sound waves hitting your eardrum. Oscillating waves form the basis of all communications, both human and electrical. The most common form of oscillating wave is the sine wave, so named because of the geometric pattern it produces. When sine waves are produced by a musical instrument or a vocal cord, we call them audio waves, and they vibrate (or oscillate) at frequencies ranging from 20 to 20,000 cycles per second (or hertz, abbreviated Hz). Sine waves have two major properties: their frequency (the number of times they oscillate per second) and their amplitude (or loudness, with regard to audio waves).

Radio waves are also sine waves, produced electromagnetically by the interaction of magnetic and electric fields. Typically, they are generated by a radio transmitter and radiated into the atmosphere using an antenna.

Sine waves. (Courtesy of James Lentz)

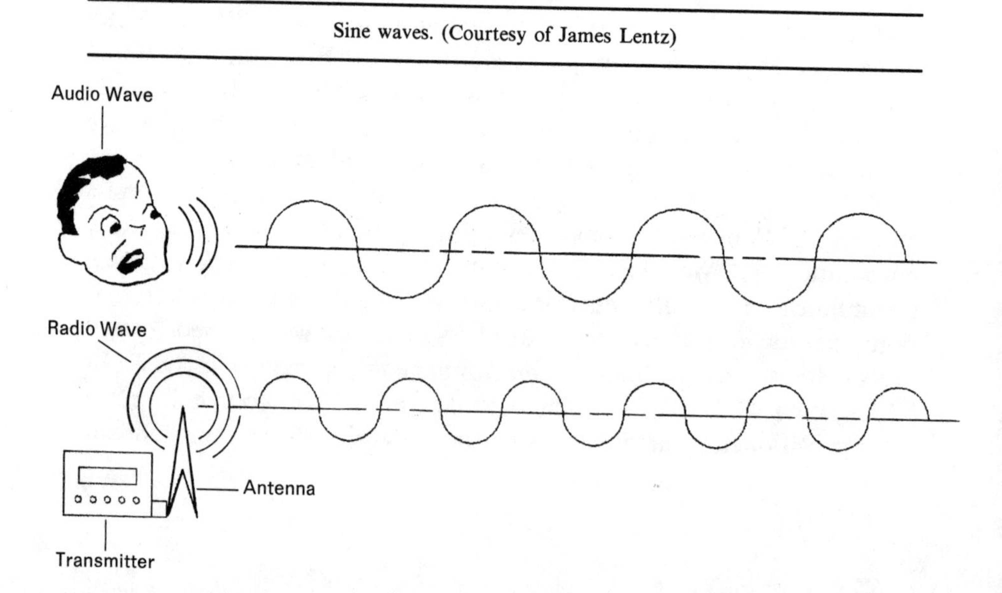

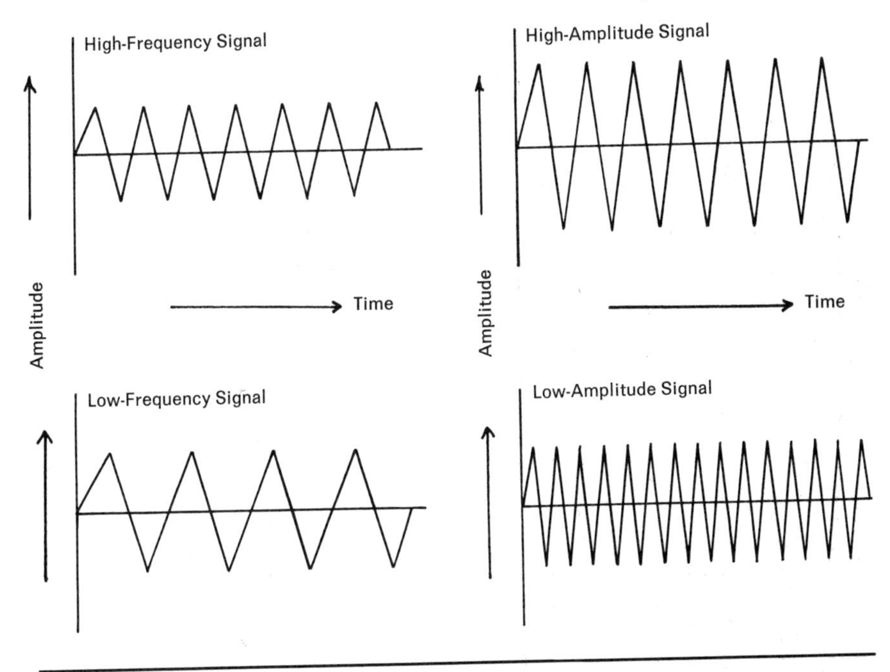

Properties of sine waves used in communications: amplitude and frequency.

Very-low-frequency radio waves, of the order of 50,000 Hz or so, are used to communicate with submarines while they are submerged, as these kinds of low-frequency waves have the ability to penetrate below the surface of the water. Medium-frequency radio waves are used to carry AM broadcasting station signals. The AM transmitters vary the amplitude of the sine-wave carrier in relation to the announcer's voice or the record's music. Very-high-frequency radio waves (VHF) are used worldwide to carry both FM radio broadcast stations and VHF television (Channels 2 to 13). The FM radio stations' transmitters shift the frequency of the carrier up and down in relation to the audio signal coming from the studio. This process of manipulating a radio station's carrier by varying its frequency is called frequency modulation.

Ultrahigh-frequency (UHF) radio waves are used to carry a second set of television channels, designated 14 to 70 on the television dial.

Interestingly, terrestrial television broadcast stations vary both the amplitude and the frequency of the transmitted carrier wave, using amplitude modulation to carry the picture information and frequency modulation to carry the audio sound track. FM is particularly immune to certain kinds of atmospheric noise; AM is not. This is why the TV's sound (FM) will be static-free during an electrical storm, but the picture (AM) will break up and display interference each time a bolt of lightning strikes.

AM radio operates on frequencies from 535,000 Hz (or 535 kilo-

TABLE 1–1.
The Electromagnetic Spectrum

VLF	LF	MF	HF	VHF	UHF	MW	Light
Frequency							
50,000–100,000 Hz	100,000 Hz–535 kHz	535–1600 kHz	1.6–50 MHz	60–216 MHz	470–806 MHz	1–20 GHz	100–10,000 GHz
Submarine communications	Navigational aids	AM radio	Short-wave radio	VHF TV Channels 2–13 FM radio 88–108 MHz	UHF TV Channels 14–70	Satellite communications, C-band 3.5–6 GHz	Infrared and light

The RCA SATCOM F-1 satellite in space—an artist's conception. (Courtesy of RCA Americom, Inc.)

hertz) to 1600 kHz. FM radio uses the 88–108-MHz band. VHF television runs from 54 MHz to several hundred million megahertz (the FM band is smack in the middle between TV Channels 6 and 7!). UHF channels start at 470 MHz and run up to 806 MHz.

Microwave frequencies begin at about 1000 MHz (or 1 gigahertz) and run up in frequency until they turn into visible light waves. At these superhigh frequencies, radio signals take on the characteristics of light waves. They can be reflected from mirrors and tend to travel by line of sight only. Communications satellites use microwaves operating in the frequency band of 3½ to 6 GHz, known as the C-band. Traveling 22,300 miles down from space, microwaves will not make it past the roof of a house; however, they can be picked up by a small receive-only earth station sitting in the backyard.

But microwave frequencies are not used only for satellite communications. Microwave ovens, medical diathermy equipment, and radar systems all utilize radio frequencies operating in the microwave portion of the electromagnetic spectrum. By and large, terrestrial telephone communications are also carried over microwave circuits using telephone company microwave transmission repeaters scattered at intervals of 40 miles or so across the country.

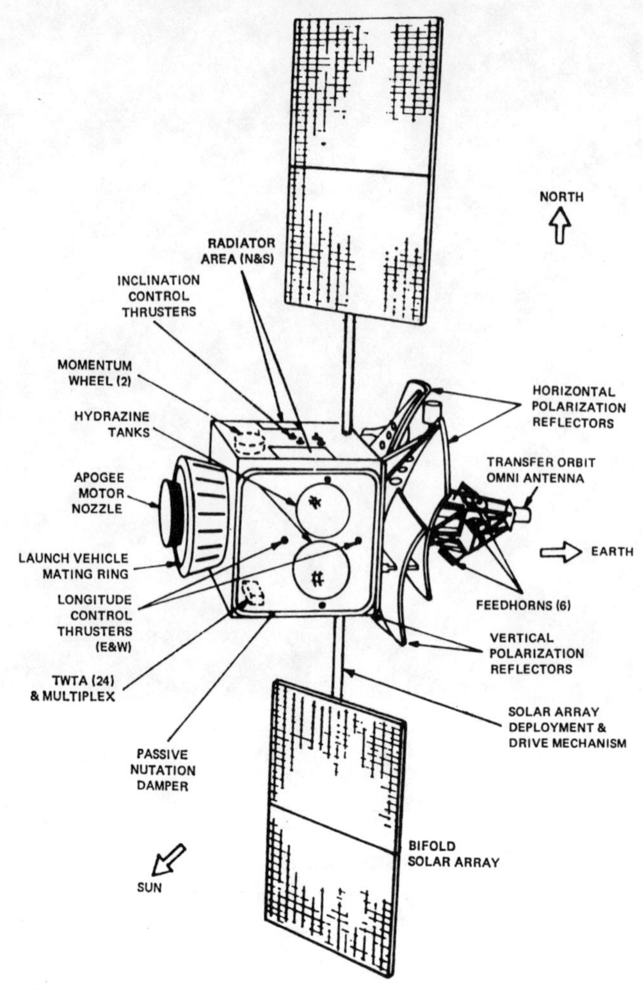

Diagram showing RCA Satcom I component locations.

(Courtesy of RCA Americom, Inc.)

Networks and Satellites

The major television networks use the telephone company's backbone microwave network to carry TV programs from network control in New York to their nationwide television station affiliates. By installing elaborate automated switching centers throughout the United States, the Bell Telephone Company can connect any television station to or disconnect it from any network feed in a matter of seconds. During the fall, when the National Football League games are in season, as many as a dozen or more regional television network feeds are

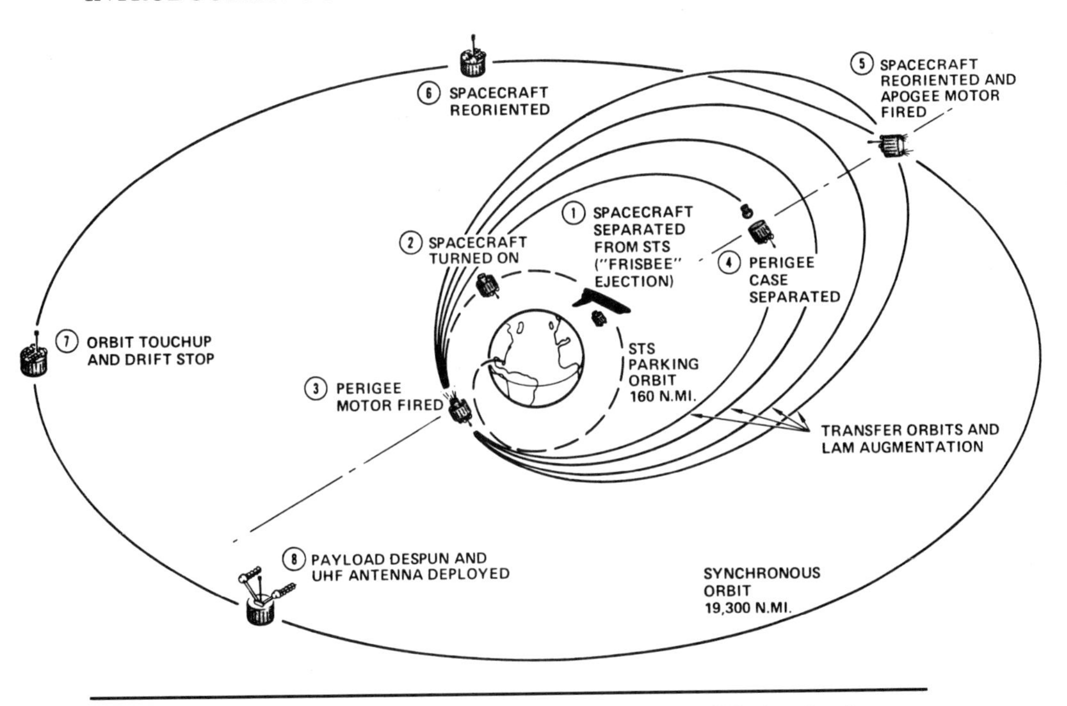

The launch of a typical communications satellite. (Courtesy of Hughes Corp.)

delivered to groups of nearby television stations under control of the master control centers. Of course, this kind of extensive terrestrial communications network is extraordinarily expensive. One television channel has the equivalent capacity to carry 2000 telephone conversations simultaneously, and it is not unusual to see a network spending upward of $100 million per year on communications channel charges. But until the launch of the domestic communications satellite systems in the mid-1970s, no other options were available!

Recognizing the need for domestic satellite communications to augment the expensive terrestrial microwave systems, the U.S. common carriers RCA and Western Union decided to launch their own SATCOM and WESTAR satellites to be used for voice, data, and television communications. COMSAT, the U.S. communications company that built the international satellite network, followed suit and leased its COMSTAR domestic satellite exclusively to AT&T and GTE for telephone company use.

With the clever use of nationwide and spot transmitter downlink beams, and the careful reuse of frequencies, each of the newer satellites would have the capability of relaying 24 separate television pictures—one per transponder—throughout the United States.

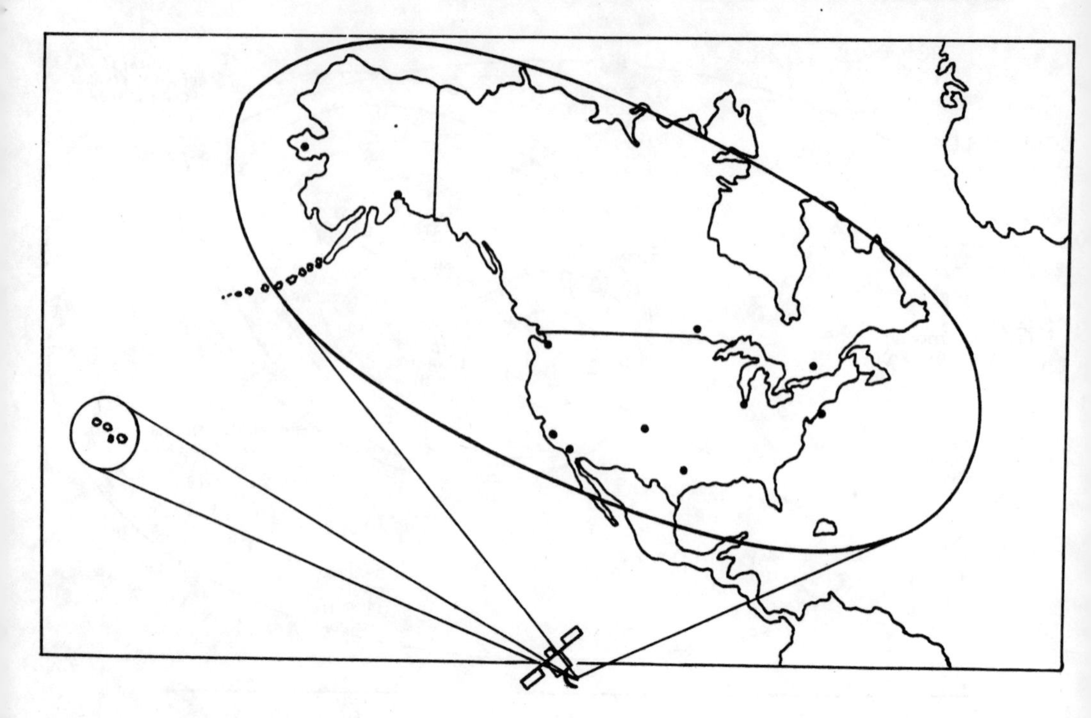

RCA SATCOM system. (Courtesy of RCA)

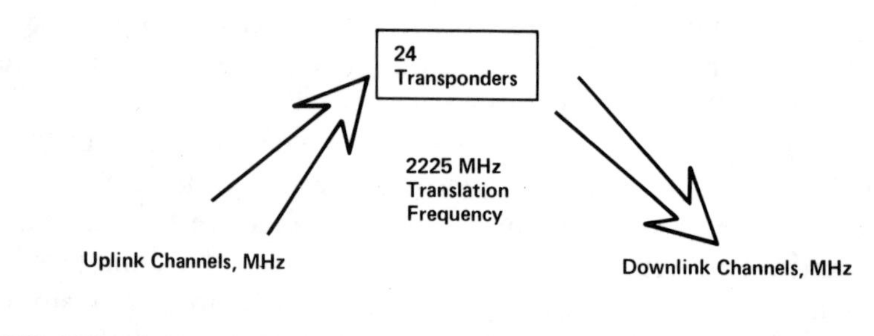

Polarization/Frequency Plan

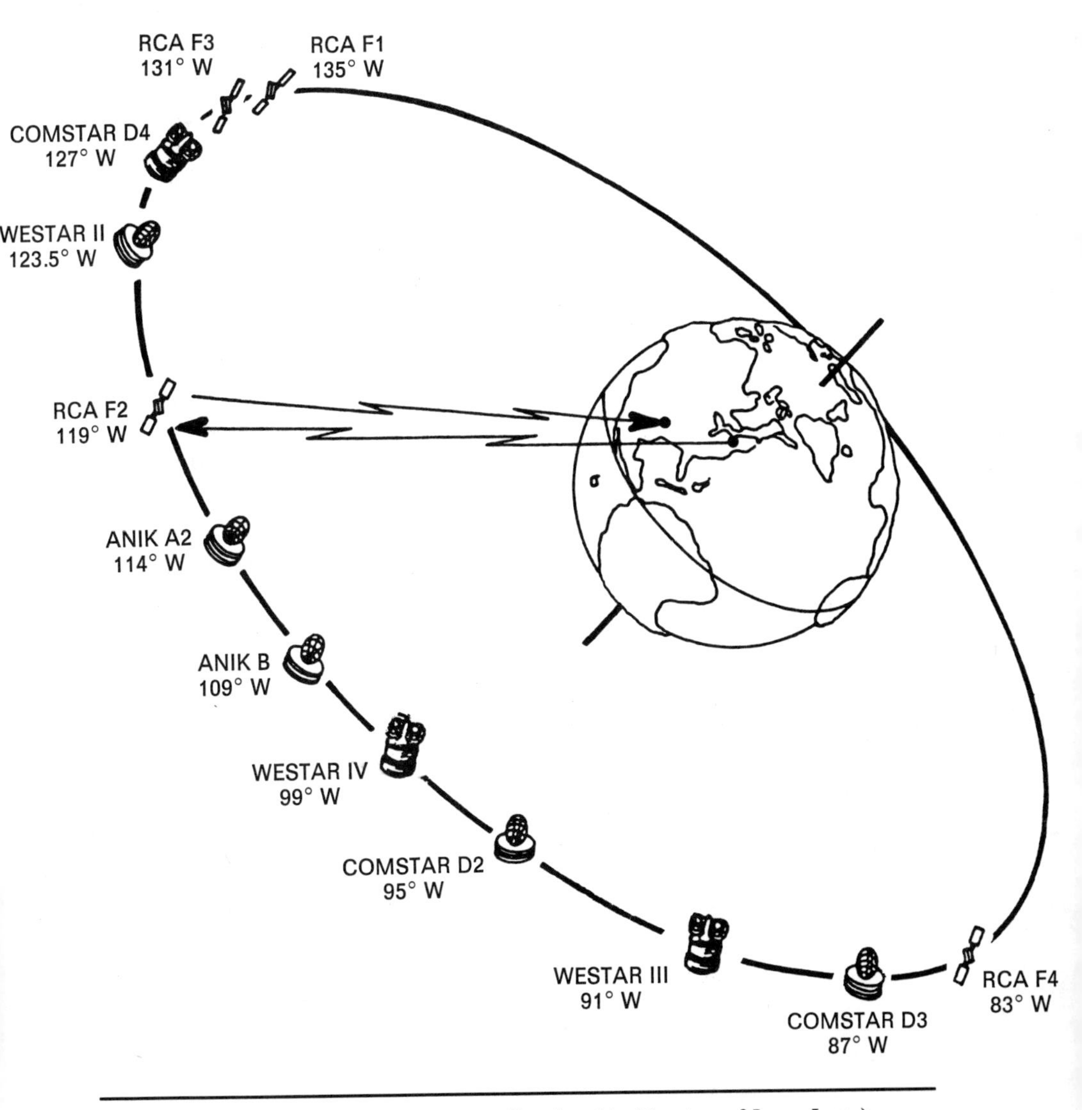

Major North American television satellites in orbit. (Courtesy of James Lentz)

The Canadians also launched three of their own domestic ANIK birds, and these, together with the INTELSAT international satellites, meant that by the end of the seventies, over 20 communications satellites had been put into space to carry domestic and overseas television feeds. However, the major television networks were cautious about using this new satellite technology to relay their programs. Bell's microwave network was expensive, but it was also absolutely reliable.

The rebirth of the cable television industry in the seventies was soon to change all that. Although community-antenna television (CATV) systems had been around since the early fifties, it wasn't until the mid-1970s that enough cities and towns had been wired for cable for it to make a significant impact. By then the new communications satellites had been launched, and several innovative programmers such as Home Box Office (HBO) hit upon the idea of using the satellites to feed their pay TV first-run-movies service directly to the CATV companies via satellite, thus eliminating the need to "bicycle" tapes and films around the country via the U.S. mails.

Distribution was far cheaper by satellite. One satellite channel or transponder rented for about $1 million per year, or just about 1 percent of the cost of the TV networks' terrestrial microwave cost. Because of its low delivery cost, HBO could keep its pay TV prices down, and millions of viewers quickly signed up through their local CATV systems to take the HBO movie service. Other television programmers were soon to follow HBO's lead in creating satellite networks geared to the CATV market. CATV had become a new force for the old-line television networks to reckon with.

Chapter 2
Cable Television and You

Lee De Forrest invented the vacuum-tube amplifier in the 1920s. Shortly thereafter, researchers at Bell Labs and elsewhere began to experiment in earnest with the process of transmitting a picture electronically by using radio waves. By the 1930s several experimental television stations were in operation, and at the 1936 World's Exposition, held in San Francisco, television pictures were the hit of the Fair. It was projected that within a few months millions of families would have their own "visual radios" in their living rooms. Then World War II broke out, and the nation's scientists had to focus their energies on the more pressing matters of national defense. Television sat on the back shelf for almost a decade.

After the war ended, major technological breakthroughs accomplished during the war helped to spark a renewed interest in television at the major radio networks—the Columbia Broadcasting System and the National Broadcasting Corporation. By 1949 several experimental television stations were operating in East Coast cities, and pioneering television-set manufacturers such as Dumont, Bendix, and Philco had begun to sell their new-fangled picture boxes in Sears, Roebuck and other department stores. The old radio networks were quick to establish television stations of their own, and to use the new AT&T nationwide microwave system, just being built, to carry the live television programs from their New York studios.

These were exciting days. A year later major cities such as New York had two or three television stations operating on a daily broadcast schedule, and the status symbol of the American family was to be the first on one's block to own a television set. The television craze had swept the country, and it was not limited to the larger cities. Small towns and communities 40 or 50 miles away from these urban centers were equally interested in having their own TV reception. In areas around New York and Pennsylvania, however, many of these smaller towns were located in valleys, with surrounding mountains that blocked the reception from the distant cities.

At about this time, a number of enterprising engineers and businesspeople began to focus their talents on the television industry, believing it to have tremendous growth potential. One of their first challenges was to bring television to the smaller communities. Recognizing the fact that television signals essentially travel by line of sight from the transmitter to the television receivers, an enterprising engineer named Irving Kahn solved the reception problem by installing a large television receiving antenna on a hilltop and bringing the signal down to the town at the foot of the hill via coaxial cable. The cable was strung on telephone poles and attached to the television set's antenna terminals in each home. In this way, the entire community could share a common master television receive-only antenna system, and the concept of community-antenna televison (or CATV) was born. Mr. Kahn's company, Teleprompter Corporation, grew over the years to become a major

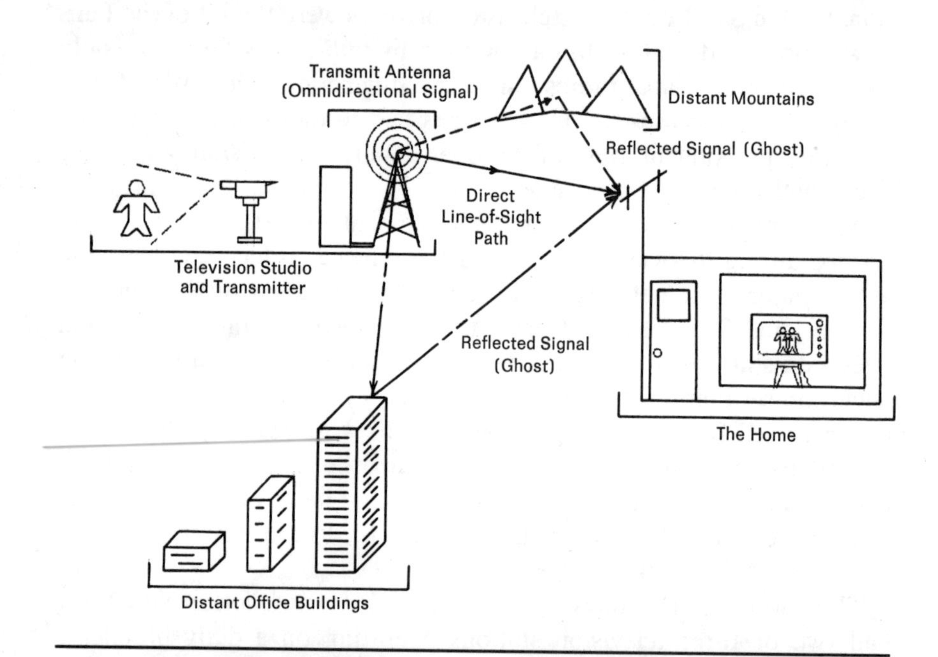

Television signals travel by line of sight, and are easily reflected by large flat objects such as buildings, water towers, mountains, and even statues. Ghosts result when the television receiver picks up both the original line-of-sight signal and one or more of these undesirable reflected signals. Rotating the TV antenna can minimize but won't eliminate ghosts. (Courtesy of James Lentz)

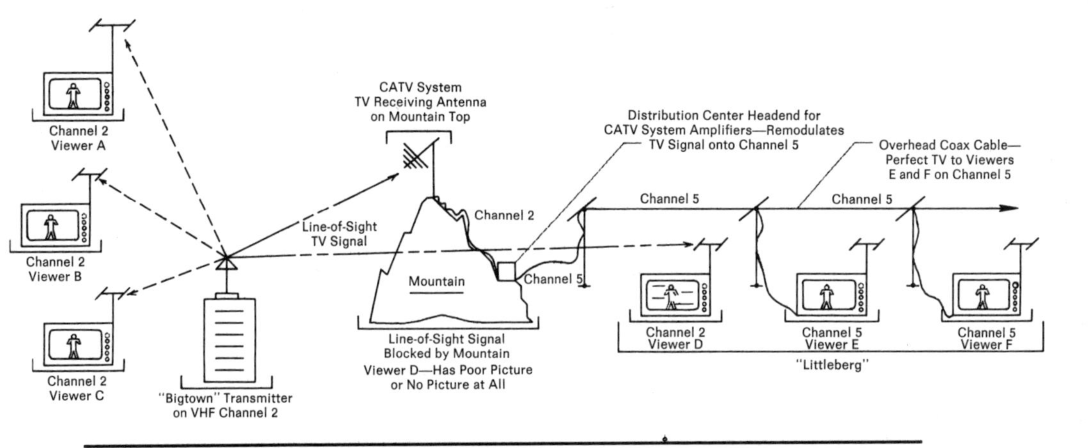

In the 1950s, the CATV companies originally constructed hilltop "antenna farms" of television receiving antennas to pick up distant television station signals. These signals were then remodulated onto new television channels (to eliminate interference or ghosting with the original broadcast signal), and sent along a coaxial cable trunk circuit from house to house. The CATV system both strengthens weak TV signals and eliminates ghosts. (Courtesy of James Lentz)

force in the CATV industry. Today it is one of the five largest multiple system operators (MSOs) of CATV systems in the United States.

On May 18, 1950, Benjamin Tongue and Isaac Blonder founded the Blonder–Tongue Laboratories to provide electronic equipment to the infant CATV industry. With an initial capitalization of $5000, they set to work designing and experimenting in a tiny loft in Yonkers, New York, and soon introduced the first commercially successful broadband TV booster amplifier, designed to increase the weak TV signals and improve fringe-area reception in distant communities. Working with the new cable television companies, manufacturing firms such as Blonder-Tongue began to wire up the hinterland with cable, providing excellent television reception that was often superior to that received by over-the-air broadcasting in the big city itself. With CATV reception, the subscriber could tune in TV stations in distant cities, each one of which was clear and ghost-free.

Government and the CATV System

Historically, receive-only antenna systems have not been licensed or regulated by the federal government. In the United States, the Federal Communications Commission (FCC) exercises statutory control over telecommunications and broadcasting facilities that operate transmitters. "Over-the-air" broadcast transmission and reception are exempt from state and local controls. Unlike their European counterparts, U.S.

television viewers have never had to pay for annual television receiver licenses, a practice followed to this day in Great Britain. As the television stations were privately owned, the government had no need to levy taxes or establish excessive regulations over the broadcasting industry. Thus the United States has almost 10,000 television stations and television translators whereas Great Britain, with a quarter of the population, has only several hundred.

Cable television systems, unlike broadcasting stations, did not initially fall under the FCC's jurisdiction and were relatively free to expand, but at the local level a CATV company did require permission to operate. Its cables crossed over public territory, and involved right-of-way permission to use government facilities such as underground conduits and utility poles. Thus cable television has always been a politically sensitive local issue. By 1965, when many CATV systems were beginning to receive or "import" their distant-city television signals by microwave transmission, the FCC had established rules governing this aspect of the CATV operation. In March 1966 the FCC set up regulations covering all cable systems, whether or not served by microwave. These rules required cable systems to carry all local television stations; prohibited systems from duplicating on the same day, via signals originated in another city, a program broadcast by a local station; and prohibited them from bringing distant signals into the 100 major television markets without a hearing to consider the probable effect on local broadcasting. In June 1968 the U.S. Supreme Court affirmed the Commission's jurisdiction over cable, thus recognizing the powerful force that the CATV systems had become.

Today, cable television systems are viewed by many as utilities and are treated similarly. A public hearing must be held whenever a new action is taken by the CATV operator, and the profits of the CATV company are tightly regulated according to the utility concept of rate of return on actual plant investment (cables, headend equipment, etc.). CATV has become a big business, and the old "mom and pop" companies have mostly been acquired by Fortune 500 corporations, which have the ability to deal with the increasingly complex regulations governing this industry. Practically overnight cable television has grown from being the stepchild of the broadcasting industry to become a multibillion-dollar business of its own.

How Cable Television Works

Broadcast television in North America was originally provided with 13 channels of frequency spectrum, beginning at 50 MHz and extending up to 216 MHz. The channels were designated 1 through 13. Almost immediately problems arose in partitioning the electromagnetic spectrum according to this plan, and the first 4 MHz of bandwidth (corresponding to Channel 1 on the television dial) were reallocated to other nontelevision communications services, leaving Channels 2 through 13 remaining on the television set. Channels 2–6, the VHF low-band channels, ran from 60 to 88 MHz. Channels 7–13, the VHF high-band channels, began at 174 MHz and ran to 216 MHz. Other new and existing services were squeezed between the low- and high-band UHF channels, including the commercial FM radio band, running from 88 to 106 MHz.

By the late 1950s it had become evident that 12 channels were insufficient to accommodate the growth of television throughout the United States. An additional 70 channels, designated 14 through 83,

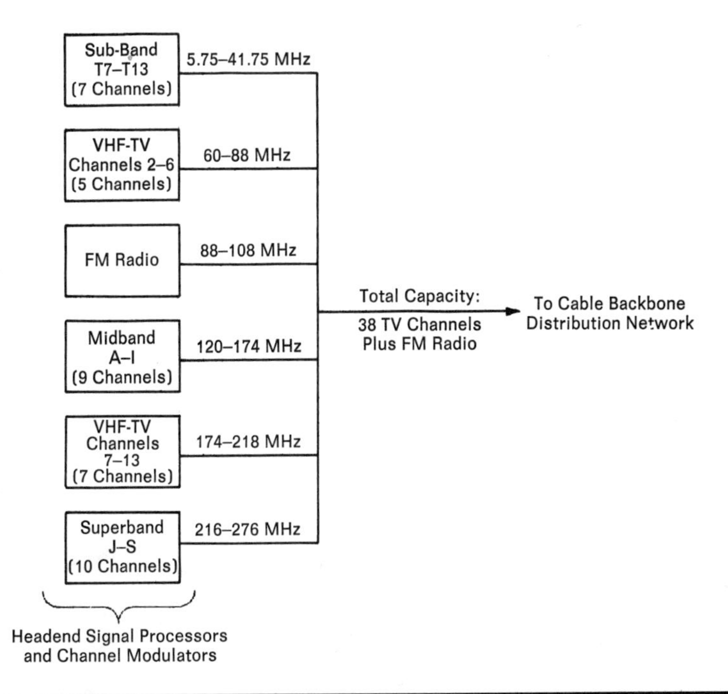

Note 1: New 400 MHz supercable can carry an additional 21 channels from 276 MHz, bringing the total cable capacity up to 59 separate TV channels.
Note 2: Conventional over-the-air UHF TV Channels 14 through 70 run from 470 to 806 MHz (broadcast only) and cannot be carried over a CATV system directly due to their ultrahigh frequency. This is why the UHF–TV channels are always converted to other channels when carried on a CATV system. (Courtesy of James Lentz)

were set aside in the ultrahigh-frequency band, beginning at 470 MHz and extending to 806 MHz. The television-set manufacturers were ordered by the FCC in the 1960s to provide both VHF and UHF tuners on their television sets, thus ensuring the rapid growth of UHF, which previously had been plagued by the "chicken and egg" problem: Too few television sets were equipped to receive the signals. (Recently the FCC reallocated the lightly used frequencies corresponding to Channels 70 through 83 to mobile telephone service. New television sets no longer need show these unused channels on their dials.)

In a cable television system the older limited-capacity twin lead wiring could only carry the 12 VHF channels successfully. But these older systems could also squeeze seven extra channels into frequencies located below the VHF television band. These channels were known as sub-band channels. Nine other midband channels, labeled A through I, could also be squeezed between the low-band Channel 6 and the high-band Channel 7, extending from 120 to 174 MHz. Few of these older CATV systems actually used these channels. Most carried only the standard 12 VHF channels, which could be tuned in by an ordinary television set.

With the advent of coaxial cable, an additional ten superband channels, labeled J through S, could be added, beginning at 216 MHz and running through 276 MHz. New superhigh-frequency cables allow modern CATV systems to carry yet an additional 20 television channels, from 282 MHz up to 400 MHz, thus allowing 58 different television programs and the entire FM radio band to be brought into one's home on a quarter-inch-thick cable!

To receive all these channels, including the sub-band, midband, and superband channels not ordinarily tunable on a television set, a special CATV convertor or decoder is installed in the subscriber's home. This convertor replaces the function of the television tuner, taking the incoming television channels from the cable, instead of through the atmosphere, and converting the signal of the desired program to be watched to VHF Channel 3 (typically), which is then fed into the antenna of the television set and received by permanently tuning the television's selector dial to that channel.

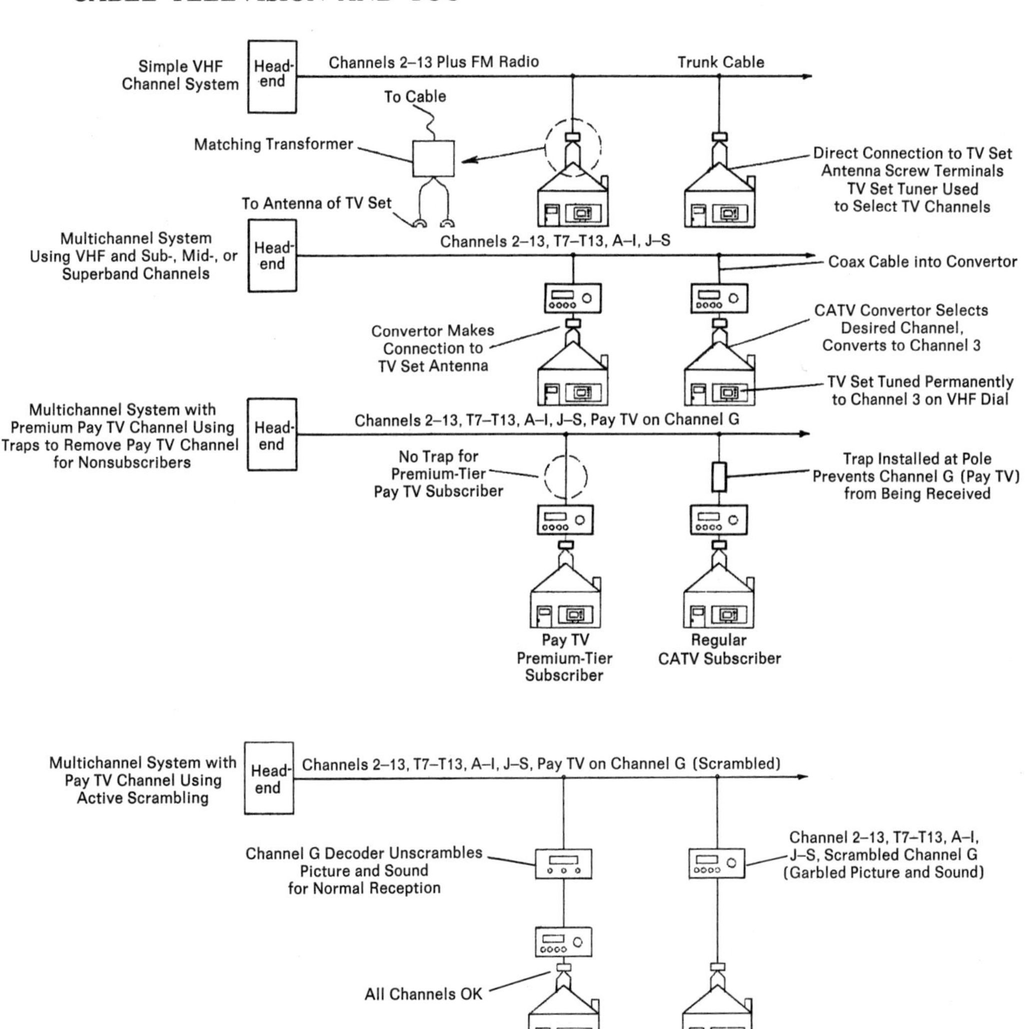

Types of CATV systems. (Courtesy of James Lentz)

The Modern CATV System

As rapidly as the cable television company could add channels, the subscribers' demands increased for yet additional channels. The CATV firms built larger and larger "antenna farms" of tall television receiver antennas aimed at distant-city television broadcasting stations. Eventually, independent common carrier organizations sprang up across the country to bring these distant signals directly to the CATV headend, or master distribution center, over private microwave circuits. Other CATV companies banded together to form their own regional networks to exchange distant-city television signals with each other. Many of these networks are now several hundred miles long, enabling one television station to have many viewers in distant states.

CATV companies also obtained direct video feeds from the television studios, thereby guaranteeing an almost perfect studio-to-home viewer signal. The broadcasting stations now saw CATV as a means of delivering the highest quality signal directly to the subscriber's home, eliminating the problems of over-the-air ghosts and fringe-area reception. CATV companies began to construct their own community access studio facilities and to operate their own videotape playback machines in many cities, and by the early 1970s, several organizations, including Time–Life Films and Viacom International, had begun to supply CATV companies with first-run motion pictures that the CATV system could carry on a special premium-service pay TV channel, using the headend videotape players.

To prevent a basic services subscriber (one who received only the over-the-air broadcast television stations via cable) from picking up these premium channels, the CATV companies used a variety of techniques to isolate or scramble the pay TV channels. Several methods are in common use. The least expensive approach is to send the premium or pay TV program down the cable on a sub-band, midband, or superband channel, one that cannot ordinarily be picked up by a conventional television set. For the subscriber who pays the additional premium-tier pay TV fee, the CATV installs the appropriate sub-, mid-, or superband decoder, which, when turned on, extracts the channel from the cable and converts it into an ordinary VHF TV channel to which the subscriber's television set can be tuned.

Another popular technique consists of trapping the signal at an individual subscriber's cable drop coming in from the utility pole or underground conduit. With this approach, a small cylindrical filter, or

A microwave pickup receiving distant-city television feeds via a terrestrial network. (Courtesy of Scientific-Atlanta Inc.)

A VHF receiving antenna array used by a CATV system. (Courtesy of Scientific-Atlanta Inc.)

trap, is inserted in series with the customer's drop wire at the trunk line connection point, or tap. The filter successfully attenuates or removes the pay TV channel for those subscribers who do not wish to receive it. When a CATV customer elects to obtain the premium-tier service, a cable installer is dispatched to the premises to remove the small filter device.

At the headend, all of these various channels of television signals coming from direct studio feeds, distant station "off-the-air" receive antennas, tape machines, and microwave networks are amplified, processed, and converted to standard baseband video signals. Each video signal is then placed on a specific channel carrier using a channel modulator, and the separate channels are combined into a multiple channel signal by a signal combiner. Depending on the age of the cable system, the multichannel signal may contain only VHF channels, or VHF channels and some combination of sub-band, midband, and superband channels. The signal is then set into the cable distribution network through a signal booster or trunk amplifier.

The Satellite Comes to CATV

Satellite development was well under way by the early 1970s. The first man had landed on the moon, and the international communications satellite network had been operating for years. Both RCA and Western Union decided to construct and launch domestic communications satellite systems to relay telephone and data communication conversations. Western Union was first in space with three 12-transponder satellites; each transponder could carry 2000 simultaneous telephone conversations (or one television channel). The birds were named WESTAR I, II, and III. RCA soon followed with the successful launch of two slightly newer 24-transponder satellites, called SATCOM I and II. For the first several years the satellites were used only slightly. The business recession of the mid-1970s prevented industry from expanding as rapidly as had been projected, and capacity far exceeded demand.

Then, in 1976, Time–Life Films, through its Home Box Office CATV pay television distribution company, hit upon the idea of using the satellites to feed first-run movies directly from the HBO master tape facilities in New York, thus eliminating the expensive and time-consuming process of bicycling videotapes around the country from cable system to cable system. Satellite television had been born!

Typical CATV selector/decoder. (Courtesy of Oak Communications Inc.)

The first CATV satellite TVRO (television receive only) earth stations were very expensive. FCC rules required the cable companies to install unnecessarily large satellite antennas, 30 feet or more in size. The earth station components themselves were expensive because of the limited number produced, and because they were originally manufactured for military use and to expensive "mil-spec" standards. It was not

Close-up of a line-mounted subscriber tap. (Courtesy of Magnavox CATV Systems, Inc.)

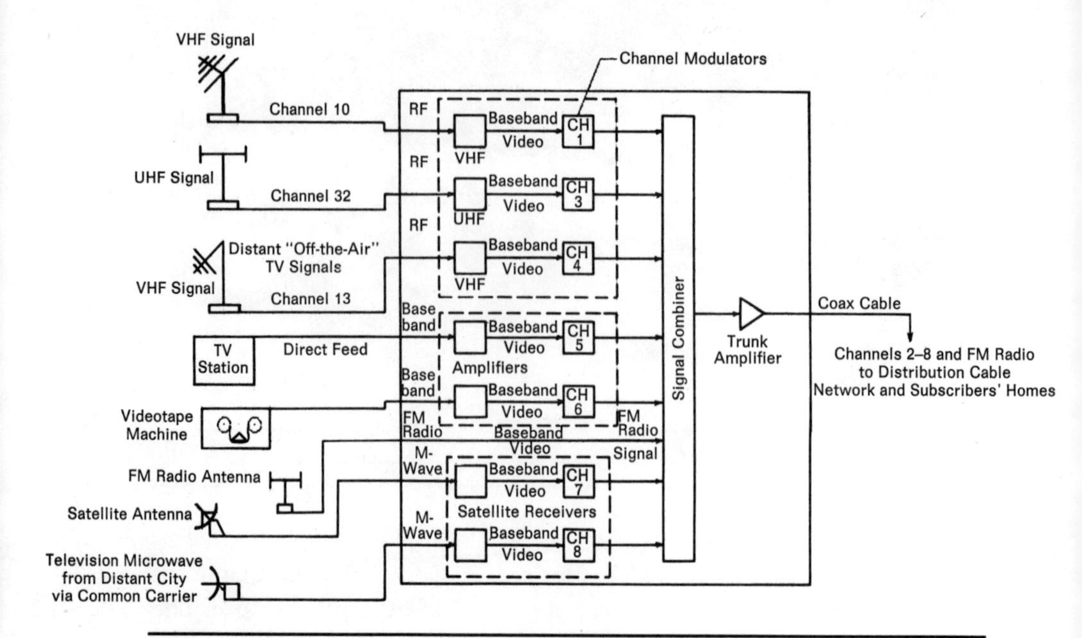

A modern CATV system has many incoming TV signals from TV receiver "antenna farms," direct studio feeds, satellite antennas, microwave TV distribution system feeds, tape machines and local automated weather and news computer terminals (not shown). These various feeds are then fed down the coax cable to the subscribers. (Courtesy of James Lentz)

CATV distribution network. (Courtesy of James Lentz)

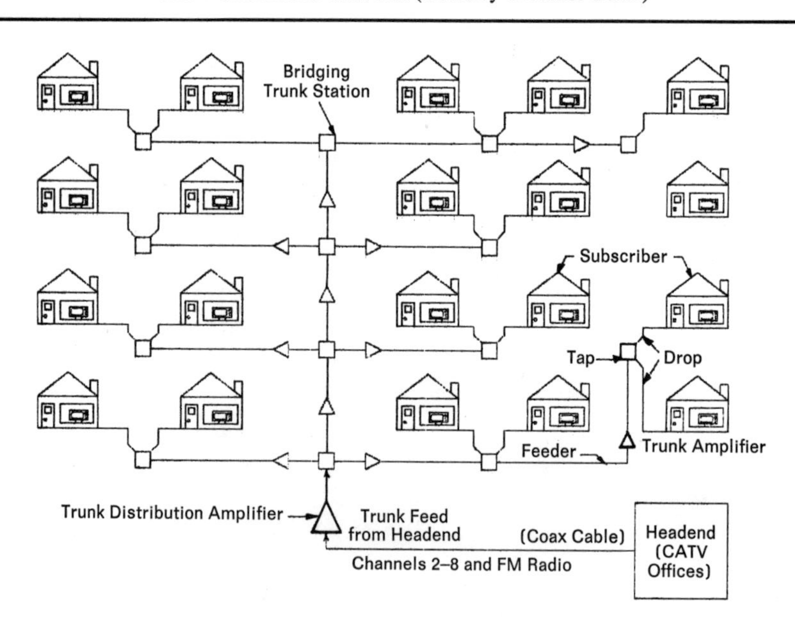

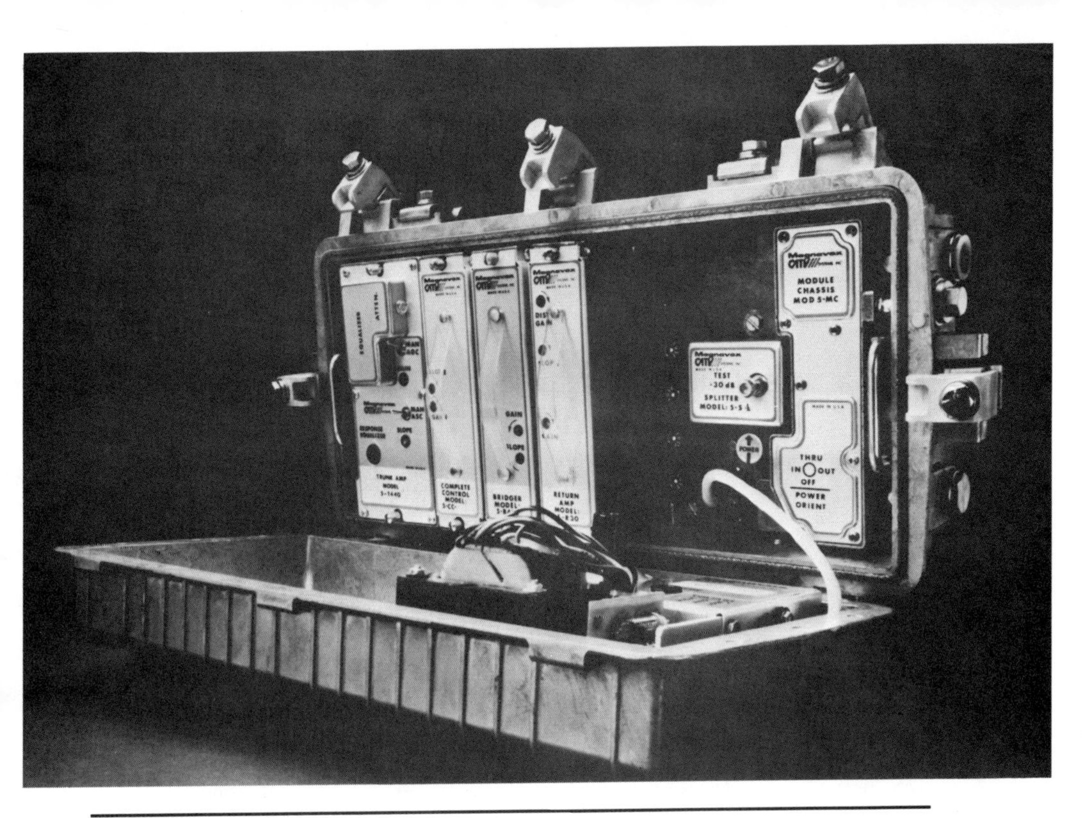

CATV trunk distribution amplifier—opened. (Courtesy of Magnavox CATV Systems, Inc.)

unusual for a CATV company in 1970 to spend between $50,000 and $100,000 for a satellite antenna system. Still, one facility could receive as many as 24 simultaneous television channels from space, and troublesome mechanical videotape recorders and costly operating personnel could be eliminated. Soon satellite earth stations were popping up like mushrooms across the countryside.

Other people were quick to recognize the potential of the satellite television distribution mechanism, and HBO was soon joined by Viacom International's Showtime service and dozens of other satellite programming suppliers and networks. As the number of TVRO systems increased, the TVRO antenna construction costs continued to fall, and the savings were passed on to the cable television companies. Then, in the late 1970s the FCC modified its TV antenna requirements to allow cable television companies to use a satellite dish as small as 15 feet in diameter. The satellite TV business boomed, and the prices of earth station equipment plummeted. Today a complete backyard home satellite TV earth station can be purchased for as little as $3000, and a skilled electronics experimenter can construct one for under $1500.

Chapter 3
The Television Satellites

Today just about everyone knows that Home Box Office is delivered to the cable television companies around the United States by satellite. As the granddaddy of all satellite television programmers, the Time–Life subsidiary has over 6 million subscribers, who are paying from $10 to $25 per month to watch first-run, commercial-free movies in their homes. And there are literally dozens and dozens of other program suppliers and television networks whose signals are carried daily by the television satellites in the United States, Canada, and throughout the world. The United States now has the capacity to carry almost 200 simultaneous channels of programming by satellite. By 1985 this figure will have increased to over 600 satellite transponders, each one of which is capable of carrying one television picture. Moreover, new multiple-picture-per-transponder diplexing technology, direct-to-home television broadcasting satellites, and higher frequency satellites promise to expand our satellite television capabilities to almost 1000 TV channels by the end of the 1980s! This does not imply, of course, that all these transponders will be used exclusively for television signals. More and more the tendency is to use satellites for ordinary telephone and telex communications as well as for new high-speed data applications being developed by U.S. business.

Similar trends are under way in Canada and in countries overseas. The Canadians are presently using two 12-transponder satellites to carry more than 15 television channels. The INTELSAT international satellite network relays dozens of global television news feeds daily. Other INTELSAT transponders are leased on a full-time basis to provide domestic television feeds from region to region within a specific country.

In addition, the Russians and Indonesians operate their own satellite television network, and many other countries are planning to lease INTELSAT transponders or to launch their own satellites over the next few years. The ARABSAT consortium of Middle Eastern nations will provide daily Arabic television feeds both within the Middle East and to other parts of the world by 1984.

The major satellite television usage, however, remains firmly centered in North America. Currently, more than 60 different satellite networks can be picked up from just about anywhere in the United States or Canada.

TV on the Birds

The organizations feeding their television programs via satellite can be divided into 11 major groups, according to the type of programming they provide and their usage of the satellite. They include:

- pay television programmers (including general, adult, and ethnic-oriented movies)
- religious organizations
- regional and nationwide television "superstations"
- sports networks
- all-news networks
- commercially sponsored networks
- public affairs programmers
- educational organizations
- the "big four" conventional television networks (NBC, ABC, CBS, PBS)
- teleconferencing organizations
- private communications networks (for example, NASA's Space Shuttle video)

Pay TV Programmers

The best known pay TV programmer is Home Box Office, a Time–Life subsidiary. HBO has more than six transponders on the RCA SATCOM III "CableNet I" satellite and the COMSTAR D-2 "CableNet II" satellite. These transponders are used to carry HBO's basic movie service and the Cinemax complimentary entertainment service to separate East and West Coast audiences.

Showtime Entertainment, Inc., a subsidiary of Viacom International, the giant cable TV owner, operates a second first-run movie service also simultaneously transmitted on both SATCOM F-3 and COMSTAR D-2 satellites. Both HBO and Showtime feature special entertainment events, Hollywood musicals, and Broadway shows, in addition to their

Home Box Office. (Courtesy of HBO)

Showtime. (Courtesy of Showtime Entertainment, Inc.)

full-length movie features. Like HBO, Showtime operates East and West Coast feeds. The Movie Channel, owned by the Warner Amex Corporation, specializes primarily in feature-length movies, and operates 24 hours a day. East and West Coast feeds are carried on the SATCOM F-3 satellite. The Home Theatre Network, a subsidiary of Westinghouse, provides first-run G- and PG-rated movies on a single SATCOM F-3 transponder.

In addition to these general-audience pay TV networks, several other organizations specialize in providing specific audiences with pay television movie services programmed to their tastes. These include the Black Entertainment Television network, which presents a variety of programming featuring black performers in dominant or leading roles, including sports, music specials, and films. BET is carried on the SATCOM F-3 satellite. Galavision is the National Spanish Network's Spanish-language service; it shows movies, specials, and shows from Mexico,

MAY

				1 NORTH AVENUE IRREGULARS G 2:06	2 HERO AT LARGE PG 1:37 Final Showing
4 CHAPTER TWO PG 2:04 Premiere	5 CHAPTER TWO PG 2:04	6 CHAPTER TWO PG 2:04	7 CLOSE ENCOUNTERS PG 2:12	8 ELECTRIC HORSEMAN PG 2:01	9 DIE LAUGHING PG 1:50
11 BEDKNOBS AND BROOMSTICKS G 2:05 Premiere	12 BEDKNOBS AND BROOMSTICKS G 2:05	13 BEDKNOBS AND BROOMSTICKS G 2:05	14 ELECTRIC HORSEMAN PG 2:01	15 HIDE IN PLAIN SIGHT PG 1:32	16 NORTH AVENUE IRREGULARS G 2:06
18 FFOLKES PG 1:39 Premiere	19 FFOLKES PG 1:39	20 WHEN TIME RAN OUT PG 1:50	21 WHEN TIME RAN OUT PG 1:50 Final Showing	22 MUPPET MOVIE G 1:34	23 MUPPET MOVIE G 1:34 Final Showing
25 STAR TREK G 2:12 Premiere	26 STAR TREK G 2:12	27 STAR TREK G 2:12	28 HIDE IN PLAIN SIGHT PG 1:32 Final Showing	29 BEDKNOBS AND BROOMSTICKS G 2:05	30 CHAPTER TWO PG 2:04

JUNE

All movies start at 8 PM Eastern Time, 7 Central, 6 Mountain and 5 Pacific, unless otherwise noted.

1 URBAN COWBOY PG 2:15 Premiere	2 URBAN COWBOY PG 2:15	3 URBAN COWBOY PG 2:15	4 DAYS OF HEAVEN PG 1:34	5 GLACIER FOX G 1:30 Final Showing	6 SNEAK PREVIEW!
8 HONEY-SUCKLE ROSE PG 1:58 Premiere	9 HONEY-SUCKLE ROSE PG 1:58	10 FIENDISH PLOT OF DR. FU MANCHU PG 1:41 Premiere	11 FIENDISH PLOT OF DR. FU MANCHU PG 1:41	12 STAR TREK G 2:12	13 BEDKNOBS AND BROOMSTICKS G 2:05 Final Showing
15 BLACK HOLE PG 1:45 Premiere	16 BLACK HOLE PG 1:45	17 BLACK HOLE PG 1:45	18 STAR TREK G 2:12	19 URBAN COWBOY PG 2:15	20 DAYS OF HEAVEN PG 1:34 Final Showing
22 RESURREC-TION PG 1:43 Premiere	23 RESURREC-TION PG 1:43	24 DIE LAUGHING PG 1:50	25 DIE LAUGHING PG 1:50 Final Showing	26 HONEY-SUCKLE ROSE PG 1:58	27 BLACK STALLION G 1:58
29 BLACK STALLION G 1:58	30 ELECTRIC HORSEMAN PG 2:01	1 COAL MINER'S DAUGHTER PG 2:05 JULY Premiere	2 COAL MINER'S DAUGHTER PG 2:05	3 COAL MINER'S DAUGHTER PG 2:05	4 BLACK HOLE PG 1:45

The Home Theater Network schedule. (Courtesy of HTN)

THE YOUNG PEOPLE'S CHANNEL NICKELODEON™

PROGRAM SCHEDULE: JUNE 1981

	DUSTY'S TREEHOUSE	Pinwheel	VEGETABLE SOUP	MATT AND JENNY	FIRST ROW FEATURES	Adventures in Rainbow Country	Studio See	what will they think of next?	Livewire	VIDEO COMICS
	MON.–FRI.			TUES. & THURS	MON., WED. & FRI.	TUES. & THURS.	MON.–FRI.			
E	8:00am, 1:30pm	8:30am	2:00pm	2:30pm, 6:00pm	2:30pm, 6:00pm	3:00pm, 6:30pm	3:30pm, 7:00pm	4:00pm, 8:30pm	4:30pm, 7:30pm	5:30pm
C	7:00am, 12:30pm	7:30am	1:00pm	1:30pm, 5:00pm	1:30pm, 5:00pm	2:00pm, 5:30pm	2:30pm, 6:00pm	3:00pm, 7:30pm	3:30pm, 6:30pm	4:30pm
M	6:00am, 11:30am	6:30am	12:00n	12:30pm, 4:00pm	12:30pm, 4:00pm	1:00pm, 4:30pm	1:30pm, 5:00pm	2:00pm, 6:30pm	2:30pm, 5:30pm	3:30pm
P	5:00am, 10:30am	5:30am	11:00am	11:30am, 3:00pm	11:30am, 3:00pm	12:00n, 3:30pm	12:30pm, 4:00pm	1:00pm, 5:30pm	1:30pm, 4:30pm	2:30pm

	VIDEO COMICS	Pinwheel	DUSTY'S TREEHOUSE	MATT AND JENNY	Adventures in Rainbow Country	Studio See	Livewire	FIRST ROW FEATURES	what will they think of next?	NICK'S FAMILY PICKS
	SATURDAYS									SUNDAYS
E	8:00am	8:30am	1:30pm	2:00pm, 6:00pm	2:30pm, 6:30pm	3:00pm, 7:00pm	3:30pm, 7:30pm	4:30pm	5:30pm, 8:30pm	EASTERN 6:00 pm
C	7:00am	7:30am	12:30pm	1:00pm, 5:00pm	1:30pm, 5:30pm	2:00pm, 6:00pm	2:30pm, 6:30pm	3:30pm	4:30pm, 7:30pm	CENTRAL 5:00 pm
M	6:00am	6:30am	11:30am	12:00n, 4:00pm	12:30pm, 4:30pm	1:00pm, 5:00pm	1:30pm, 5:30pm	2:30pm	3:30pm, 6:30pm	MOUNTAIN 4:00 pm
P	5:00am	5:30am	10:30am	11:00am, 3:00pm	11:30am, 3:30pm	12:00n, 4:00pm	12:30pm, 4:30pm	1:30pm	2:30pm, 5:30pm	PACIFIC 3:00 pm
	SUNDAYS									
E	8:00am	8:30am	1:30pm	2:00pm, 4:30pm	2:30pm, 5:00pm	3:00pm	3:30pm, 8:00pm		5:30pm	
C	7:00am	7:30am	12:30pm	1:00pm, 3:30pm	1:30pm, 4:00pm	2:00pm	2:30pm, 7:00pm		4:30pm	
M	6:00am	6:30am	11:30am	12:00n, 2:30pm	12:30pm, 3:00pm	1:00pm	1:30pm, 6:00pm		3:30pm	
P	5:00am	5:30am	10:30am	11:00am, 1:30pm	11:30am, 2:00pm	12:00n	12:30pm, 5:00pm		2:30pm	

ARTS ON NICKELODEON
EASTERN 9:00 pm
CENTRAL 8:00 pm
MOUNTAIN 7:00 pm
PACIFIC 9:00 pm

SPECIAL DELIVERY SCHEDULE: JUNE 1981

	WED., JUNE 3	FRI., JUNE 5	WED., JUNE 10		FRI., JUNE 19	SUN., JUNE 21 (FATHER'S DAY)	FRI., JUNE 26	MON., JUNE 29
E	8:30-9:00pm	7:30-8:30pm	5:30-6:30pm	6:30-7:30pm	7:30-8:30pm	3:00- 3:30pm	5:30-6:00pm	6:00-7:00pm
C	7:30-8:00pm	6:30-7:30pm	4:30-5:30pm	5:30-6:30pm	6:30-7:30pm	2:00- 2:30pm	4:30-5:00pm	5:00-6:00pm
M	6:30-7:00pm	5:30-6:30pm	3:30-4:30pm	4:30-5:30pm	5:30-6:30pm	1:00- 1:30pm	3:30-4:00pm	4:00-5:00pm
P	5:30-6:00pm	4:30-5:30pm	2:30-3:30pm	3:30-4:30pm	4:30-5:30pm	12:00-12:30pm	2:30-3:00pm	3:00-4:00pm
	INTERNATIONAL CHILDREN'S FESTIVAL AT WOLF TRAP	RICK DERRINGER	INCREDIBLE, INDELIBLE, MAGICAL, PHYSICAL, MYSTERY TRIP	MAGICAL, MYSTERY TRIP THROUGH LITTLE RED'S HEAD	STAR CITY ROLL OUT	PORTRAIT OF GRANDPA DOC	INTERNATIONAL CHILDREN'S FESTIVAL AT WOLF TRAP	MICHEL'S MIXED-UP MUSICAL BIRD
	pre-empts What Will They Think of Next?	pre-empts Livewire	pre-empts Video Comics, and First Row Features	pre-empts First Row Features and Studio See	pre-empts Livewire	pre-empts Studio See	pre-empts Video Comics	pre-empts First Row Features

Nickelodeon program schedule. (Courtesy of Nickelodeon)

South America, and Spain. Galavision is also carried on the SATCOM F-3 satellite.

Several children's networks are offered on satellite, including the Nickelodeon Emmy-award-winning service operating on its own exclusive transponder on SATCOM F-3. Nickelodeon includes old-time adventure serials, comic strips, and new and original stories. The USA Network, also on SATCOM F-3, although not usually considered a pay TV service, does provide programming for children with its Calliope children's film series. The USA Network is usually carried by the CATV company in its basic service package. Commercials are included throughout the programming.

A number of R-rated adult movie programmers have gone on the birds, including Satori Productions—whose late-night Private Screenings service is carried on WESTAR III. Escapade, a similar late-night offering, is fed by Rainbow Programming Services over the COMSTAR D-2 satellite. As yet, no X-rated movie services are fed to cable companies over satellite circuits; there appears to be some question as to whether the federal government's pornography laws would prohibit such an action.

Religious Networks

Turning from the Devil to God, the fastest growing services, after the R-rated movie programmers, are the religious organizations. The granddaddy of the satellite users is the Christian Broadcasting Network, recently reorganizing to become the Continental Broadcasting Network. CBN features Christian programming 24 hours a day, including the *700 Club* and other enormously popular entertainment shows and musical events. CBN is carried via SATCOM F-3, and like all the religious services it is provided completely free of charge to any CATV company, organization, or individual who wishes to pick up the signal via satellite. People That Love Television Network (PTL) provides a similar 24-hour-per-day Christian programming service also carried by SATCOM F-3. PTL includes talk and variety shows and children's programming with a Christian orientation. The Trinity Broadcasting Network (TBN) originates from a Los Angeles UHF television station. Offering a broadly based Christian programming and inspirational service 24 hours per day, TBN is carried by COMSTAR D-2. The National Christian Network (NCN) offers somewhat greater nondenominational programming, primarily during the daytime hours. NCN is carried via the COMSTAR D-2 satellite. Although the nondenominational Christian services are the oldest religious programs on satellite, National Jewish Television and other religious organizations also sublease blocks of air time on various satellite transponders to distribute their programming, primarily via WESTAR and COMSTAR satellites.

The National Christian Network. (Courtesy of NCN)

People That Love Television Network. (Courtesy of PTL)

The Superstations

The superstations are regional television stations that have "gone national" by having their broadcast signal carried by cable television companies nationwide via satellite retransmission. The cost for a CATV company to carry a superstation's signal is usually very low, of the order of 10 to 15 cents per month. The superstations make their profit by charging their advertisers much higher rates for commercials, which are viewed by a far larger audience than is true for the regional stations.

The leading entrepreneur, second only to the folks at HBO, in developing satellite television is Ted Turner, owner of the Atlanta Braves and Atlanta Hawks, and a successful defender in the America's Cup yacht race. Turner's WTBS–Atlanta Channel 17 superstation boasts a fabulous film library acquired at bargain basement prices just before the station went on the bird. WTBS also provides full sports coverage of the Turner-owned teams and operates 24 hours a day on the SATCOM F-3 satellite. WOR-TV, New York City's independent television station, features a variety of entertainment, news, sports, and movies on a 24-hour schedule. WOR brings New York City to America via the SATCOM F-3 satellite. WGN-TV, Chicago's independent superstation, operates in a similar format and originates the popular Phil Donahue show live from its Chicago studios. WGN is carried 24 hours per day by a SATCOM F-3 transponder.

In addition to these well-known superstations, two other television stations feed their programming via satellite. XEW-TV, the Mexico City television station, delivers its Spanish-language programming to U.S. television stations under a special Western Union tariff. XEW can be found on the WESTAR III satellite. KUSK-TV, a new country and western television station in Prescott, Arizona, plans to rebroadcast its

Programming subject to change without notice.

WGN the ONE from UNITED VIDEO!

Time	MON	TUE	WED	THU	FRI	SAT	SUN
6	FIRST REPORT TOP OF THE MORNING						SUPERMAN
:30	BULLWINKLE CARTOONS					BUYER'S FORUM 3 SCORE/1st REPRT	CARTOONS 6:45 1st REPRT
7	BOZO SHOW					US FARM REPORT	FAITH 20
:30						BULLWINKLE	COMM. CALENDAR WHAT'S NU?
8						REX HUMBARD	MASS FOR SHUT-INS
:30	BEWITCHED					SATURDAY MORNING MOVIE I	CHURCH HOUR
9	MY THREE SONS						HOUR OF POWER
:30	YOUR NEW DAY						SERGEANT PRESTON
10	MIKE DOUGLAS					ABBOTT, COSTELLO AND FRIENDS 10:45 IRS	SPACE: 1999
:30							
11	DONAHUE					ISSUES UNLTD.	THE CISCO KID
:30						CHARLANDO	LONE RANGER
12	$50,000 PYRAMID					SATURDAY MATINEE	SHERLOCK HOLMES
:30	HOLLYWOOD SQUARES					SATURDAY MATINEE II	CHARLIE CHAN
1P	AFTERNOON MOVIE						
:30							MOVIE GREATS
2	Five Minute News (Approx. 2 pm)						
:30							
3	BUGS BUNNY						
:30	SCOOBY DOO					AMERICA'S TOP TEN	FAMILY CLASSICS
4	FLINTSTONES					SOUL TRAIN	
:30	I DREAM OF JEANNIE						
5	GOOD TIMES						
:30	WELCOME BACK, KOTTER						KUNG FU
6	BARNEY MILLER						
:30	CAROL BURNETT						SUNDAY EVENING MOVIE
7	SOLID GOLD	PRIME TIME MOVIE				WILD KINGDOM	
:30						IN SEARCH OF ...	
8	MONTE CARLO					PEOPLE TO PEOPLE	LAWRENCE WELK
:30						MAUDE	
9	THE 9 O'CLOCK NEWS						
:30							
10	PRISONER CELL BLOCK H					SOLID GOLD (ENCORE)	MORECAMBE & WISE
:30	THE ODD COUPLE						
11	WGN PRESENTS						
12:30							NIGHTBEAT
1A	NIGHTBEAT						CROMIE CIRCLE
1:30	ALL NIGHT PROGRAMMING						
3:00							

WGN's program schedule—Central time zone. (Courtesy of WGN)

unique country programming to over 130 new low-power television stations scattered throughout the United States. KUSK and the LPTV network project are backed by the Allstate Venture Capital Group of Sears. The WESTAR I satellite is planned as the primary feed.

Sports Programming

The Entertainment and Sports Programming Network (ESPN) brings more than 1200 annual hockey, basketball, NCAA, and collegiate games and other sports events to the cable companies 24 hours a day. Owned primarily by the Getty Oil Company, ESPN feeds the

The USA Network—sports programming. (Courtesy of USA Network)

Monthly subscription guide and schedule. (Courtesy of Entertainment and Sports Programming Network, Inc.)

Saturday, June 27

A.M.
- 9:30 **Bowling:** National Collegiate Championship
- 11:30 **Rugby:** National Club Championship

P.M.
- 1:30 **Budweiser Presents Top Rank Boxing from Las Vegas**
- 4:00 **SportsCenter**
- 5:00 **Golf:** 1971 British Open Highlights
- 6:00 **NASL Soccer:** Portland at Tulsa (L)
- 8:00 **SportsCenter**
- 8:30 **Supercross from Detroit:** Part 1
- 10:00 **Golf:** 1971 British Open Highlights
- 11:00 **SportsCenter**
- 11:30 **NASL Soccer:** Portland at Tulsa

Sunday, June 28

A.M.
- 1:30 **Formula I Belgium Grand Prix**
- 4:00 **SportsCenter**
- 4:30 **To Be Announced**
- 5:00 **Rugby:** National Club Championship
- 7:00 **SportsCenter**
- 7:30 **Superstar Volleyball Cup:** Ontario vs. Saskatchewan
- 8:00 **NASL Soccer:** Portland at Tulsa
- 10:00 **Wrestling:** National Senior Greco-Roman Championships

P.M.
- 12:00 **Diving:** USA vs. Peoples Republic of China
- 2:00 **Formula I Spanish Grand Prix**
- 4:30 **SportsCenter**
- 5:00 **Grand Slam of Horseshow Jumping:** Part 2
- 7:00 **Golf:** 1972 British Open Highlights
- 8:00 **SportsCenter**
- 9:00 **Formula I Spanish Grand Prix**
- 11:30 **SportsCenter**

Monday, June 29

A.M.
- 12:00 **Rugby:** National Club Championship
- 2:00 **Wrestling:** National Senior Greco-Roman Championships
- 4:00 **SportsCenter**
- 5:00 **NASL Soccer:** Portland at Tulsa
- 7:00 **SportsCenter**
- 8:00 **1981 Top Ace Handball Championship —** Final
- 8:30 **Polo:** World Cup Semifinal #2
- 10:00 **Auto Racing '81**

P.M.
- 12:30 **Professional Team Rodeo:** Tulsa vs. Amarillo
- 2:00 **Rugby:** National Club Championship
- 4:00 **SportsCenter**
- 5:00 **F.A. Soccer:** British Home Championships — England vs. Scotland
- 7:00 **Supercross from Detroit:** Part 2
- 8:30 **SportsCenter**
- 9:00 **Diving:** USA vs. Peoples Republic of China
- 11:00 **SportsCenter**
- 11:30 **Golf:** 1972 British Open Highlights

Tuesday, June 30

A.M.
- 12:30 **Auto Racing '81**
- 3:00 **NCAA Golf:** Division II Championship
- 4:00 **SportsCenter**
- 5:00 **Australian Rules Football:** Teams To Be Announced
- 6:30 **To Be Announced**
- 7:00 **SportsCenter**
- 8:00 **NCAA Lacrosse:** Division I Championship
- 10:30 **Wrestling:** National Senior Greco-Roman Championships

P.M.
- 12:30 **Grand Slam of Horseshow Jumping:** Part 2
- 2:30 **Australian Rules Football:** Teams To Be Announced
- 4:00 **SportsCenter**
- 5:00 **NASL Soccer:** Week in Review
- 5:30 **Golf:** 1973 British Open Highlights
- 6:30 **PKA Full Contact Karate:** World Welterweight Championship from Atlanta, Georgia
- 8:30 **SportsCenter**
- 9:00 **NASL Soccer:** Week in Review
- 9:30 **F.A. Soccer:** British Home Championships — England vs. Scotland
- 11:30 **SportsCenter**

ESPN's Daily Program Schedule can be viewed at 5:00 a.m., 8:00 a.m. and 11:00 p.m. Every Weekday.

country via the SATCOM F-3 satellite. The Hughes Television Network supplies its multidimensional sports programs to independent television stations nationwide. Hughes is noted for its big-name events, including major golf tournaments and other sports extravaganzas often carried by network television affiliates. The Hughes Television Network is retransmitted by WESTAR III. The USA Network features Thursday night pro baseball, Madison Square Garden sports, and NHL, NBA, and NASL games. USA is oriented to CATV systems, and feeds its programming via SATCOM F-3. The Video Sports Network concentrates primarily on southeast U.S. sports events in a format similar to the Entertainment and Sports Programming Network. VSN shares a transponder with the Appalachian Community Service Network on the RCA SATCOM F-3 satellite.

News Networks

The Cable News Network is the second creation of the ubiquitous Ted Turner. From its world headquarters in Atlanta, Georgia, the CNN organization provides 24 hours a day of news, weather, and sports information to over 1000 U.S. cable networks. News bureaus in Washington, D.C., New York City, and Los Angeles provide live feeds hourly. A dozen other bureaus worldwide keep the Cable News Network in immediate touch with late-breaking stories. CNN is carried on the RCA SATCOM F-3 satellite. The Independent News Network, operating on WESTAR I, feeds dozens of independent VHF and UHF television stations over the nation with nightly news programs originating from New York City and other flagship stations.

Both NBC and CBS also lease private contract channels on the RCA

Spanish International Network Hispanic programming. (Courtesy of SIN)

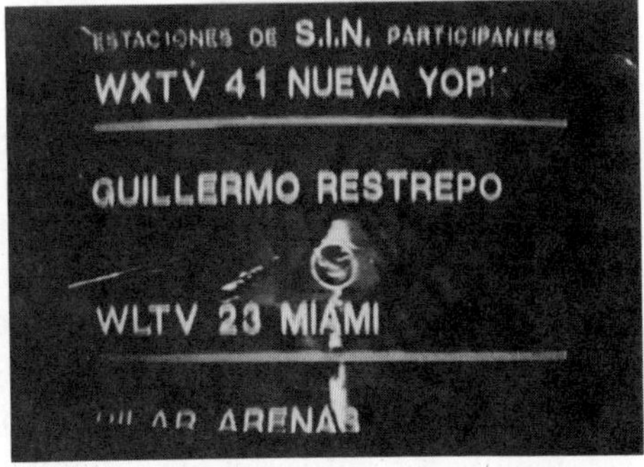

WESTAR III and COMSTAR D-3 satellites. National networks use these transponders for feeding the nightly news across the country, handling news prefeed, and relaying certain network programming, including daytime soap operas and late evening entertainment. The WESTAR III and COMSTAR D-3 transponders are used for news bureau coordination as well, and it is not unusual to see the nightly news anchor people conversing with each other over the circuits about the events of the day. Interesting news items and tidbits of information not normally aired on the news programs sometimes are aired on these private news video circuits, and make for interesting viewing an hour or two before the local news feeds are scheduled to occur.

Special Television Networks

A number of programming organizations have specialized in creating network services that cater to a specific element of the general audience and that are supported by conventional commercial advertising.

Among these organizations, the Spanish International Network (SIN) is probably the oldest. SIN is America's national Spanish television network, furnishing Spanish-language variety shows, drama, nightly news, sports, and live feeds from Puerto Rico, Spain, Mexico, and South America to several dozen independent television stations in U.S. cities that have large Spanish-speaking communities. SIN's programming is carried via the WESTAR III satellite.

The Modern Satellite Network, a division of Modern Talking Picture Service, Inc., carries entertainment and informational programs oriented toward the general consumer during the daytime. These half-hour television shows are sponsored through commercials and by the program producers themselves, often major U.S. corporations. Howard Ruff's *Ruff House* television show is aired over MSN. The network is carried to CATV companies via SATCOM F-3.

Southern Satellite Systems provides the Satellite Programming Network (SPN) with a similar general-viewership, commercially sponsored network delivered to CATV companies via the WESTAR III satellite. SPN carries classic movies, the TeleFrance USA French programming service, and a number of other interesting talk shows, corporate public relations events, and syndicated programs 24 hours a day.

The USA Network (which originally began as a sports feed for Madison Square Garden) has broadened its base of programming to include children's shows and The English Channel, which shows British-made movies and television series sponsored by paid advertisements. Recently the USA Network entered into a contractual agree-

MONDAY, JUNE 15

PM
12:00 NETWORK ON TONE
BUSINESS VIEW – Host Bob Jones with "Man's Material Welfare," on the American free enterprise system, and "The Innovators," on designing and building auto bodies.
1:00 **AEROBIC DANCING** – The Janet Sloane Aerobic Dancercize Program.
1:30 **FAMILY GUIDE TO BOATING FUN** – New products in the areas of power boating, sailing and water safety.
2:00 **TELEFRANCE U.S.A.** – "The **Gentleman Burglar,"** starring Georges Descrieres as detective Arsene Lupin.
Chroniques de France: Aspects of France and French life.
Three Star Cooking Series: A royal recipe that pleased Madame de Pompadour.
Cinefrance: "Marriage on the Rocks," a wife, her husband, their lovers, and an incident that changes their relationships.
NETWORK BREAK (4:58:50)
NETWORK OFF TONE (4:59:58)

TUESDAY, JUNE 16

PM
12:00 NETWORK ON TONE
MODERN LIFE – Host Suzanne Leamer with "Challenger," a look at race driver Janet Guthrie, "Wherever You Are," coping with arthritis, and "Put Wings on Your Career" in aviation maintenance.
1:00 **AEROBIC DANCING** – The Janet Sloane Aerobic Dancercize Program.
1:30 **BUSINESS VIEW** – Presentations from the world of business and industry, hosted by Bob Jones.
2:30 **CONSUMER INQUIRY** – Host Debbie Durham with features of consumer interest.
3:30 **MODERN LIFE** – Host Suzanne Leamer presents "Everyone," man's survival through mobility, "The Travels of Timothy Trent" on safety packaging, and "Shelley, Pete and Carol," a film on teenage pregnancy. Includes a "Viewpoint" interview with Dr. Carol Garvey of the U.S. Health Services Administration, on the subject of teenage pregnancy.
4:30 **THE HOME SHOPPING SHOW** – Demonstrations of how to select a vacuum cleaner; free barbecue recipes; a gallon of pure drinking water for pennies.
NETWORK BREAK (4:58:50)
NETWORK OFF TONE (4:59:58)

WEDNESDAY, JUNE 17

PM
12:00 NETWORK ON TONE
BUSINESS VIEW – Host Bob Jones with "Challenger," race driver Janet Guthrie for Kelly Services, "The Next Step." from Breeder Reactor Corp., "Report to Consumers" from Montgomery Ward, and "Business for the Birds."
1:00 **AEROBIC DANCING** – The Janet Sloane Aerobic Dancercize Program.
1:30 **WHAT'S COOKING?** – Norinne Cole and Marlene Cummins offer recipes and tips from the kitchens of R. T. French.
2:00 **THE HOME SHOPPING SHOW** – Discussions and demonstrations of interesting products and services that may be ordered by telephone and mail.
2:30 **CONSUMER INQUIRY** – Host Debbie Durham with "Saved by the Bell," on home fire safety, "How to Buy a Used Car," "Always Tip the Fat Lady," on sound financial planning, and "Play It Safe" with power lawn tools.
3:30 **BUSINESS VIEW** – Bob Jones presents features from the world of business.

7

(Courtesy of Modern Satellite Network)

ment with Procter and Gamble to furnish several hundred original hours of programming a year, thus significantly expanding the USA Network's activities.

Both CBS and ABC television have entered the cable TV race by establishing cable programming subsidiaries. CBS Cable features fine arts programming from New York City, including "up market" plays, cultural events, and musical presentations. CBS can be found on the COMSTAR D-2 satellite. ABC's new offering—Alpha—the arts channel of cultural events and first-run operas, shows, and orchestral performances, can be found on the RCA F-3 satellite.

PROGRAM GRID

EASTERN TIME	PACIFIC TIME	MONDAY	TUESDAY	WEDNESDAY	THURSDAY	FRIDAY	SATURDAY	SUNDAY
7:00	4:00	INTERNATIONAL BYLINE					COWBOY FLICKS	INTERNATIONAL BYLINE
7:30	4:30	MEDICINE MAN	CELEBRITY	FROM NEW YORK	RUFF HOUSE	BUCKY DENT		WOMEN'S CHANNEL
8:00	5:00	BUCKY DENT		THE GOURMET	WALL STREET TODAY	GOOD LIVING		J. L. BROWN
8:30	5:30	WOMEN'S CHANNEL					PLANT GROOM	PHYSICIAN IN HIS SERVICE
9:00	6:00	SUSAN NOON					CELEBRITY	WOMEN'S CHANNEL
9:30	6:30	FRAN CARLTON						BILLY JAMES HARGIS
10:00	7:00	MOVIETOWN					CYCLE-AMERICA BMX	KENNETH COPELAND
10:30	7:30						BUCKY DENT	
11:00	8:00						JIMMY HOUSTON OUTDOORS	JOE BURTON JAZZ
11:30	8:30	DON KENNEDY'S SPOTLIGHT					WOMEN'S CHANNEL	
NOON	9:00	THE GOURMET	KALEIDOSCOPE	RUFF HOUSE	CONNIE MARTINSON	MEDICINE MAN	WORLD LEAGUE WRESTLING	CONNIE MARTINSON
12:30	9:30	JOAN FONTAINE						PLANT GROOM
1:00	10:00	CHEF SECRETS					WOMEN'S CHANNEL	KALEIDOSCOPE
1:30	10:30	DAVID GRUEN					AMERICAN ANGLER	FINANCIAL INQUIRY
2:00	11:00						MONSTER FLICKS	BILL DANCE OUTDOORS
2:30	11:30	SEW WHAT'S NEW	GOOD LIVING	GARDENING AT HOME	REAL MONEY	COPING		FROM NEW YORK
3:00	NOON	GREAT IDEAS						SPN MOVIE
3:30	12:30	WOMEN'S CHANNEL					CHAMPIONSHIP WRESTLING	
4:00	1:00	FRAN CARLTON						
4:30	1:30	MOVIETOWN					CHAMPIONSHIP FISHING	
5:00	2:00						AMERICA SINGS	BUCKY DENT
5:30	2:30						WOMEN'S CHANNEL	THE GOURMET
6:00	3:00	JOAN FONTAINE					ROCKWORLD	HISTORY OF SPACE
6:30	3:30	HISTORY OF SPACE	RUFF HOUSE	MEDICINE MAN	WALLSTREET TODAY	BILL DANCE OUTDOORS		MEDICINE MAN
7:00	4:00	CHAMPIONSHIP FISHING	FISHIN' WITH MIKE & LARRY	AMERICAN ANGLER	JIMMY HOUSTON OUTDOORS	CYCLE—AMERICA BMX	GARDENING AT HOME	FISHIN' WITH MIKE & LARRY
7:30	4:30	GOOD LIVING	SEW WHAT'S NEW	CONNIE MARTINSON	THE GOURMET	REAL MONEY	RUFF HOUSE	AMERICA STILL
8:00	5:00	GREAT IDEAS					DON KENNEDY'S SPOTLIGHT	
8:30	5:30	WOMEN'S CHANNEL					WOMEN'S CHANNEL	
9:00	6:00	TELEFRANCE — U.S.A.						
9:30	6:30							
10:00	7:00							
10:30	7:30							
11:00	8:00							
11:30	8:30							
12:00	9:00	DON KENNEDY'S SPOTLIGHT					REAL MONEY	JOE BURTON JAZZ
12:30	9:30	PAUL RYAN					PAUL RYAN	
1:00	10:00	ALL NIGHT AT THE MOVIES						
7:AM	4:00							

*Local advertising availability for cable systems during network breaks (average 2 minutes per hour)

SPN - A DIVISION OF SATELLITE SYNDICATED SYSTEMS INC.
P.O. BOX 45684 TULSA, OK. 74145 (918) 481-0881

SSS TM

(Courtesy of SPN, a division of Satellite Syndicated Systems)

(Courtesy of CBS Cable)

The Robert Wold organization, a company that for years has specialized in sports and other special event feeds for the national television networks, has created a new satellite television network called Wold Communications. Wold's programming is designed nationally, for broadcast by independent television stations, and its major drawing card is the daily Merv Griffin show, uplinked from Hollywood. Wold also uses its facilities to feed private teleconferences and special event gatherings such as political fund-raising shows and corporate multicity public stockholder annual meetings, and uplinks Entertainment Tonight live from Hollywood each evening to television stations nationwide.

Public Affairs and Educational Channels

C-SPAN, the Cable Satellite Public Affairs Network headquartered in Washington, D.C., transmits daily live coverage of the U.S. House of Representatives via the SATCOM F-3 satellite. The Canadian Broadcasting Corporation also feeds both English- and French-language coverage of the Canadian House of Commons from Ottawa on the ANIK A-3 satellite.

The Appalachian Community Service Network (ACSN) offers educational programming for adults and postsecondary groups, including college-level courses and continuing education seminars in a variety of fields. ACSN appears on RCA SATCOM F-3.

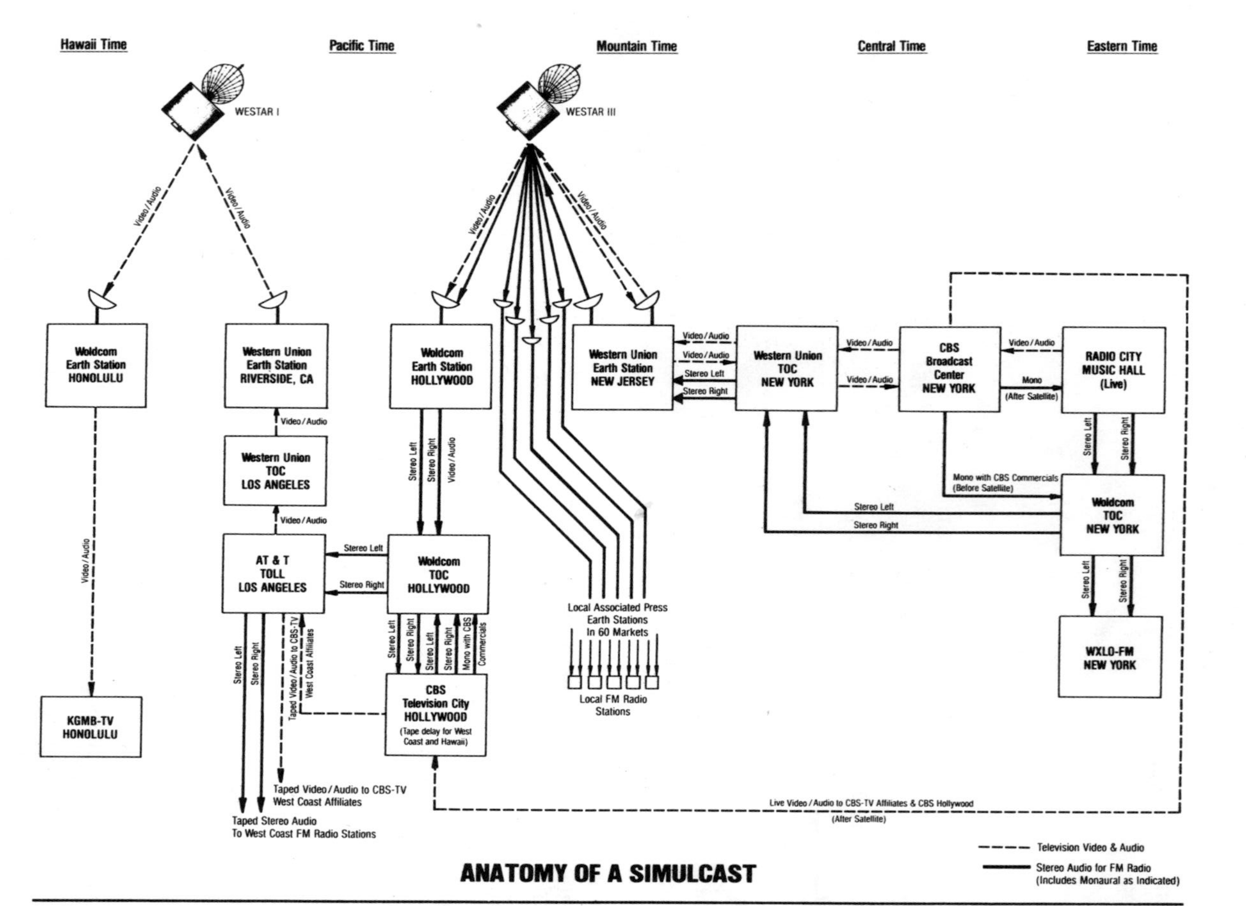

ANATOMY OF A SIMULCAST

On Feb. 25, Wold Communications transmitted a stereo simulcast of the Grammy Awards from New York's Radio City Music Hall to some 60 FM stations across the country. The intricate hook-up, illustrated above, successfully circumvented the one-third-of-a-second satellite delay and matched the audio with the video broadcast by CBS-TV. Distributor for the FM simulcast was N.K.R. Productions.

(Courtesy of Wold Communications)

A rather unusual "educational service" is provided by United Video on SATCOM F-3 in the form of a video shopping guide. By watching this shopping channel, a cable television viewer can see the latest gadgets, products, and services on the market. The manufacturers pay to present this continuously running commercial, uninterrupted by annoying programming!

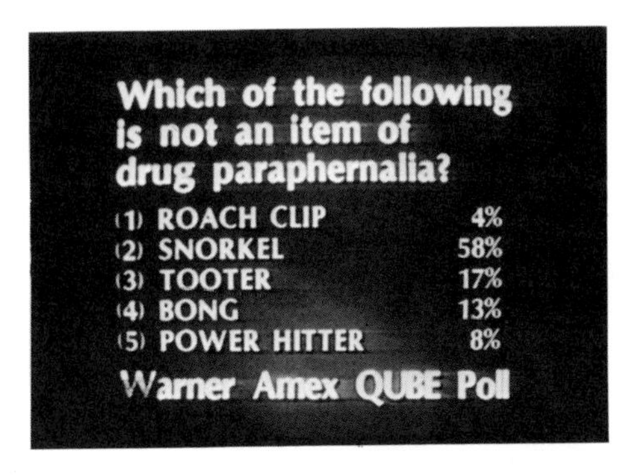

Warner Amex QUBE Poll. (Courtesy of Warner Amex)

The American Educational Television Network (AETN) supplies continuing-education courses for various professionals. The AETN seminars are often sponsored by professional associations, including medical, legal, and accounting societies, and its programming is carried on RCA F-3.

The Conventional TV Networks

As noted, both NBC and CBS use satellite feeds to deliver news programs and unedited news prefeeds between their East and West Coast network master control centers. In addition, all three networks —NBC, CBS, and ABC—carry many of their regular shows over the

ACSN—The Appalachian Community Service Network. (Courtesy of ACSN)

Canadian House of Commons daily coverage. (Courtesy of Dept. of Communications, Ottawa, Canada)

Public Broadcasting Service uses three full-time WESTAR satellites. (Courtesy of Paraframe Inc.)

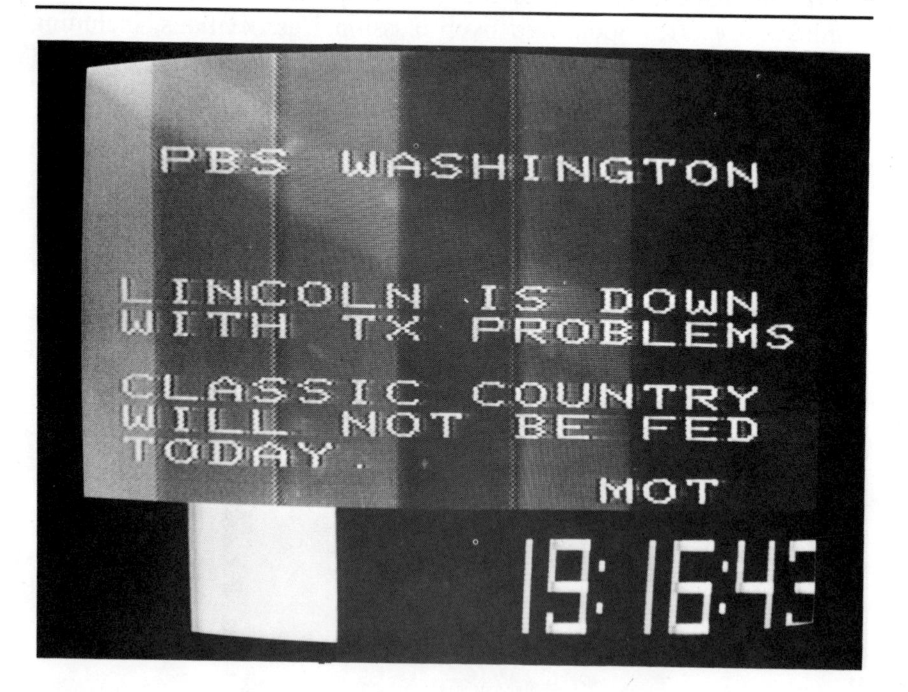

birds, both between their network centers and to the Alaskan television station network, hours or even days before these programs are actually aired. The satellites most frequently used for these feeds are the WESTAR III and the COMSTAR D-3. In addition, both the Canadian Broadcasting Corporation (CBC) and the Canadian Television Corporation (CTV) broadcast their entire programming day via the ANIK B-2 and ANIK A-3 Canadian TELSAT satellites. Many U.S. situation comedies and other network shows are aired in Canada the same week they were originally presented via the CBC and CTV networks. The armed forces satellite network uses SATCOM F-2 to feed U.S. network programming to U.S. military television stations in Alaska and Cuba, and this occasionally includes live news events and sports coverage.

The Public Broadcasting Service (PBS) operates three full-time transponders and one occasional-use transponder on the WESTAR I satellite. If you live on the West Coast of the United States, for example, you can watch *Wall Street Week* live at 5:30 p.m. Pacific Time, when it is uplinked at 8:30 p.m. Eastern Time from the Maryland Center for Public Broadcasting Studios.

Private and Special Networks

Perhaps the most fascinating of all satellite television are the "secret" feeds, those channels that contain data and program signals other than television pictures. Most of these require some form of additional data decoder to unscramble their signal successfully, but the audio network channels can often be received by using a conventional short-wave multiband receiver whose antenna is connected to the *video* output (not audio!) of the television satellite receiver.

These private networks include the Reuters "Monitor" high-speed business data service, which is fed to desk-top display terminals via the SATCOM F-3 satellite. United Press International and the Associated Press news services transmit several channels of teletypewriter wire copy, audio news feeds, and facsimile photographs over subcarriers of regular video channels on the SATCOM F-3 satellite. Several dozen audio networks are also delivered to radio stations across the nation via the WESTAR III satellite, including sports, Spanish-language, and black network news feeds; national public radio; and a number of independent radio chain feeds.

The National Aeronautics and Space Administration (NASA) operates a contract channel on the SATCOM F-2 satellite to provide video

TABLE 3–1.
Subscriber Counts—Satellite-Fed Networks Subscriber Counts as of May 1, 1981

Basic Services (Television)	Systems	Homes	Percent Change in Homes from March 31
WTBS (total)	3,006	11,758,391	+2.35
WTBS (part-time)	176	1,525,626	−7.24
WTBS (full-time)	2,830	10,232,765	+3.95
CBN	2,758	9,642,114	+2.89
ESPN	1,840	9,486,331	+6.41
USA Network	1,350	7,500,000	+7.14
C-SPAN	1,010	7,500,000	+5.63
BET	643	6,645,998	+1.63
CNN (as of 4/22)	1,074	5,693,655	+5.95
WGN	1,586	5,490,821	+2.52
Nickelodeon/ARTS (ARTS new this month)	975	4,500,000	NC
PTL	319	3,850,000	+1.32
WOR	751	3,500,000	+6.71
Modern Satellite Network	470	3,475,000	−.71
SIN (estimated Spanish-speaking households)	103	2,652,100	NC
SPN (as of 4/15)	220	2,275,133	+9.35
Video Concert Hall	131	2,208,310	NC
Trinity	80	938,745	+87.75
North American Newstime (formerly UPI Newstime)	67	650,000	−20.95
Appalachian Community Service Network	135	600,000	+1.69
Basic Services (audio)			
WFMT	60	434,357	+19.14
Lifestyle	14	41,275	+892.90
Pay Services			
Home Box Office	2,500	6,000,000	NC
Showtime	1,000	1,600,000	NC
The Movie Channel	950	875,000	NC
Cinemax	200	200,000	NC
Home Theater Network	160	160,000	NC
Rainbow	32	80,000	+26.98
Private Screenings (includes STV systems)	5	78,000	NC
GalaVision	73	70,000	+7.69

NC designates no change.

coordination between Houston, Cape Canaveral, and Washington, D.C. Space Shuttle video occasionally can be seen on this transponder.

Table 3-1 presents a summary of the leading satellite-fed network services based on their CATV viewership figures.

Who Owns the Channels?

The domestic U.S. and Canadian satellites are owned and operated by five major common carrier groups: RCA Americom, Inc., a subsidiary of the RCA Corporation; Western Union Telegraph Corporation; The Communications Satellite Corporation; American Telephone and Telegraph Company; and TeleSat Canada. These organizations operate the SATCOM, WESTAR, COMSTAR, TELSTAR, and ANIK satellite systems respectively. In addition, several other U.S. companies have been approved to own and operate domestic satellite systems, including Hughes Satellite, General Telephone and Electronics, and the Southern Pacific Communications Corporation. (See Table 3-2.)

COMSAT is also the U.S. operator of the INTELSAT Global Satellite System, and in partnership with IBM has launched the Satellite Business Systems' all-digital private network satellite.

Although the common carrier organizations own and operate their communications satellite systems (consisting of the satellite space segment and the telemetery, tracking, and command control ground stations), the satellites normally are built by other organizations, such as the Hughes Satellite Corporation and the RCA Satellite Corporation. U.S. and Canadian satellites are launched by NASA for a multimillion-dollar launching fee.

Common carriers are usually prohibited from providing their own network programming services, and the satellite transponders are treated by the FCC from a regulatory standpoint as if they were ordinary terrestrial video channels. Therefore, the common carrier satellite owners must rent their transponders to other organizations that provide the original programming, usually through their own 30-foot uplink earth stations. Organizations such as the Christian Broadcasting Network, PBS, and the Cable News Network own their own satellite transmit earth stations. The TVRO satellite downlink receive-only earth stations are usually owned independently by the cable television companies and television broadcasting stations.

Under present FCC regulations, WTBS, WOR, and WGN, the U.S.

TABLE 3–2.
American TV Satellites (North and South America)

Satellite	Common Carrier	Orbital Slot °West Longitude	Frequency	Turn-on Date
Hughes I	Hughes	74	4/6 GHz	1983
SPCC I	Southern Pacific Communications	70	Dual Band 4/6 & 12/14 GHz	1984
SatCol	Columbian Satellite	75	4/6 GHz	1983
Advanced WESTAR	Western Union & AMSAT	79	Dual band 4/6 & 12/14 GHz	1983
SATCOM IV	RCA Americom	83	4/6 GHz	1982
SATMEX	Mexico	85	4/6 GHz	1983
TELSTAR	AT&T	87	4/6 GHz	1983
COMSTAR D-3	Comsat General (for AT&T and GTE)	87	4/6 GHz	In service
WESTAR III	Western Union	91	4/6 GHz	In service
Advanced WESTAR	Western Union & AMSAT	91	Dual band 4/6 & 12/14 GHz	1983
SBS III	SBS	94	12/14 GHz	1983
COMSTAR D-2	Comsat General (for AT&T and GTE)	95	4/6 GHz	In service
TELSTAR II	AT&T	95	4/6 GHz	1984
SBS II	SBS	97	12/14 GHz	In service
WESTAR I	Western Union	99	4/6 GHz	In service
WESTAR IV	Western Union	99	4/6 GHz	1982
GSTAR	GTE	100	12/14 GHz	1984
GSTAR	GTE	103	12/14 GHz	1984
ANIK A-1	Canadian Satellite	104	4/6 GHz	In service
ANIK D-1	Canadian Satellite	104	4/6 GHz	1982
SBS I	SBS	106	12/14 GHz	In service
ANIK B-1	Canadian Satellite	109	Dual Band 4/6 & 12/14 GHz	In service
ANIK C-3	Canadian Satellite	109	12/14 GHz	1983
ANIK D-3	Canadian Satellite	109	4/6 GHz	1984
ANIK C-1	Canadian Satellite	112.5	12/14 GHz	1983
ANIK A-3	Canadian Satellite	114	4/6 GHz	In service
ANIK D-2	Canadian Satellite	114	4/6 GHz	1983
ANIK C-2	Canadian Satellite	116	12/14 GHz	1984
SATCOM II	RCA Americom	119	4/6 GHz	In service
SPCC II	Southern Pacific Communications	119	Dual band 4/6 & 12/14 GHz	1984
WESTAR II	Western Union	123.5	4/6 GHz	In service
WESTAR V	Western Union	123.5	4/6 GHz	1982
COMSTAR D-4	Comsat General (for AT&T and GTE)	127	4/6 GHz	In service
COMSTAR D-1	Comsat General (for AT&T and GTE)	128	4/6 GHz	In service
SATCOM III R	RCA Americom	131	4/6 GHz	In service
SATCOM I	RCA Americom	135	4/6 GHz	In service
Hughes II	Hughes	135	4/6 GHz	1983
SATCOM I R	RCA Americom	139	4/6 GHz	1983
SATCOM II R	RCA Americom	143	4/6 GHz	1983

GTE and SBS are each authorized two orbital slots. Other locations include: GTE—106°; SBS—100°, 122°, 125°, 128°.

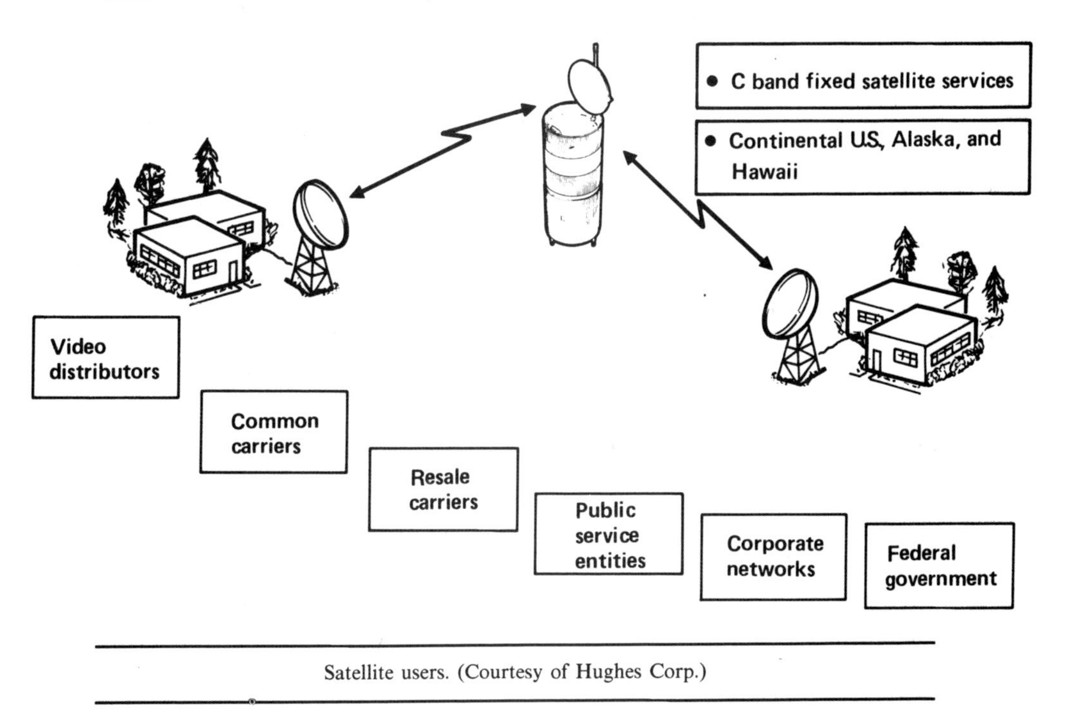

Satellite users. (Courtesy of Hughes Corp.)

superstations, are not permitted to lease satellite transponders directly, so several independent organizations have sprung up to pick up the superstation signals locally and retransmit them via leased transponders. Called resale common carriers, Eastern Microwave, Syndicated Satellite Systems, and United Video deliver the programs of WOR, WTBS, and WGN, respectively, to cable companies for a tariff charge of 10 to 15 cents per subscriber per month. This nominal charge covers the cost of distribution and allows the resale carriers to realize a small profit. The superstations, of course, benefit indirectly by obtaining significantly larger viewing audiences, thereby allowing them to increase their air-time advertising rates substantially.

The International Connection

Canada operates its own domestic satellite system, whose signals can easily be picked up in the United States. Canada's ANIK satellites are owned by TeleSat Canada, an independent corporation formed by the government of Canada and owned by the Canadian Treasury. TeleSat Canada provides the Canadian telephone companies with telephone circuits via satellite, and interconnects the National Telex Network operated by CN–CP Telecommunications. TeleSat Canada also provides leased transponders to the two Canadian television networks, and plans to provide a multiprogramming service to Canadian CATV companies similar to the RCA SATCOM F-3 cable network. In addition, four new Canadian superstations—located in Toronto, Vancouver, Hamilton, and Edmonton—have begun transmitting their programming to smaller CATV systems under the joint Cancom programming package. Finally, the French Quebec independent television network is now carrying its shows to all parts of Canada via ANIK. The Canadian satellites also can be viewed from any point in the United States, although the farther south one travels, the larger the antenna required.

The Pacific and Atlantic INTELSAT satellites carry international network feeds between continents, and can be viewed on the respective coasts of the United States. The ABC nightly news programs and other special network news events make heavy use of the INTELSAT system. Often the INTELSAT satellites will feed outbound programs headed to Japan, Australia, South America, and Europe, as well as relay programs from continent to continent by way of the U.S. connection.

In addition, a number of nations lease INTELSAT transponders for exclusive use internally. Thus the Brazilian national television network can be found beaming its own network shows and news programs from Rio de Janeiro using an Atlantic INTELSAT satellite.

The Russians operate several different and incompatible communications satellite networks whose television programming can be seen in many areas of the United States. One satellite system, INTERSPUTNIK, is geosynchronous, with Russian satellites positioned over the Atlantic acting as relays between the Soviet Union, Cuba, and other nations affiliated with the U.S.S.R. The Russian domestic satellite system, MOLYNIA, is not, however, a geosynchronous satellite facility. Rather, a series of satellites have been placed in unusual polar orbits, which require earth station antennas in the United States to be pointed northward toward the north pole instead of toward the equator in order to receive them. Because the MOLYNIA satellites are continually moving, the TVRO earth station must track them by using a motorized antenna. These satellites are placed in these nonstationary orbits so that they can be seen from the far-north earth stations located throughout the seven time zones of the Soviet Union. As a result of time-zone differences, one can watch the 6:00 o'clock evening news broadcast from Moscow at 11:00 in the morning in New York. It is quite fascinating to see a map of the United States projected behind a Russian announcer, as is frequently done on their nightly news shows! Interestingly, because of their extreme polar orbits, Russian domestic satellites begin their transmissions to the Soviet Union when positioned almost directly over the United States!

Chapter 4

Home Satellite TV Reception: Direct from Space to You

The satellite television system consists of four major components: the programmer's television studio and master control facility where the television feeds originate; the uplink earth station; the space segment —a geostationary communications satellite; and the downlink system —a television receive-only (TVRO) earth station terminal. In addition, a separate satellite command station (known as TT & C stations in the industry) is maintained by the satellite common carrier owner to monitor and control the overall operation of the satellite itself.

Depending on the programmer, the studio and master control facilities can be part of and affiliated with an existing terrestrial television station (as is the case with the Trinity Broadcasting Network); the program can originate from a separate studio and control facility created specifically for this purpose (as with the Cable News Network); or the programs can be uplinked by the common carrier itself using videotape machines to play back prerecorded and stock movies and shows (as with Home Box Office, which utilizes the RCA satellite uplink control facilities in New Jersey to transmit its first-run pay television movies delivered to the RCA control center by HBO in the form of videotapes).

The uplink earth station facility consists of, typically, a 30-foot parabolic antenna and associated transmitter. This uplink is often owned by the common carrier that owns the satellite. In this case the common carrier receives the incoming television signal from the satellite pro-

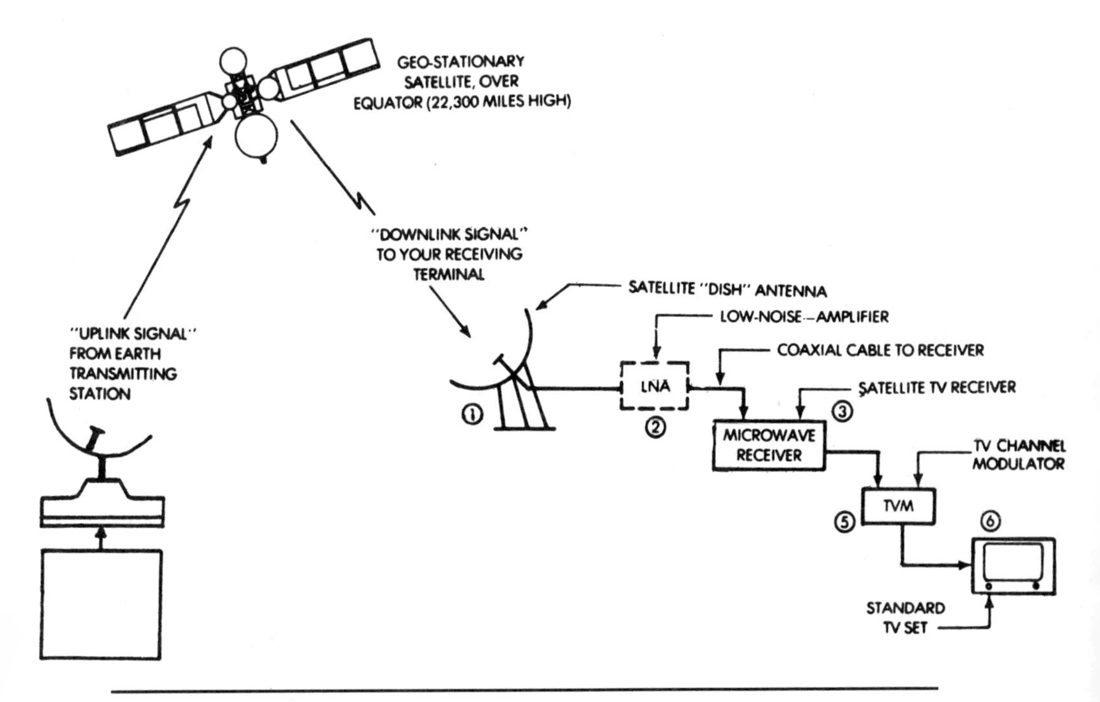

The home satellite TV system. (Courtesy of Robert Luly Associates)

grammer's studio/master control facility via AT&T's terrestrial television circuits. Optionally, the satellite programmer owns and operates its own uplink earth station, usually located in the parking lot adjacent to the programmer's master control and studio facility. As more and more satellite programmers are installing their own uplink facilities, it is not unusual to have ten or more uplink antennas throughout the United States beaming their television programs to the same satellite.

The space segment, or satellite, consists of, typically, a 12- or 24-transponder geosynchronous satellite operating in the C-band 3.7-GHz microwave-frequency spectrum. These satellites are owned by such national and international common carrier organizations as RCA, Western Union, and COMSAT (U.S. satellites), TeleSat Canada (the Canadian ANIK system), and INTELSAT (the international satellite network).

The downlink TVRO facility, located in one's backyard, or at the headend of a cable television system, or in the parking lot of a modern hotel, consists of five major components: the satellite dish antenna, a low-noise preamplifier (which strengthens the extremely weak satellite TV signal), a satellite TV receiver, a television VHF modulator (which converts the satellite receiver's video signal to an ordinary VHF televi-

Master control center for TeleFrance USA. (Courtesy of TeleFrance USA)

A typical installation. (Courtesy of Channel One Inc.)

A home TVRO earth station. (Courtesy of James Lentz)

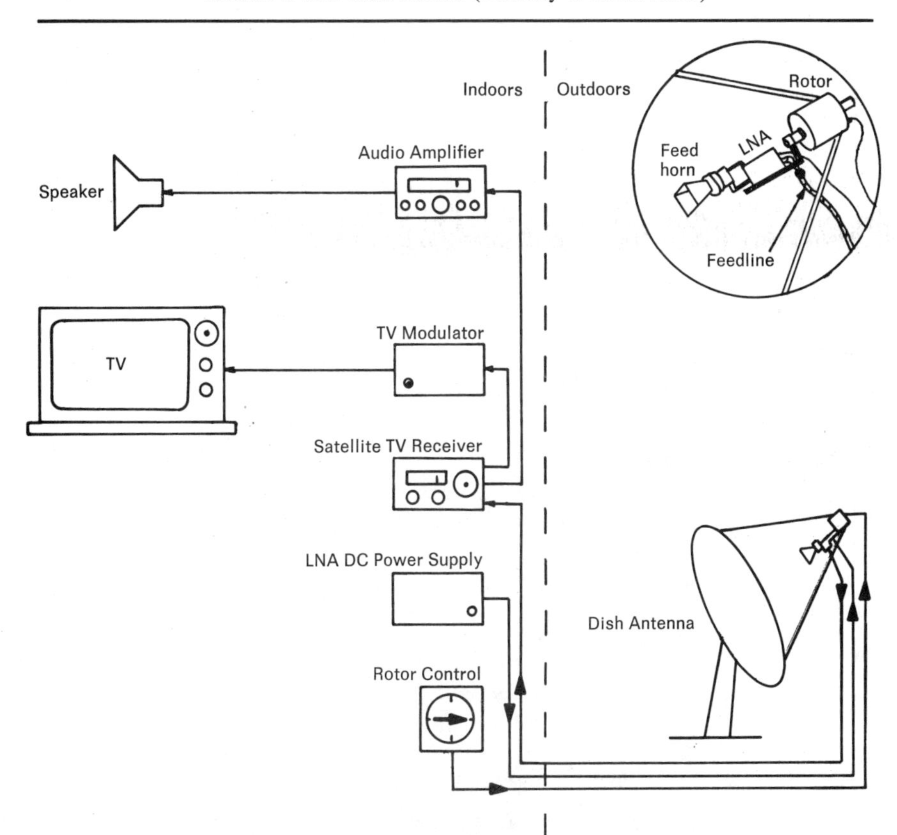

sion channel), and the television set itself. In addition, several other components may be found in the TVRO satellite terminal, depending upon its configuration.

The Home Satellite TVRO Terminal

The typical private earth station usually feeds a home television set. The audio portion of the television programming is often fed into the stereo hi-fi amplifier as well.

Many people think of the satellite terminal as the dish antenna itself but this is but one component, although certainly the largest! Satellite TVRO antennas come in two popular types: the parabolic dish and the spherical antenna, so named because of their unique shapes.

A 4.85-meter petalized parabolic antenna using an inexpensive wood frame. Note prime forces feedhorn with LNA can be rotated by small TV rotor to change from vertical to horizontal satellite transponders—top left of picture. (Courtesy of Paraframe Inc.)

Downlink's new Skyview II building-block antenna goes just about anywhere, like this rural Connecticut roof. Eight-footer, shown here, can be broken down to 4 × 4-foot modules, and reassembled in hours into a 12-, 16-, or 20-foot antenna. (Courtesy of Downlink Inc.)

Typical home TVRO parabolic antennas vary from 8 to 15 feet in diameter, and are constructed of metallicized fiberglass or lightweight spun aluminum. They weigh between 150 and 500 pounds. One of the most novel parabolic antennas is the Luly umbrella, named for its inventor, Bob Luly, a southern California home-TV satellite manufacturer. The Luly antenna unfolds like a giant 12-foot beach umbrella, and consists of a special Mylar conductive surface supported by a precisely machined rib structure. It weighs less than 40 pounds, and can be set up in a matter of minutes using an ordinary heavy-duty 35-mm camera tripod as its stand. (Because of its extreme light weight, the Luly antenna should be enclosed in a portable tent or permanent geodesic dome structure to prevent it from being blown off the bird by a gust of wind.)

The other popular design is the spherical antenna, noted for its ability to receive television signals from several satellites simultaneously. Unlike the parabolic antenna, which focuses the energy of a single satellite on a single focal point located slightly in front it, the spherical antenna focuses an arc of five or six satellites on five or six separate focal points. By locating a television detector, known as a feedhorn, at each of these independent focal points, the owner of a spherical antenna literally can pick up directly from space almost 100 television channels *simultaneously.* The spherical antenna is inexpensive to construct, and popular units often use wood strips and window screening mesh as their major

One novel antenna, the Luly, is a 12-foot parabolic which simply opens up like an umbrella! (Courtesy of Robert Luly Associates)

Details of the multisatellite advantage of the spherical antenna. (Courtesy of James Lentz)

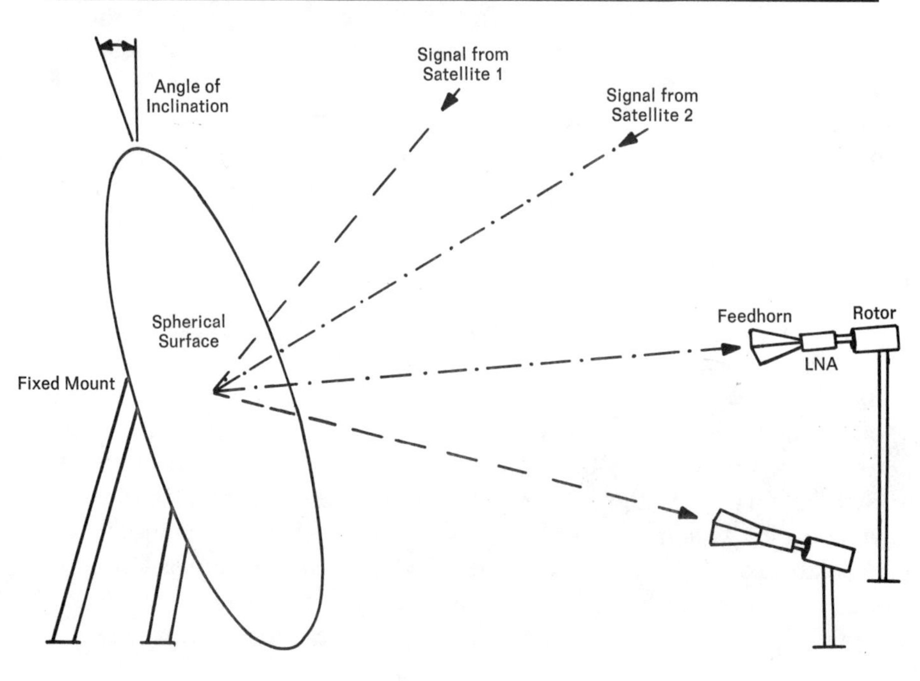

components. Spherical-antenna kits sell for under $800, and a "do-it-yourself" hobbyist can build one for less than $200 worth of parts.

In addition to its low cost, a spherical antenna also minimizes the problem of high winds and snow conditions; its reflector surface utilizes a wire-mesh screen that presents a minimal surface to the wind and that helps to prevent the accumulation of snow and ice.

TVRO antennas must be positioned properly to point at the correct location in space where the satellite is parked. To accomplish this, the antenna is rotated left to right, and tilted up and down. The angle of left-to-right rotation is known as the antenna azimuth. Due north has an azimuth of zero (or 360) degrees. East is 90 degrees, south represents 180 degrees, and west points to 270 degrees. When the antenna is aimed at the horizon, it has an elevation of zero degrees. Pointed straight up, its elevation angle is said to be 90 degrees.

To position a spherical antenna, it is necessary only to rotate its base structure and to raise or lower its rear support beams. Because a spherical antenna can receive two or more satellites simultaneously, once it has been positioned properly its azimuth and elevation angles usually need not be changed.

There are three popular types of parabolic antenna mounts: the fixed mount, the AZ/EL mount, and the polar mount. In each of these structures, adjustments allow for the correct positioning of the antenna. Although the AZ/EL mount corresponds most readily to the azimuth and elevation angles normally supplied by antenna manufacturers as an aid to positioning the antenna for any geographic location (see Appen-

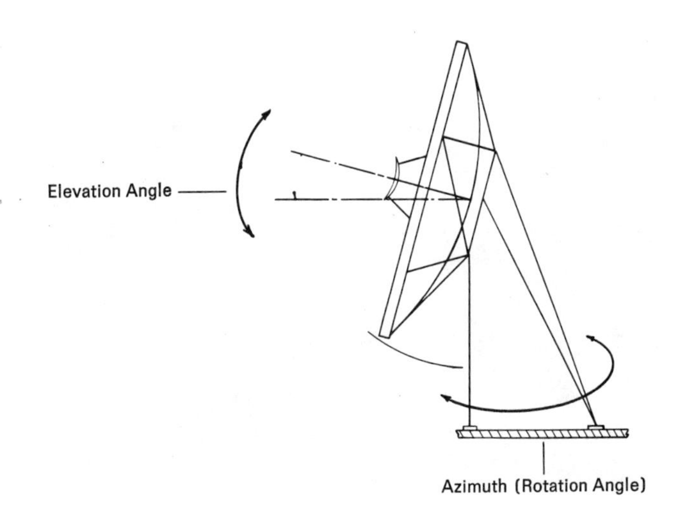

dix 6: How to Find the Birds), the polar mount is recommended. Once the polar mount has been properly positioned by adjusting its declination axis angle, the antenna will point approximately at any satellite in the geosynchronous orbit arc by simply rotating its hour axis from left to right. Although minor adjustment of the declination axis will still be necessary (Appendix 6 explains why), the ability to swing the antenna easily through an arc to pick up the various satellites is a tremendous feature indeed! Some home TVRO manufacturers make motorized antenna mounting systems that permit the user to control the antenna position remotely by dialing in the correct azimuth and elevation angles for the desired satellite on a small living-room antenna console. Even fancier systems allow the user simply to type the desired satellite's name

Example of fixed-mount parabolic antenna (with prime-focus feed). (Courtesy of Cayson Electronics)

Scientific-Atlanta 4.6-meter commercial antenna showing polar mount with cassegrain feed. Notice handle at base of antenna that changes antenna tilt angle. (Courtesy of Scientific-Atlanta Inc.)

AZ/EL mount for Paraframe antenna. Bolts can be loosened at base and on central wood brace to allow for left-right rotation and up-down tilt. (Courtesy of Paraframe Inc.)

Some antenna mounts have motors that will allow the antenna's azimuth and elevation angles to be changed by remote control. Here is a picture of one such control center, a combination satellite receiver and remote-controlled motorized antenna mount. Closeup of deluxe-mount receiver/antenna steering-control system and Advent Videobeam projection television system console. (Courtesy of Microwave General)

Two types of satellite antenna feeds. (Courtesy of Robert Luly Associates)

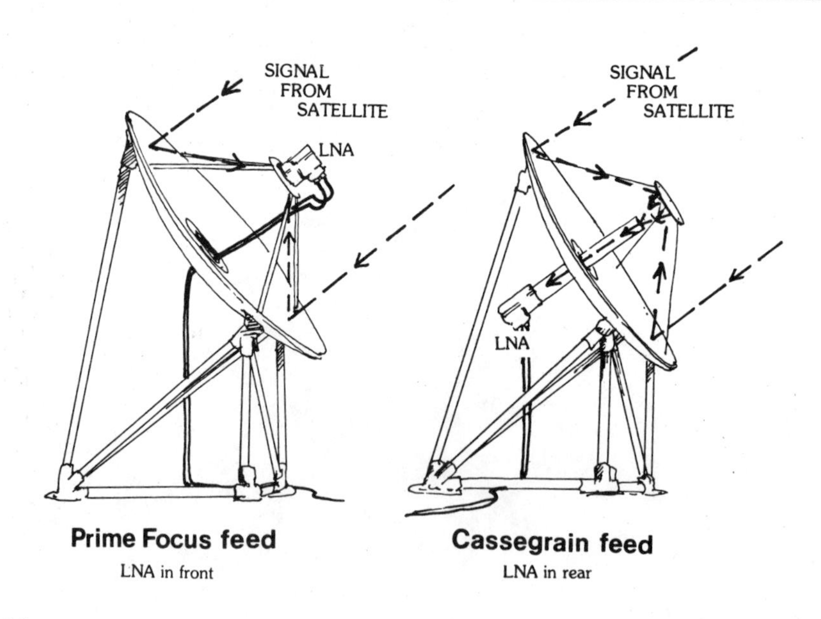

into an Apple or Radio Shack home computer, which then computes the proper azimuth and elevation angles, and directs the antenna to aim itself at that point in the heavens.

At the focal point of the antenna, a feedhorn is positioned to pick up the weak signals from space, which the antenna has amplified, reflected, and focused. The feedhorn is a metallic signal reflector and transmission waveguide that carries the microwave signals into the low-noise amplifier for initial preamplification. There are two popular types of feeds: the prime focus feed, in which the feedhorn is mounted directly on the low-noise amplifier (LNA) at the focal point in front of the antenna; and the Cassegrain feed, in which the feedhorn routes the microwave signal through a metal funnel to the LNA located on the rear of the antenna. Both feeds are equally popular in TVRO antenna design today, although the prime-focus feed has obtained somewhat greater popularity in the home TVRO terminal. One of the most popular feeds is the circular feedhorn, which provides a slightly better ability to collect all the stray microwave signals successfully. In marginal TVRO systems the use of a circular feedhorn will often make the difference between obtaining a poor television pictures that is full of snow and sparklies and one that is nearly perfect.

The LNA is connected to the end of the feedhorn. This device consists of an extremely sensitive high-frequency transistor circuit, which can amplify the extraordinarily weak signals from space without swamping them with an excessive amount of noise produced by the amplifier itself. Because the signal from a television transponder is so very weak (a typical satellite transmits with an output power of only 5 watts), and is coming from so far away (the signal is less than one millionth as strong as an ordinary television broadcasting station's signal!), substantial preamplification is necessary. At these very low signal levels, even the natural noise produced by the earth radiating its heat energy into space can interfere with the distant satellite's signal. For this reason the LNAs are typically rated in terms of degrees Kelvin of noise figure, that is, the absolute amount of noise the LNA itself generates compared with that of the device operating at zero degrees Kelvin (−273°C). The amplified output signal from the LNA is fed through a special large-diameter low-loss coaxial cable known as Heliax into a down-convertor, a device that converts the microwave-frequency television signal into an intermediate frequency (or IF) signal, which can be easily carried over less expensive conventional coaxial television cable.

Until very recently the down-convertor was located inside the satellite receiver, thus requiring the use of expensive, thick, and unsightly Heliax cable to bring the microwave-frequency television signals from

Example of multisatellite low-cost spherical antenna. To change satellites, one must simply reposition the feedhorn located in the foreground. Alternatively, several feedhorn and LNA combinations can be installed, allowing the user to switch instantly from satellite to satellite. Block Island, RI, resident William Bendokas (left), owner, New England Home Video store, gets a briefing on satellite locations from Downlink VP Mark Kulaga before "cherry picking" the satellite TV channels. (Courtesy of Downlink Inc.)

Picture of the Chaparral Superfeed high-efficiency feedhorn bolted to 120-degree LNA. Use of the Superfeed can increase picture quality significantly in marginal installations. (Courtesy of Chaparral Communications Co.)

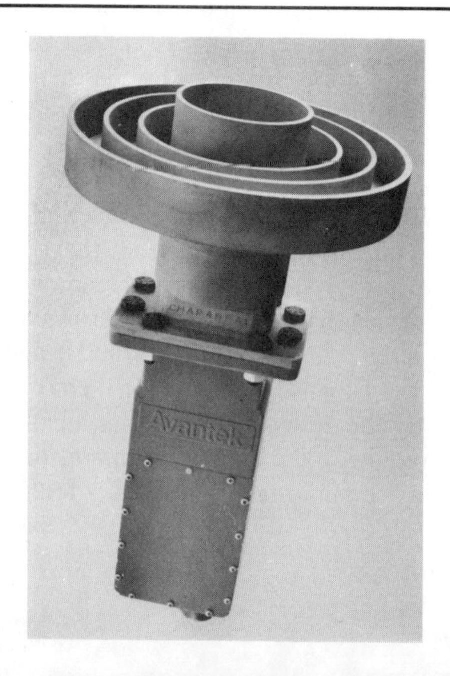

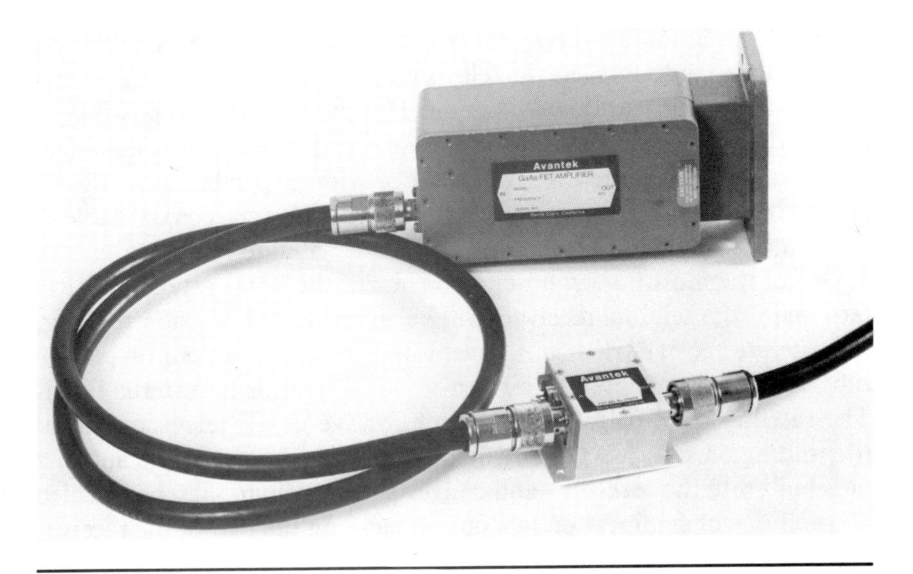

LNA (low-noise amplifier) mounted on feedhorn via flange at right. Signal is amplified and passes out of LNA via heliax coaxial cable (at left) to down-convertor and satellite receiver. (Courtesy of Avantek)

Gillaspie combined their down-convertor with other image-fitting circuitry, calling it the Image Reject Mixer. Home-style satellite receiver and down-convertor and Image Reject Mixer located in separate antenna-mounted box. (Courtesy of Gillaspie & Associates, Inc.)

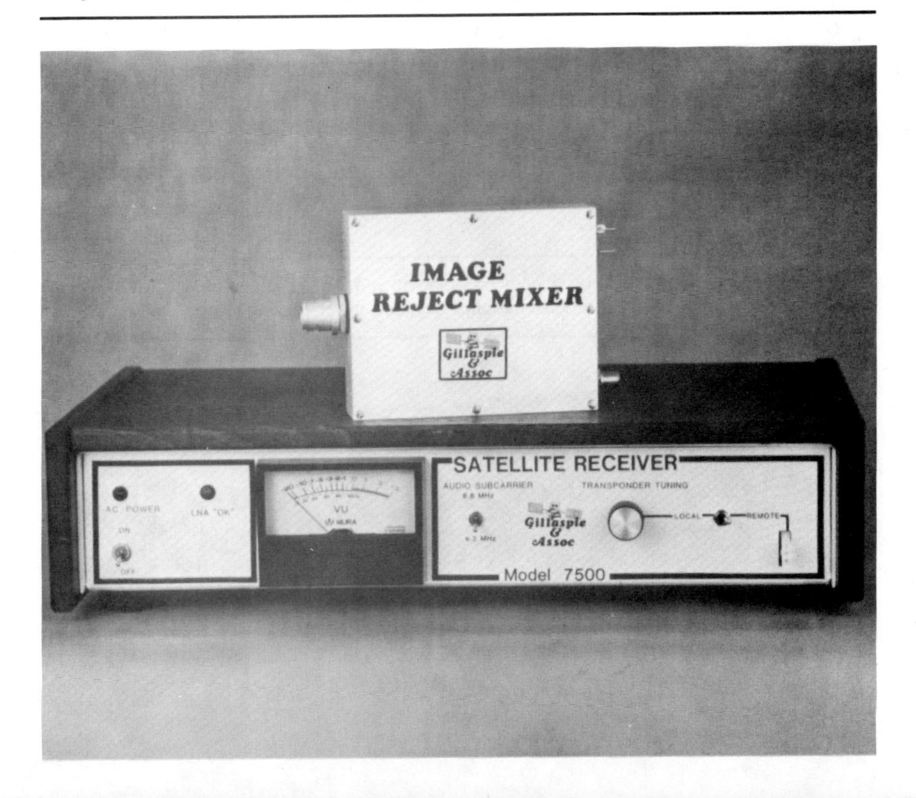

the LNA mounted on the antenna outside to the TVRO satellite receiver located indoors near the television set. Indeed, most cable television systems still use this technique. However, it is now possible to mount the down-convertor at the antenna, and several LNA manufacturers (notably Avantek and Dexcel) provide a combination LNA/down-convertor that allows an inexpensive CATV-type coaxial cable to be used between the antenna and the home satellite receiver.

One of the most important components of the home satellite TVRO terminal is the satellite receiver. Unlike an ordinary television receiver, the satellite receiver does not have a built-in audio system or picture tube. Instead, it is more like a short-wave receiver or hi-fi stereo tuner. The satellite receiver takes in the microwave or IF television signal (depending on whether the down-convertor is mounted at the antenna or is built into the receiver), and converts it to ordinary baseband video and audio, which appear on two output jacks on the rear of the receiver cabinet. The baseband video signal cannot be fed directly into an ordinary home television set; it is similar to the signal that comes from a television camera. To display the video on a TV set, the signal must be fed into an RF modulator, which converts it into an ordinary VHF television channel that can be tuned in by the set. The output of the modulator appears on a wire that is simply connected to the VHF antenna terminals of the television set, which is typically tuned to Channel 3 or 4, depending on the modulator used. Alternatively, the output of the satellite TVRO receiver can be fed into a professional color-television monitor or projection television system directly, or into the video input of a video-cassette recorder. The audio output of the satellite television receiver can be fed into the VHF modulator, which will allow the sound to be heard over the television set's speaker, or into a hi-fi stereo amplifier for outstanding high-fidelity bass and treble response.

The original home satellite TV receiver, which was introduced en masse in 1979. One of the best-selling receivers. (Courtesy of International Crystal Manufacturing Co.)

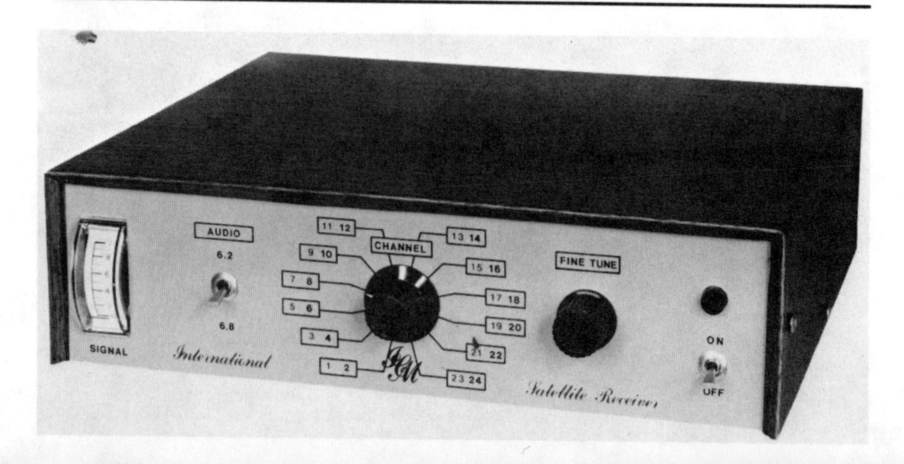

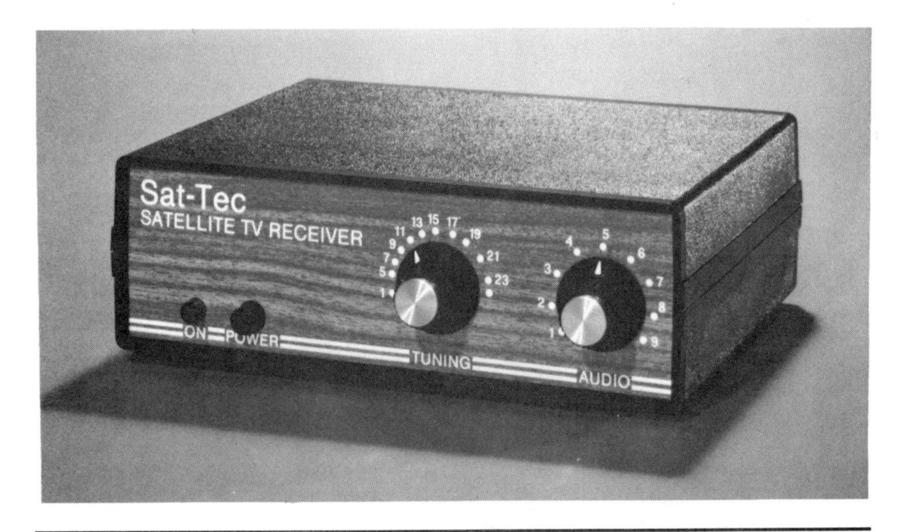

A very popular home satellite TV receiver features low cost. (Courtesy of SAT-TEC)

Many manufacturers currently produce home satellite television receivers of various shapes and sizes and with various features. Perhaps the granddaddy of these is the International Crystal Manufacturer's model, first produced in 1979. The ICM unit tunes the 24 different transponders (12 vertical, 12 horizontal) and allows the user to select between the audio subcarrier scheme used on the RCA-type satellites and the audio scheme on the Western Union satellites. The receiver also works with 12-transponder satellites.

Popular Coleman-designed home satellite receiver with remote control. (Courtesy of Earth Terminals, Inc.)

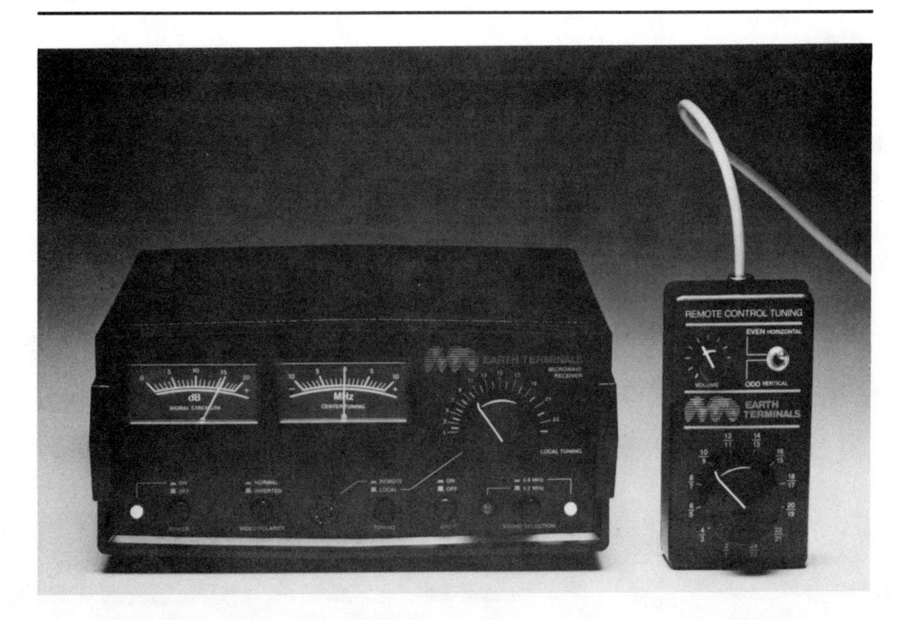

Downlink's new receiver is quite small, and boasts a polished wood cabinet. (Courtesy of Downlink Inc.)

Once your home satellite system is installed, the picture from space can be tuned in from your easy chair. (Courtesy of Downlink Inc.)

The 24-transponder satellites utilize a frequency-reuse technique known as polarization. In this arrangement 12 channels are transmitted to earth using microwave signals whose waves are polarized horizontally with respect to the position of the satellite. The other 12 channels are transmitted using microwave signals whose waves are polarized vertically with respect to the position of the satellite. Therefore, these waves are orthogonal, or at 90-degree angles to each other, and will not interfere with each other even though the channels may be overlapping in frequency. To pick up all 24 channels, either the feedhorn/LNA combination must be rotated by 90 degrees (typically by using a small TV antenna rotor), or two separate LNAs and their corresponding coaxial cable feeds must be mounted on a special feedhorn that splits the received signal into its corresponding vertical and horizontal components. In either case all home satellite TV receivers are capable of tuning in all 24 satellite transponders if the particular transponder's signal is fed into the satellite receiver's RF input jack on its rear.

Many satellite TV receivers offer a remote-control feature that allows the viewer to change channels from an easy chair, without having to walk over to the receiver itself.

The very best example of the new wave of home satellite receivers, Dexcel's DXR 1000 is only three inches high! A combination LNA/down-convertor, also made by Dexcel, converts satellite TV signals to intermediate 70 MHz frequency, allowing an inexpensive coax cable to be fed from the antenna to this receiver. Associated line of Dexcel hi-fi components fits in with DX 1000 for a perfect match. (Courtesy of Dexcel Inc.)

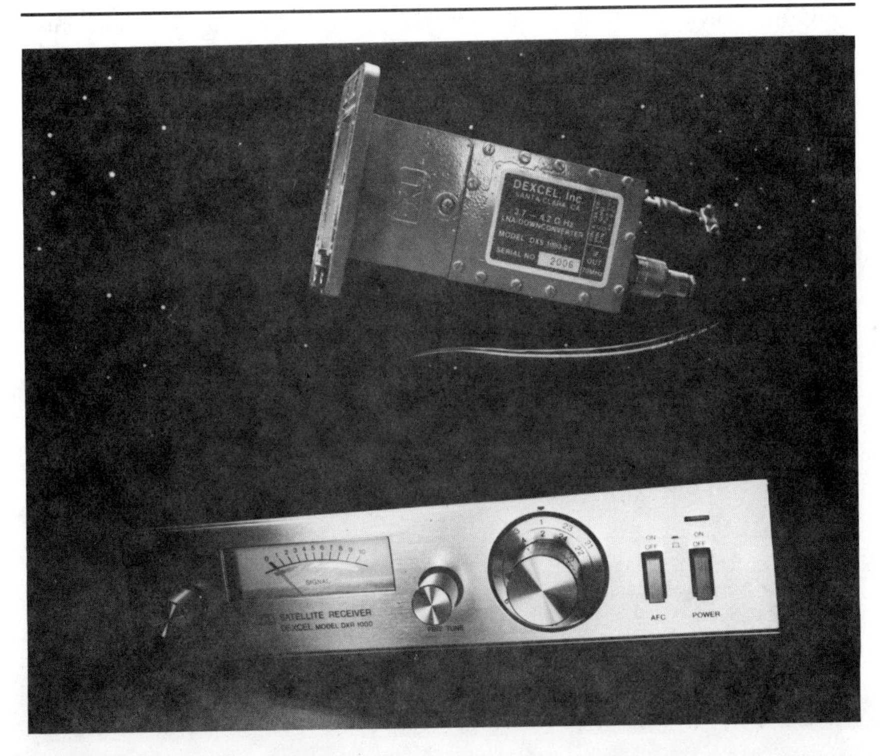

Home System Television Configurations: How to Plug 'Em Together

The home satellite TVRO terminal can be thought of simply as another television source, similar to the video-cassette recorder, video-disk player, cable television feed, or STV pay television service. The outputs of the satellite TVRO receiver are ordinary video and audio feeds. In the simplest configuration these outputs are connected through cables to a small VHF television channel modulator whose FR signal is then delivered to the television set's antenna terminals.

With the addition of a home video-cassette recorder, the requirement for a separate VHF channel modulator is eliminated, as the VCR has a built-in television channel modulator. In this configuration the satellite TVRO audio and video outputs need only be connected to the audio and video input jacks on the VCR. When the VCR is placed in the record mode, the satellite TV picture will be displayed (with some VCRs it is unnecessary to activate the record mode; and so the instruction manual should be checked for clarification).

More than 20 percent of today's homes are equipped with cable television service. Although the ownership of a home satellite TVRO terminal will eliminate the need for CATV service to pick up the distant signals, in many instances the user will wish to retain the service to bring in local television signals ghost-free. In this case the satellite TVRO terminal easily can be used in conjunction with the CATV system, either by tuning the outboard VHF channel RF modulator associated with the satellite feed to an unused channel different from the one used by the CATV decoder/convertor unit, or by connecting the output of the CATV decoder/convertor to the VHF television RF input on the video-cassette recorder, when one is present. The VCR then becomes the master video control board for the entire television system.

For the video aficionado who has created a multiple-television-source system, it will probably be easiest to switch the various video devices at their VHF channel RF signal feed, using an inexpensive RF switcher box. With this configuration is is possible to handle two or more video-cassette recorders, allowing the various video signals to be routed either to the television set or to one or both recorders simultaneously.

A. Home system TV configuration: simple straight-through connections.
B. Using a video-cassette recorder with the satellite system.
C. TVRO with VCR and VHF/UHF television and CATV television.
D. TVRO with two VCRs and other video sources.
E. TVRO with TV monitor and/or projection TV system.
(Courtesy of James Lentz)

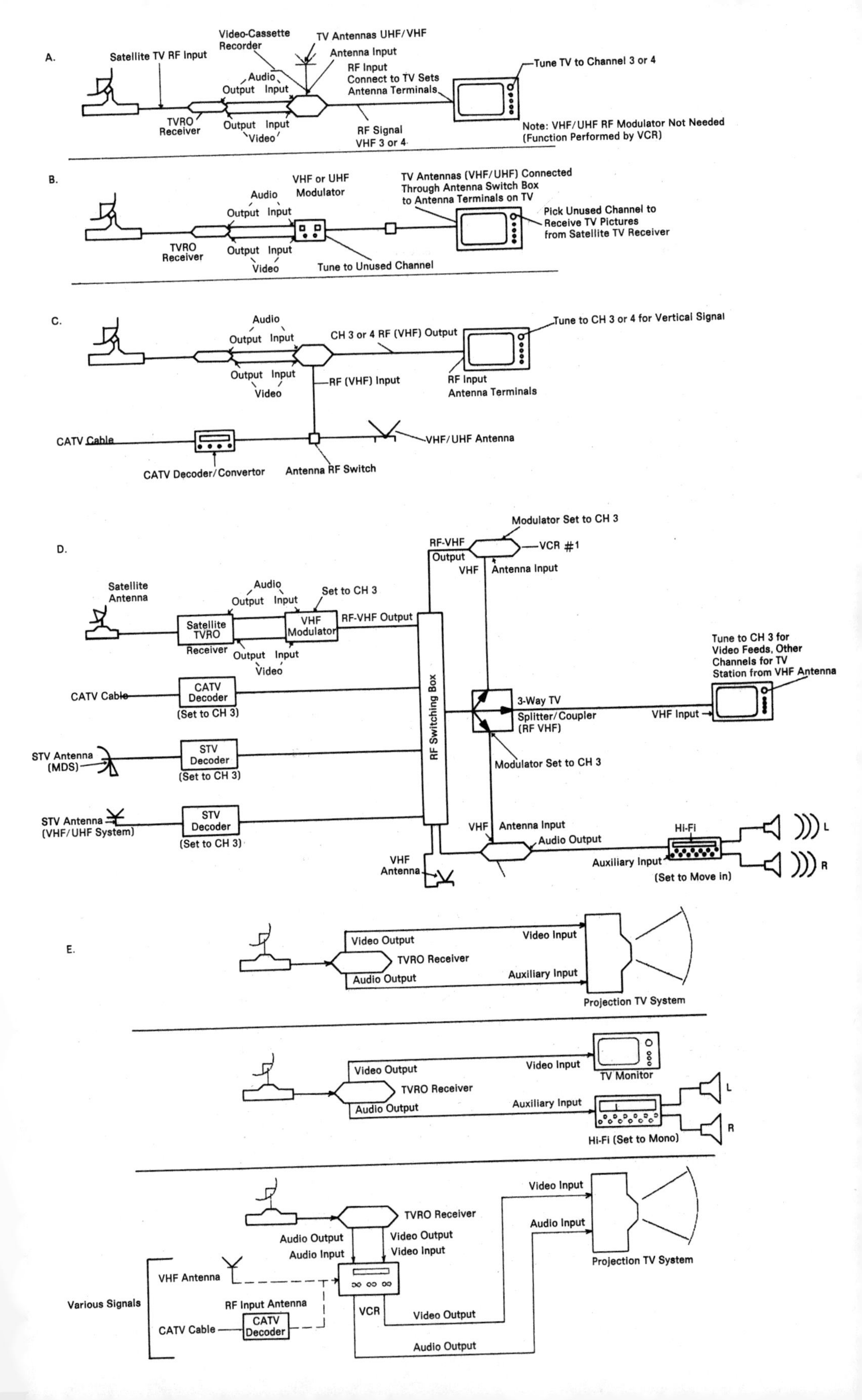
A.
Satellite TV RF Input
Video-Cassette Recorder
TV Antennas UHF/VHF
Antenna Input
Audio
Output Input
RF Input Connect to TV Sets Antenna Terminals
Tune TV to Channel 3 or 4
TVRO Receiver
Output Input
Video
RF Signal VHF 3 or 4
Note: VHF/UHF RF Modulator Not Needed (Function Performed by VCR)
B.
VHF or UHF Modulator
Audio
Output Input
TV Antennas (VHF/UHF) Connected Through Antenna Switch Box to Antenna Terminals on TV
Pick Unused Channel to Receive TV Pictures from Satellite TV Receiver
TVRO Receiver
Output Input
Video
Tune to Unused Channel
C.
Audio
Output Input
CH 3 or 4 RF (VHF) Output
Tune to CH 3 or 4 for Vertical Signal
Output Input
Video
RF (VHF) Input
RF Input Antenna Terminals
CATV Cable
VHF/UHF Antenna
CATV Decoder/Convertor
Antenna RF Switch
D.
Modulator Set to CH 3
RF-VHF Output
VCR #1
VHF
Antenna Input
Satellite Antenna
Audio
Output Input
Set to CH 3
Satellite TVRO Receiver
VHF Modulator
RF-VHF Output
Output Input
Video
Tune to CH 3 for Video Feeds, Other Channels for TV Station from VHF Antenna
CATV Cable
CATV Decoder
(Set to CH 3)
RF Switching Box
3-Way TV Splitter/Coupler (RF VHF)
VHF Input
STV Antenna (MDS)
STV Decoder
(Set to CH 3)
Modulator Set to CH 3
STV Antenna (VHF/UHF System)
STV Decoder
(Set to CH 3)
VHF
Antenna Input
Hi-Fi
Audio Output
L
R
Auxiliary Input
(Set to Move in)
VHF Antenna
E.
Video Output
Video Input
TVRO Receiver
Audio Output
Auxiliary Input
Projection TV System
Video Output
Video Input
TV Monitor
TVRO Receiver
Audio Output
Auxiliary Input
L
R
Hi-Fi (Set to Mono)
TVRO Receiver
Video Input
Audio Input
Audio Output
Audio Input
Video Output
Video Input
Projection TV System
VHF Antenna
Various Signals
RF Input Antenna
VCR
CATV Cable
CATV Decoder
Video Output
Audio Output

The home satellite TVRO terminal works exceptionally well with projection television systems and studio-quality television monitors, as both of these readily accept baseband video and audio signal inputs directly. This eliminates the need for an external VHF channel RF modulator, which can degrade the overall picture quality slightly (the same is true of a video-cassette recorder). When connected to a projection television system, satellite television pictures have a network-broadcast quality, comparable only to the pictures on the studio's color monitors themselves. The color fidelity and hi-fi sound obtainable through such an arrangement must be experienced to appreciate the outstanding capabilities of satellite television reception. Received by line of sight directly from the satellite television space relay, satellite TV pictures are totally free of ghosts and other interference normally associated with terrestrial television reception. The best and simplest configuration of all consists of a home satellite TVRO terminal in conjunction with a video-cassette recorder, projection television system, and hi-fi stereo, and videotapes recorded from satellite usually produce higher quality pictures than those obtained from commercial video-cassette recordings.

An "everything box" ties all the video equipment together. (Courtesy of AMCO Electronics)

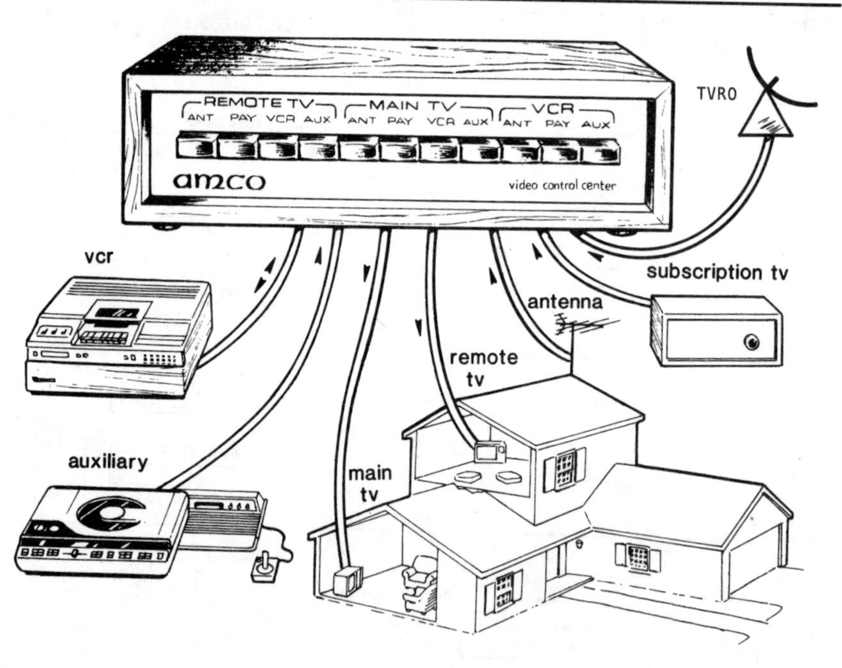

Fully-integrated satellite receiver and remote-control antenna arming system. (Courtesy of Microwave General.)

Chapter 5

Starting Your Own Mini-CATV System

Twenty years ago the cable television industry was still mostly a "mom and pop" affair. Mom would run the front office and take care of books and pop would go out on the pole to string the cables from house to house. Cable television systems existed to pick up weak television broadcast signals from distant cities by using a common antenna system mounted on a tall tower or a nearby mountaintop. Thousands of these mini-CATV firms still exist today, with hundreds of new systems going into business annually. Many are members of the Community Antenna Television Association, an Oklahoma-based organization of mini-CATV companies. Although the corporate giants of the communications industry have bought into the CATV business over the last five years, there still remains tremendous opportunity for the small entrepreneur to provide a service and make an excellent income. And the creation of the new television satellites makes it that much easier to enter the mini-CATV business.

Mini-CATV systems can be found throughout the United States. Many consist of only a few dozen homes tied together by a common coaxial TV cable. Other mini-CATV companies have wired up entire housing developments, and are often owned by the nonprofit housing associations themselves. Still other mini-CATV systems confine their activities to a single apartment building or condominium complex. These cable TV arrangements are usually known as MATV systems, for master antenna television distribution.

There is often a blurring of distinction between the mini-CATV and MATV system concepts, but generally a mini-CATV system is owned by an organization that is independent of the apartment complex or housing development management. In addition, mini-CATV systems charge a fee for their services whereas most apartment MATV systems are provided by the management as part of the monthly rental. And mini-CATV systems' cables often cross city or state property lines, thus requiring a license from the Public Utilities Commission to operate.

Although residential communities historically have provided the base for cable television operations, high-rise downtown office buildings are now being wired up by the business-oriented mini-CATV operators as well. These entrepreneurs see the specialized satellite television business services such as Reuters Monitor system, Dow Jones, UPI and AP's newswire feeds, and the Cable News Network as valuable business communications tools. Satellite video teleconferencing can also be brought to the office building via a mini-CATV system. Most of the new CATV distribution system hardware (such as line amplifiers or couplers) allow for two-way communications, both from the headend to the subscriber's premises and back, thus enabling a firm's computers in the basement of an office building to talk with data terminals throughout the building, employing the same coaxial cable that carries the television channels. Several major cable companies, including Viacom International and Manhattan Cable (a subsidiary of Time-Life), have entered into agreements with one of the data communications common carriers—Tymnet—to allow computers in New York and San Francisco to talk with each other at high speeds, by utilizing the local city distribution capabilities of the respective cable television companies. This trend of carrying data and television pictures over the same cable will reach a peak in the 1980s. The mini-CATV and MATV entrepreneur can capitalize on the situation by plugging into the communications satellites via his or her own TVRO earth terminal.

Technical Considerations for the Mini-CATV System

The commercial mini-CATV system is similar to the home satellite TV terminal. A receiving antenna dish is still required, along with a down-convertor, LNA, and receiver hardware. In addition, however, a downstream distribution system must be constructed to deliver the signal to each of the subscribers. In the case of an existing apartment building MATV system, the addition of the new satellite TVRO terminal may simply consist of feeding the satellite receiver's (or receivers') television output into the existing cable TV network. In most cases,

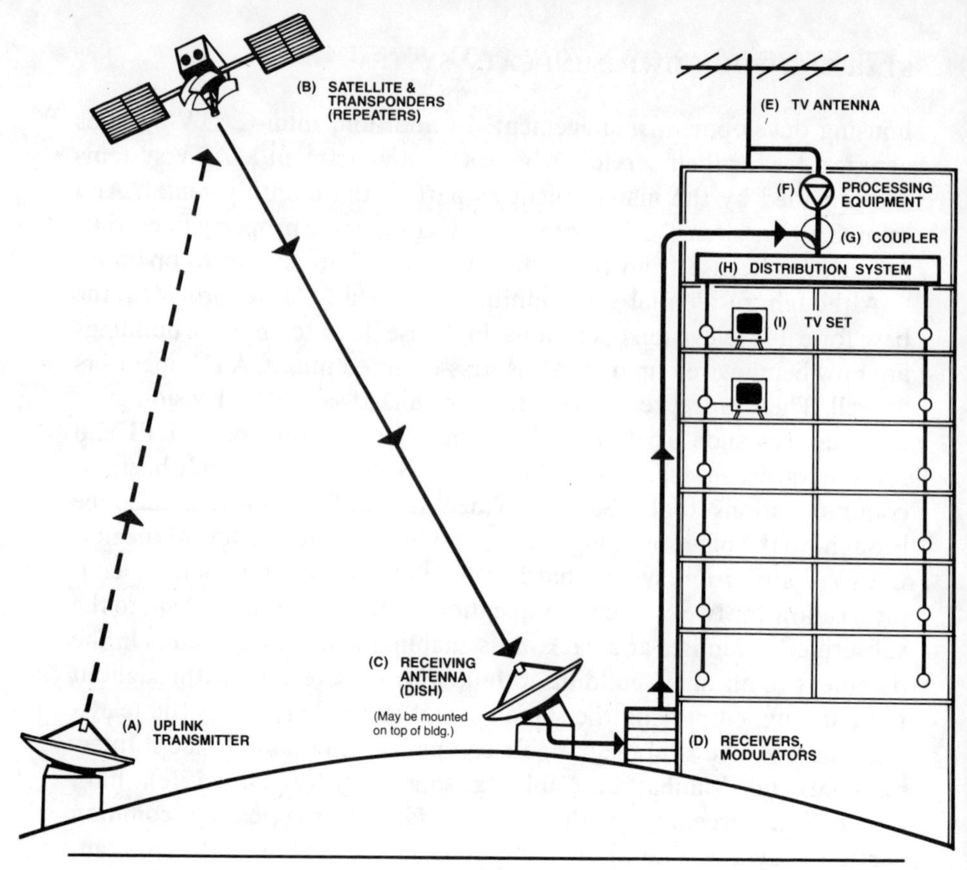

A mini-CATV system for an apartment building. (Courtesy of Anixter Mark, Inc.)

however, further hardware will have to be provided, and the satellite terminal should be installed according to commercial standards and specifications.

The first consideration is to ensure that the proper antenna size is used. (Appendix 7 discusses this in more detail.) In general, where a 10-foot satellite TVRO antenna would be quite acceptable for home use, a 4.6-meter or 15-foot antenna is the one most frequently chosen by cable companies. The larger size antenna ensures that sufficient signal will be available to feed a multiple set of satellite receivers with topnotch pictures. Although major cable companies have used parabolic antennas for their installations almost exclusively, the spherical antenna has the advantage of being able to receive several satellites simultaneously, thereby allowing the mini-CATV operator to broaden the selection of programming available to subscribers. Moreover, the spherical antenna, once installed, is relatively maintenance-free, and provides less resistance to the wind, a factor to be considered when placing an antenna on a roof.

TYPICAL RECEIVE TERMINAL CONFIGURATIONS

BASIC DUAL POLARIZED/TWO RECEIVER SYSTEM

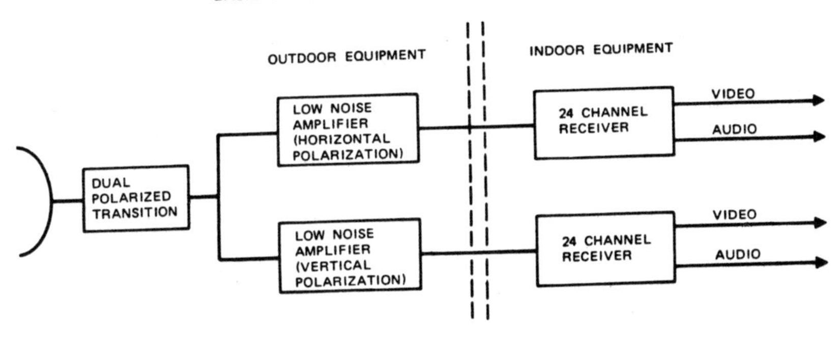

A typical mini-CATV system will require dual LNAs to allow simultaneous reception of both vertical and horizontal transponders. (Courtesy of Robert Luly Associates)

Antenna location is a prime consideration. The antenna must be positioned in such a way as to minimize potential interference from existing or future terrestrial microwave systems (such as the telephone company's). One way of minimizing this problem is to obtain a computerized frequency search of possible interference from one of the engineering firms specializing in this area, such as Compucon of Dallas, Texas, or Comsearch, Inc., of Falls Church, Virginia (see Appendix 2 for addresses).

Using power dividers or splitters, multiple receivers can be plugged in to pick up many channels simultaneously. (Courtesy of Robert Luly Associates)

DUAL POLARIZED/MULTIPLE RECEIVER SYSTEM

DUAL POLARIZED TRANSITION
LOW NOISE AMPLIFIER (HORIZONTAL POLARIZATION)
LOW NOISE AMPLIFIER (VERTICAL POLARIZATION)
POWER DIVIDER
POWER DIVIDER
1 24 CHANNEL RECEIVER
2 24 CHANNEL RECEIVER
N ADDITIONAL RECEIVERS
1 24 CHANNEL RECEIVER
2 24 CHANNEL RECEIVER
N ADDITIONAL RECEIVERS
VIDEO
AUDIO

A mini-CATV TVRO 10-foot antenna installed in a housing development. (Courtesy of Rigel Systems)

Fred Hopengarten, president of Channel One, standing in a 5-meter satellite TV dish antenna, a popular size for commercial mini-CATV use. (Courtesy Channel One Inc.)

A mini-CATV operator does not require an FCC license to construct or use a receive-only satellite antenna. However, by obtaining such a license the company is guaranteed that no future interference will occur, thus preventing the telephone company, for example, from constructing a microwave tower several blocks away that might unintentionally beam its signals directly past the TVRO terminal.

Whereas the home satellite terminal will usually have only a single LNA and feedhorn assembly rotated by a small motor to pick up either the vertical or horizontal satellite transponder channels, the commercial system will use a feedhorn that provides for two LNAs, allowing for simultaneous reception of both the 12 vertical and 12 horizontal transponder channels.

Separate coaxial cables are required to bring the two LNA feeds from the satellite antenna to the down-convertors, which are usually built into and are an integral part of the commercial satellite receiver. Thus, a short run of the more expensive and bulky Heliax cable is required for each LNA.

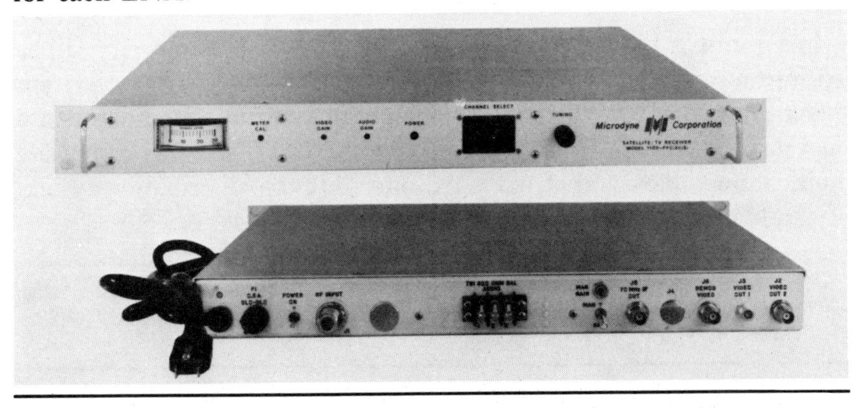

Typical miniaturized down-convertor/receiver combination used in mini-CATV systems, mounted in multiple-channel racks. Receiver sells for about $2300. (Courtesy of Microdyne Corp.)

As most mini-CATV systems will want to pick up more than one vertical and horizontal transponder channel at the same time, a bank of multiple receivers will be required, each tuned to a separate transponder channel. In this case, two-, four-, or eight-way power splitters will be connected to the incoming Heliax cable. The multiple outputs of the power splitters can then feed each receiver separately.

This headend equipment can be located in a small metal shed or enclosure adjacent to the TVRO antenna. The shed serves to protect the equipment from unauthorized access as well as from rain, snowstorms, and lightning.

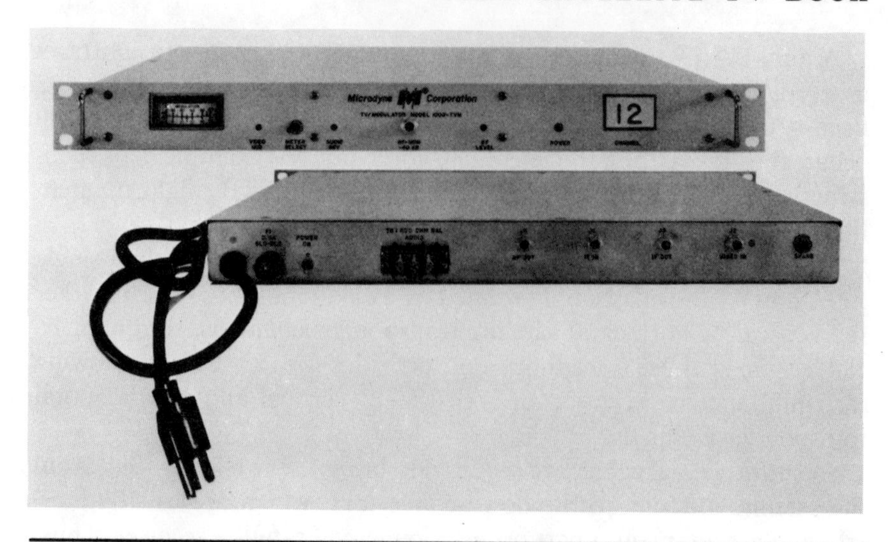

Stand-alone VHF-TV channel modulator takes video baseband signal input and converts to UHF channel for feeding onto CATV cable. (Courtesy of Microdyne Corp.)

In a mini-CATV system each of the multiple receiver outputs will be fed into a separate VHF channel modulator to convert the television signal to a regular VHF channel. The outputs of these modulators are then mixed together, using a signal combiner unit, and this composite multichannel television signal is fed into the cable distribution system of amplifiers and television set couplers.

Combination satellite receiver with built-in down-converter and TV channel modulator to transmit TV signal down the CATV cable to home subscriber's TV set for reception as a conventional VHF channel. (Courtesy of Microdyne Corp.)

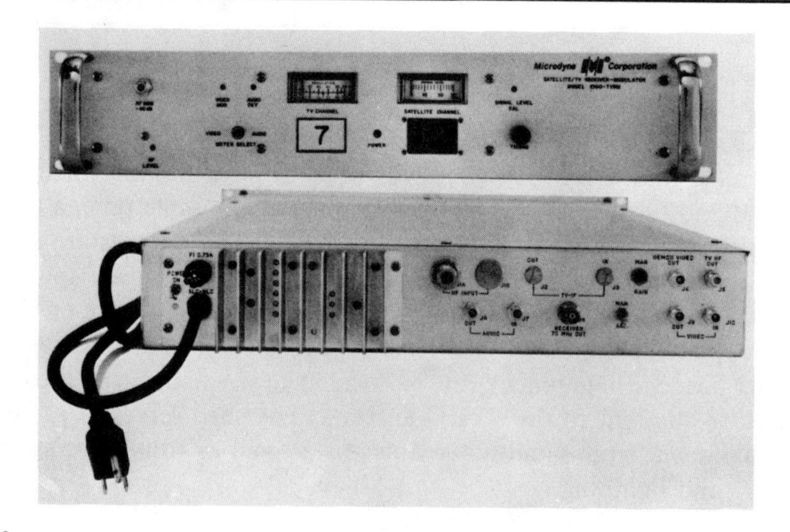

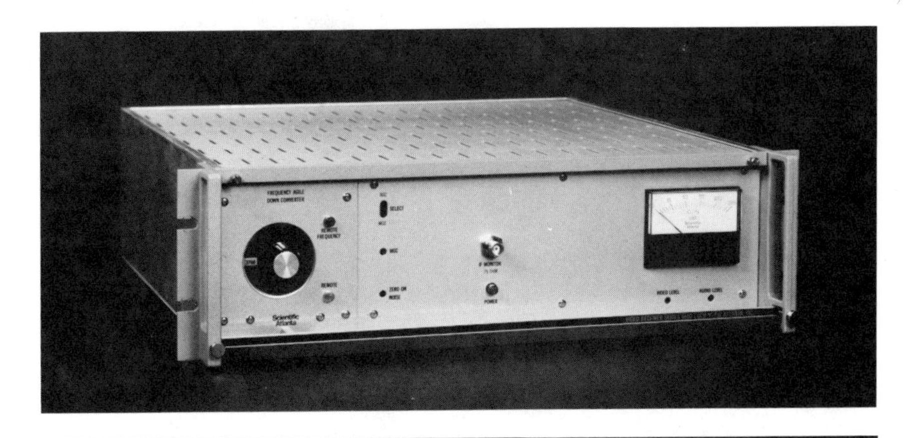

A commercial-quality combination down-converter and frequency-agile receiver with remote-control ability can be placed near antenna with remotely controlled tuner located at the TV set. (Courtesy of Scientific-Atlanta Inc.)

Many mini-CATV and MATV systems will also want to pick up ordinary television broadcasting signals using conventional UHF and VHF receive antennas. In this case a separate bank of VHF and UHF tuners or broadcast television signal-processing equipment (as it is known in the industry) is installed next to the satellite TVRO receivers. Their signals are then coupled into the cable distribution system via the signal combiner.

Many mini-CATV systems offer several tiers of programming. A basic monthly service may consist of the local television stations and one or more satellite feeds such as the Christian Broadcasting Network,

Deluxe multireceiver bank with push-button transponder/receiver channel selection can be remotely controlled by telephone to allow a mini-CATV operator to run several geographically scattered CATV systems from a central location, permitting special events and sports activities fed over different transponders to be switched "on line" instantly. (Courtesy of Avantek)

The CATA labs in Oklahoma. Not atypical of the "antenna farm" of a medium-sized CATV system. Note the satellite dish and the stack of VHF and UHF receiving antennas pulling in distant-city broadcast stations. (Courtesy of John Kinick)

superstation WTBS, and the Cable News Network. The second tier of programming may include one or more pay television first-run movie services, such as Home Box Office. Here it will be necessary to provide some simple method of electronically preventing subscribers who are paying for only the basic service from receiving the premium pay television channel(s) as well. There are several ways in which this can be accomplished.

Perhaps the simplest method consists of installing a small trap in series with the customer's television antenna lead. The trap prevents the pay TV signal from passing through to the television set. If the subscriber elects to obtain the additional service, the trap is simply removed. This process is simple and low in cost, but the trap can be easily removed by the subscriber.

Another technique consists of sending the special channels (including, if desired, the basic satellite service) down the cable using the

Inexpensive block convertor allows the mini-CATV system to carry premium programs on the hidden sub-, mid-, and superband cable channels not normally found on a TV set. Convertor sits atop subscriber's television set, changes channels to standard VHF television channels. (Courtesy of Oak Communications Inc.)

midband and superband channels not normally found on an ordinary television tuner's dial. (See Appendix 2 for sources of other special CATV equipment.) This technique requires installation of a special block convertor connected between the incoming cable and the subscriber's television set antenna terminals. When the switch on the block convertor is thrown from standard to premium television, all of the hidden channels are converted by the box into standard television channels that can be tuned in by the television set's VHF selector. Block convertors are more expensive than simple traps (a trap costs less than a dollar, a block convertor about $17). They do provide a significantly greater level of security, however.

The block conversion technique has one serious drawback. Many of the video-cassette recorders presently on the market can tune in to these hidden channels, although their instruction manuals usually do not reveal this. The midband channels are located between VHF Channels 6 and 7, and superband channels start just above Channel 13 on the VCR's channel tuning dials. If the mini-CATV entrepreneur wires up an apartment building of video enthusiasts and uses this technique to protect the pay TV channels, there probably will be few premium-channel paying subscribers left once the word gets out!

An economically priced 35-channel varactor-tuned channel-by-channel convertor, the new Econo-Line Thirty-Five, picks up all VHF, sub-, mid-, and superband cable TV channels. (Courtesy of Oak Communications Inc.)

The best way to protect the television signals electronically is through the use of a full-channel CATV decoder box that has a built-in signal-descrambling system. Of course, such a technique requires the use of corresponding channel scramblers at the system headend. The newer systems allow a special electronic serial number to be assigned to each decoder box. These boxes can be individually turned on and off remotely via a small headend control panel. Thus, once the box has been installed, future service calls to the customer's premises to upgrade or change the box from one pay service to another are eliminated. Also, if the box is stolen, its serial number can be removed from the system electronically, thereby preventing it from receiving any television pictures when it is again plugged into the mini-CATV system.

The most effective way that a mini-CATV system can solve the problem of pay TV "pirates," however, is simply to offer the subscriber a single monthly service charge that provides for all television channels, including the pay TV movie services.

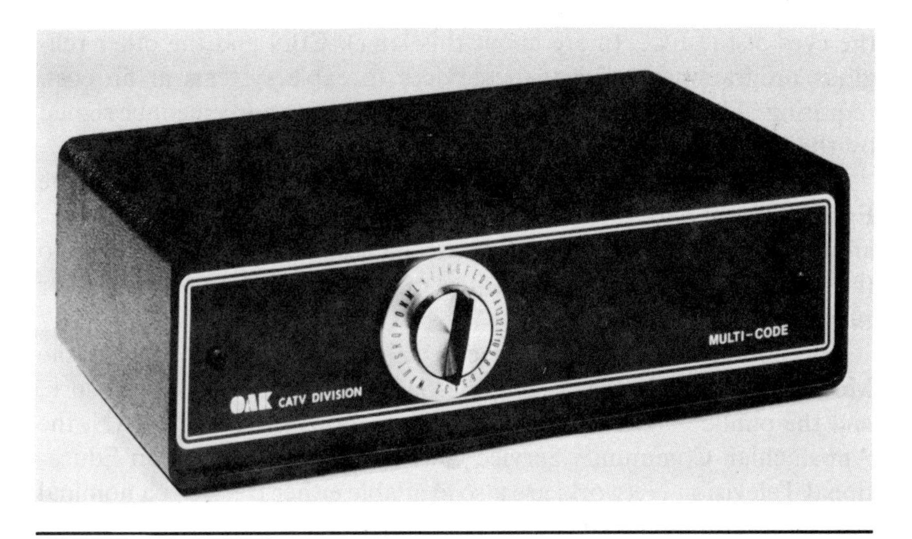

The ultimate in CATV flexibility, this decoder/selector can be remotely controlled by a headend microcomputer. This system allows multiple tiers of pay TV channels to be offered to different groups of pay subscribers. (Courtesy of Oak Communications Inc.)

Programming Considerations

The mini-CATV operator will usually elect to carry a number of different programming services, including basic (or free) satellite-supplied channels, pay TV channels, superstations, and local broadcast station pickups. Each of these services has its own advantages and disadvantages, which should be considered independently.

Basic Services

The basic services include those satellite programming feeds that are available to the mini-CATV company either free or at low cost. The most popular of these feeds are the religious programmers, such as the Christian Broadcasting Network. CBN is not considered a broadcast television station under FCC rules and regulations, and so may be carried by any cable television system without restriction. This is not true, however, of the Trinity Broadcasting Network, which, in part, rebroadcasts the signal of a Los Angeles UHF television station. Thus, it is considered by the FCC to fall under the rebroadcasting rules concerning the pickup and transmission of distant broadcast television stations. Carrying TBN automatically makes one a CATV system in the eyes of the FCC (more about this later). CBN and the other religious programmers offer their services to cable systems at no cost, requiring only that a one-page signal authorization agreement be signed by the mini-CATV system.

Other popular basic services include the Cable News Network, the Entertainment and Sports Programming Network, the USA Network, and the children's Nickelodeon service. These services are available to the mini-CATV operator for a fee ranging from 4 to 20 cents per subscriber per month.

The commercially sponsored entertainment networks such as the Modern Satellite Network and the Satellite Programming Network, and the public affairs/educational organizations such as C-SPAN, the Appalachian Community Service Network, and the American Educational Television Network, are also available either free or at a nominal cost of several pennies per viewer per month.

The advantage of offering a subscriber basic services is that their programming costs are extremely low. The disadvantage is that many people will not subscribe to the mini-CATV service if only basic services are offered. People want movies! Each basic service ties up a television channel on the cable system, and requires separate satellite TVRO receivers and channel modulators.

Movie (Pay TV) Service

For these reasons, most mini-CATV operators will offer a first-run pay television movie service such as HBO, Showtime, The Movie Channel, or the Home Theater Network to entice viewers to subscribe to the overall cable service. Although the charges vary from organization to organization, all pay TV satellite programmers charge a sliding-scale fee based on the price the minicable system charges the home subscriber and the number of total subscribers in the system. Fees typically range between $2.50 and $5 per month per premium-tier pay TV subscriber. The movie programmers base their fees on the charge that the cable television firm levies on the subscriber. To minimize these pay TV payments, cable systems that offer first-run pay TV movies often do so for a separate fee, billed to the subscriber independently of the basic cable monthly service charge.

Superstations

The third satellite program type consists of the national television superstations: WTBS (Atlanta), WOR (New York), and WGN (Chicago). The mini-CATV company can obtain these services for 10 to 15 cents per month per cable subscriber. A single-page authorization

EARTH STATION
FCC Call Sign ________

SATELLITE SERVICE REQUEST AND AGREEMENT

1. This Agreement, made and effective this ___day of _______________, 19__, is by and between United Video, Inc., a communication common carrier, with its principal place of business at 5200 S. Harvard, Suite 215, Tulsa, Oklahoma (hereinafter referred to as "Carrier" and _______________ (hereinafter referred to as "Customer").

2. To receive signals of WGN-TV, Chicago, Customer hereby orders satellite transmission service from Chicago, Illinois to Customers' receiving location(s). Customer agrees to payment of the monthly rates designated on Schedule 1 (reverse side) and further described in Carrier's Tariff #13, which is on file at the Federal Communications Commission (FCC). This Agreement and the obligations imposed hereby is and shall be governed by, subject to, and interpreted in accordance with all terms, rules and regulations, charges, practices and conditions of Carrier's tariff, including supplements or amendments thereto. Carrier and Customer shall comply with all provisions of that tariff, which is incorporated herein by reference and shall be part of this Agreement.

3. Customer shall be responsible for any necessary FCC and Copyright registration or authorization to permit Customer to utilize the service provided by the Carrier. Customer shall be responsible for construction and operation of earth station facilities necessary to receive Carriers' signal.

4. Since continued availability of this satellite transmission service is subject to the Carrier maintaining all necessary FCC, Copyright and other legal authority, as well as other factors, Carrier shall not be responsible for interruptions or discontinuance of satellite service. In the event of such interruptions or discontinuance of satellite service, the liability of the Carrier shall be limited to a pro rata refund of any service charges paid by Customer for the period during which service was not provided.

5. Customer agrees to receive this transmission service for a period of no less than three (3) years from the date service is implemented to the location(s) given on Schedule 1 (reverse side). Payment for service is due at the office of Carrier on or before the 1st day of the month for which service is provided.

WITNESS the due execution of this Agreement on the day stated above.

CUSTOMER	CARRIER
______________	United Video, Inc.
Authorized Corporate Name	Authorized Corporate Name
______________	5200 South Harvard, Suite 215
Street Address	Street Address
______________	Tulsa, Oklahoma 74135
City, State and Zip Code	City, State and Zip Code
(X)______________	(X)______________
Authorized Officer or Signer	Authorized Officer or Signer
______________	Roy L. Bliss
Typed Name of Signer	Typed Name of Signer
______________	1-800/331-4806
Area Code and Phone Number	Area Code and Phone Number

Typical contract between CATV System and United Video to authorize CATV pickup of WGN-TV. (Courtesy of United Video) agreement, similar to the religious programmer's agreement, is all that is required legally to pick up these signals from the resale common carrier. Because the superstations are regular television broadcasting stations, however, their inclusion in a mini-CATV's cable system is complicated somewhat by additional FCC and Commerce Department copyright regulations.

Under the FCC rules, any mini-CATV company that has more than 50 subscribers cannot legally carry any of these distant television broadcasting station signals unless all of the local television broadcasting stations are first carried on the cable system (certain jointly owned systems such those operated by a condo or housing development association are exempt; see Appendix 5 for a detailed discussion of these requirements). As the other programming sources such as HBO, Cable News Network, and Christian Broadcasting Network do not originate from regular television stations, they are not considered broadcasting stations under FCC rules. Likewise, the Copyright Tribunal also imposes conditions on the mini-CATV system that carries broadcasting station signals of any kind, either locally originated or imported via satellite. Depending on the semiannual gross revenue of the mini-CATV system, the organization must pay copyright usage fees to the federal government of $30 per year and up. These fees ultimately are distributed by the Copyright Tribunal to the television programs' producers, directors, writers, and actors. (See Appendix 5 again for a more thorough discussion of the copyright rules and regulations that apply to mini-CATV systems.)

Local UHF and VHF television station pickup comes under the same regulations that govern the superstations. Most mini-CATV systems will want to carry one or more such broadcasting stations, but it is interesting to note that the carriage of *only nonbroadcasting* satellite networks and programs by the mini-CATV system will *eliminate* its designation by any branch of the federal government as a CATV company, and thus exempt it from any rules and regulations at the federal level.

Network Television

Although many mini-CATV companies would like to carry the Canadian CBC and CTV networks as well as the ABC, CBS, NBC, and PBS network feeds (all of which appear on various satellites), at the present time it appears that this is not officially possible. (Home satellite TV viewers can, of course, watch whatever they wish to in the privacy of their homes.) The U.S. television networks generally will not sign authorization letters to allow a mini-CATV company to pick up their private network feeds. To do so would undermine their contractual obligations to their affiliate broadcasting stations. However, both ABC and CBS Television have introduced new made-for-cable cultural and entertainment networks, and both the CBS Cable and ABC Alpha services will be happy to sign mini-CATV agreements similar to those of the other satellite networks. Likewise, the FCC restricts U.S. cable companies from picking up signals from Canadian satellites, as there is no official agreement between the two countries allowing this to take place. However, if a mini-CATV company does not carry any broadcast television signals from the United States, there may be some dispute as to the FCC's jurisdiction. This question may be worth pursuing further, especially in those U.S. communities that have large Canadian populations.

The Canadian mini-CATV firm will discover that, officially, its satellite television feeds are limited to the Canadian ANIK satellites. However, literally hundreds of northern mini-CATV systems have been aiming their antennas south of the border for a number of years, with no consistent opposition from either the Canadian Radio–Television Telecommunications Commission, the regulatory authority for Canada, or the FCC. Because of this reality, a number of new Canadian pay TV movie channels are being developed in an attempt to wean the Canadian mini-CATV systems back to picking up Canadian satellite programming, and satellite programming choices for the Canadian mini-CATV system are now looking very bright.

A review of all the programming options for the mini-CATV company reveals that there are many available alternatives. A typical system might include two or three local television channels, the Cable News Network, a religious channel, a sports network, C-SPAN, a pay TV channel, and a superstation. If a mini-CATV company wanted to, it could carry more than 30 channels of programming just from the satellites!

Legal Considerations

As briefly discussed, mini-CATV may fall within the rules and regulations of both the FCC and the Copyright Office of the Commerce Department. If the CATV system carries television station broadcasts picked up off the air or via satellite, then the operation is considered a cable system in the eyes of the Copyright Tribunal. The cable company must complete a formal application, and make semiannual payments according to a fee schedule that depends on the company's gross yearly revenues. Similarly, if there are 50 or more subscribers on a mini-CATV system that carries broadcast television signals, another set of forms must be filed with the FCC, and annual reports must be returned to the Commission. In addition, the CATV system comes under the full jurisdiction of all federal laws concerning equal opportunity employment, and of regulations issued by other agencies that the FCC has determined shall also apply to CATV companies. Since these regulations do not apply to MATV apartment building systems that include the cable television connection as part of the monthly rental, it may be to the advantage of a mini-CATV operator actually to install a separate cable feed into each apartment or dwelling unit to deliver the satellite television services over a completely different wire. This would serve to separate the cable television operation from the nonregulated MATV operation administratively, technically, and legally. In most cases, however, compliance with both the FCC and copyright agencies is straightforward. By being listed with these agencies as a legal CATV company, the mini-CATV operator may also find it easier to be treated as a going organization when dealing with some of the program suppliers, notably the pay television movie services.

Local building codes must be considered when installing any satellite antenna for commercial use. In many jurisdictions height limitations will require that a building permit be obtained before a satellite TVRO antenna can be installed on a roof. This may involve the services of a licensed professional engineer to verify that the roof is capable of supporting an antenna system that weighs several hundred pounds. It is a good idea in any case, as insurance policies may also require such review when a 15-foot-diameter dish suddenly appears on top of a building. Locating the TVRO antenna in a nearby parking lot or a field adjacent to the apartment or housing complex may be more convenient. A small fence can be constructed around the antenna to prevent unauthorized personnel from obtaining access to it.

Many cities and states require that business permits be obtained by organizations that deal with the general public. A mini-CATV system is such a retail organization. These permits are usually issued automatically, and are concerned with the general nature of retail establishments rather than with specific laws concerning CATV companies.

Regulations governing CATV companies themselves fall within the jurisdiction of the city and state public utilities commissions. By and large, PUCs define cable television companies as those with common carrier-like organizations, whose cables are stretched across public property or under city streets. It will often be possible, therefore, for a mini-CATV firm to minimize or avoid local PUC regulation completely by restricting its operation to privately owned property. In the case of a multibuilding apartment complex, the building owner can issue the mini-CATV operator written authorization to extend the cables from the centrally located headend (perhaps a small shed adjacent to the TVRO satellite antenna located in the parking lot) to each of the apartment dwellings. A suburban housing developer might retain a strip of access land behind the buildings, for use by the housing association. The CATV cable could be buried along this route or mounted on poles or, optionally, extended underground from house to house, assuming that the individual property titles have included a CATV clause to permit future access to the buried cable.

Although there is no legal requirement to do so, many mini-CATV entrepreneurs will find it to their advantage to incorporate. Certain tax advantages can accrue with incorporation, and incorporating a company can provide some additional liability protection. One of the major advantages, however, is that an incorporated company will probably find it easier to deal with cable television programmers and equipment suppliers, which are geared to do business with corporations.

Condo associations and housing community neighborhood organizations may find it advantageous to operate a not-for-profit cable television system themselves. When the mini-CATV system is jointly owned by all of its subscribers, the FCC does not ordinarily consider the organization to be a company. Such a mini-CATV system is, therefore, exempt from the FCC regulations.

However the mini-CATV system elects to structure itself for the regulatory agencies, it should always endeavor to appear to the satellite program suppliers as an ordinary CATV company rather than as an apartment MATV complex. The satellite programmers, especially the pay TV organizations and commercially sponsored satellite networks, prefer to do business with cable television companies, and in some instances their contractual agreements with their advertisers require this. Thus a mini-CATV system, given the choice, should call itself Cadillac Community Television rather than the Cadillac Apartments MATV System.

The Business Plan: Projecting Costs and Revenues

The costs involved in putting together a mini-CATV system will vary considerably from location to location and from system to system, depending upon the number of channels carried, the number of subscribers projected, and the luxuriousness of the hardware used. In general, there are six major cost items to consider: initial equipment hardware and its installation, programming fees, sales commissions and employee salaries, indirect operating expenses (telephone service, etc.), ongoing maintenance and repair costs, and miscellaneous expenses (outside consultants, etc.).

Example 1. For purposes of illustration, let's assume that our typical mini-CATV organization, the Cadillac Community Cable Company, has been created to provide service to 150 subscribers, each of whom will pay $20 per month to be connected to the cable system. Cadillac CATV presently feeds to one 15-story apartment building, known as Cadillac Towers, in which all of the subscribers live. Cadillac's gross receipts amount to $3000 per month. Let us assume that installation fees of $25 per subscriber are also charged, bringing the total installation fee income to $3750.

Based on the assumption that six programming channels are carried —C-SPAN, Cable News Network, Christian Broadcasting Network, WTBS, Entertainment and Sports Programming Network, and a premium movie service—six separate satellite TVRO receivers and associated VHF channel modulators will be required.

Table 5-1 describes our typical costs.

The total hardware costs amount to $21,050; based on a five-year financing arrangement at 20-percent interest, this amounts to $557.70

TABLE 5–1.
Mini-CATV Example No. 1: Cadillac Community Cable Company
(Based on 150 subscribers at $20/month each)
Installation fee of $25/unit.
CATV system is located in a multistory apartment complex or condo complex.

Gross monthly receipts: $3,000
Total installation fees: $3,750

EXPENSES: Dealer Cost		
Hardware	$3,500.00	Antenna and mount (15′) (H&R or Paraframe)
(Satellite and CATV distribution equipment)	1,900.00	Dual LNAs w/cable and coupler/feedhorn (Avantek 120° LNAs)
	400.00	2 4-way power splitters
	12,600.00	6 satellite receivers w/VHF channel modulators (Comtech Model 550-A)
	300.00	1 signal combiner
	2,000.00	5 distribution amplifiers and 250 cable "taps," miscellaneous cable, clamps, parts, etc.
	350.00	1 color TV monitor
	$21,050.00	
5-year lease-purchase at 20%/year	$557.70	

Total principal and interest payback	$33,461.78	
Installation	$600.00	Antenna at $20/man-hour
	9,000.00	CATV cable distribution network
	$9,600.00	150 units at 3 hours/unit; $20/hour
5-year financing at 20%	$254.34	$15,260.48 total P & I
Programming		
150 × 0.01	$1.50	C-SPAN (1¢/viewer/month)
150 × 0.15	22.50	CNN (Note: CNN and WTBS—lower combined rate not counted)
150 × 0.10	15.00	WTBS (or other superstation) discount 20¢/month total
150 × $4.00	600.00	Premium Movie Service
Free	0.00	CBN (or other religious network)
150 × 0.04	6.00	ESPN Sports Network
	$645.00	(Pay TV service accounts for over 93% of the programming costs)
Salaries	$400.00	Part-time office assistant/bookkeeper
	400.00	Part-time technician
	$800.00	(960)
Overhead	$100.00	Rent
	20.00	Phone
	50.00	Utilities
	30.00	Miscellaneous
	$200.00	(240)
Other	$200.00	(240) Insurance, outside consultants, etc.
Total monthly costs (first 5 years)	$2,657.04	$2,557.69 if installation fee is applied to reduce initial loan costs

Assumes equipment and installation was financed. If not, costs are $2,085/month and return on initial cash of $30,650 is at rate of about 12%/year, compounded.

Net profit per month (before taxes)	$342.96	Total over 5 years—$20,577.60
After 5 years (assuming overhead and salary increases of 20%)	$2,085.00	
Gross receipts increase to: (assuming monthly rate increases to $22)	$3,300.00	
Therefore, from the 60th month onward, income becomes	$1,915.00	per month ($14,580/year)

per month. Assuming that installation costs of $9600 are also similarly financed, the monthly cost is $254.34.

Programming expenses are $645, of which over 90 percent goes to pay the premium movie service supplier.

Assuming that a part-time office assistant/bookkeeper and a part-time service technician work 10 to 20 hours per week, the salaries will run approximately $800 per month. Overhead and other costs are projected at an additional $400 a month.

The total monthly costs amount to $2657.04, producing a net profit per month (before taxes) of approximately $343. However, if the installation fees are applied to reduce the amount of initial bank financing required to cover the installation costs, then the total expenses (including financing costs) drop to under $2558 per month, increasing net monthly profit to $442.31 or a yearly income of over $5000.

If one estimates that at the end of five years operating costs will have increased by 20 percent and the monthly service charge to the subscribers raised to $22, then yearly income from year six onward will amount to $14,580.

Example 2. In the second mini-CATV example, the Suburban CATV Corporation provides service at the same rates to a 500-unit garden-type apartment or townhouse complex.

In this illustration (see Table 5-2), hardware and installation expenses are somewhat higher as additional distribution amplifiers and

TABLE 5–2.
Mini-CATV Example No. 2: The Suburban Community CATV Corporation
(Based on 500-unit garden-type apartment or townhouse complex at $20/unit)
Installation fee of $25/unit.

Gross monthly income: $10,000
Installation fee: $12,500

EXPENSES:

Hardware	$3,500.00	Antenna and mount
	500.00	Protective shield
	1,900.00	2 LNAs w/cable, coupler feedhorn
	400.00	2 4-way power splitters
	12,600.00	6 SAT receivers and modulators
	300.00	1 signal combiner
	5,000.00	20 distribution amplifiers w/cable taps, cable, etc.
	350.00	1 color TV monitor
	$24,550.00	
5-year lease purchase financing at 20%/year	$650.42	

Installation	650.00	Antenna and headend
	30,000.00	CATV cable distribution net
	$30,650.00	(500 units at 3 hours/unit; $20/hour)
5-year financing at 20%	$812.04	
Programming		
500 × 0.01	$5.00	C-SPAN
500 × 0.15	75.00	CCN
500 × 0.10	50.00	WTBS
500 × 4.00	2,000.00	Premium Movie
500 × 0	0.00	CBN
500 × 0.04	20.00	ESPN
	$2,150.00	
Salaries	$1,500.00	Full-time office staff person
	1,500.00	Full-time technician
	$3,000.00	
Overhead	$500.00	
Other	$500.00	
Total monthly costs	$7,612.46	
Net Monthly Profit (before taxes)	$2,387.54	$28,650/year
However, if installation fees are used to partially cover installation salaries instead of financing them, monthly costs drop to:	$7,281.28	($480.86)
Which increases net profit to:	$2,718.72	$32,625.59/year

Therefore, this system is profitable enough to provide, initially, over $32,000 per year of income during its first five years of operation, and over $48,600 per year after that (assuming that costs increase by 25 percent and monthly fees are raised to $22).

---

cables are required. Monthly programming costs are also increased by a prorated amount, and an office manager and a service technician are hired as full-time employees. Office overhead and other costs have increased by 130 percent.

Total monthly expenses now become $7612.46, providing a monthly profit (before taxes) of $2387.54. As in the first example, however, if the installation fees are used partially to cover the cost of installation, thereby reducing the financing requirements, the monthly costs drop to

$7281.28, increasing the monthly profit to almost $2719, or $32,625.59 per year, before taxes.

At the end of the first five years of operation, and assuming the same circumstances that applied in the first example, net yearly income would rise to $48,600. If the mini-CATV firm can install just ten similar housing complex systems, its net income will reach almost one-half million dollars per year. There are over 30,000 such condominiums, apartment buildings, housing developments, and trailer courts of this size in the United States. Another 50,000 housing developments are half as large. And yet the present level of cable television development throughout the United States is just over 20 percent, indicating that almost four out of five U.S. households do not presently have cable television service. Clearly, a tremendous opportunity exists for the mini-CATV entrepreneur.

Where to Go from Here: Using the Pros

The examples of the two mini-CATV systems are, of course, just that—examples. Labor costs vary dramatically from location to location, as does the cost of transportation and money. Also, whereas one mini-cable system may be able to charge its subscribers $20 per month, another may be able to charge only $15 per month.

The best way to minimize one's risk in beginning such a venture is to ask the advice of professionals in the business. Often, contacting a local registered professional engineer who has had experience in installing cable television equipment will be of value. This is especially true when some governmental agency requires that building codes be met in the installation of the TVRO antenna dish. The National Society of Professional Engineers in Washington, D.C., can help locate a qualified professional engineer, as can the state agency that licenses PEs.

The Community Antenna Television Association (CATA) in Oklahoma and the National Cable Television Association (NCTA) in Washington, D.C., are two nonprofit organizations whose members are cable television operators. The CATA membership consists primarily of mini-CATV companies. NCTA's membership ranges from the smallest "mom and pop" firm to the largest corporate giant. Both organizations hold annual conferences and trade shows, and sponsor administrative, marketing, and technical seminars. Many states also have regional associations of cable television companies that hold monthly meetings, and are a good source of local information and feedback on the industry. One of the biggest is California Community Television Association, headquartered in Castro Valley, California.

Bob Cooper's company, Satellite Television Technology in Oklahoma, can provide some consulting assistance, and the Society of Private and Commercial Earth Station Users (SPACE), the Washington-based private terminal user group, is also interested in having mini-CATV companies as its members.

A number of excellent cable television magazines are on the market, and the weekly *CableVision* publication has become the *Business Week* of the industry.

Of course, the CATV manufacturers themselves can provide a significant amount of assistance in both the preliminary stages of planning and on an ongoing basis. This applies to the commercial suppliers such as Microdyne, Avantek, and Comtech. The home TV satellite suppliers are usually too new and understaffed to offer the kind of support needed by a beginning cable television system.

Finally, The Satellite Center in San Francisco, California, has put together a detailed mini-CATV study package and a variety of other management services for the independent mini-CATV entrepreneur. The Satellite Center can provide engineering, financing, and, under some conditions, discount equipment buying services.

Detailed information on the equipment manufacturers and their products, the programming suppliers and their services, and the organizations mentioned throughout this chapter can be found in Appendices 2, 3, and 4. The call of Johnny Appleseed awaits: The towns and hamlets of America want cable television now!

Chapter 6

Satellite Television for Hotels, Taverns, Restaurants, and Hospitals

At first blush one might wonder what a nonprofit organization such as a hospital has in common with a commercial establishment such as a hotel, a tavern, or a restaurant. The answer is: They all have patrons. Although a patient might not be considered by the layman in the same light as a guest in a hotel, nonetheless to the hospital administrator it's pretty much the same thing. Restaurant and tavern patrons are simply shorter-term guests. And all of these establishments can potentially benefit from furnishing their patrons with television via satellite.

The Innkeeper and Satellite TV

Let's take the case of a hotel or motel. Ninety-nine percent of the time a master antenna television system (MATV) already exists, and color television sets are usually provided in each of the guests' rooms as an entertainment service of the establishment. In most cases these systems are functioning properly, and are periodically maintained. In a service-competitive business such as the hotel industry, each innkeeper must continually vie with the hotel down the street to give newer and better service.

One such service is the MDS (multipoint distribution system) pay TV movie feature now available in many hotels nationwide. This movie channel is fed to the hotel by an independent MDS common carrier over a special terrestrial microwave frequency that cannot be picked up by an ordinary television set. The hotel installs a small antenna on its roof aimed at the nearby MDS movie channel's transmitter, and feeds the signal into its internal MDS TV system. The MDS operator, which most likely picks up the movie channel using its own TVRO earth station, charges the hotel a fee for each guest who elects to watch the pay television movie on the room TV set. A small decoder box, similar to the CATV convertor, sits atop each room's television set to convert the hidden pay channel to a standard VHF channel, and in many systems to register electronically when the movie channel has been selected. (Chapter 11 looks into the MDS concept in more detail, and shows how it differs from both satellite television and mini-CATV systems.)

By installing its own satellite TVRO earth station, a hotel can obtain these movie services directly, thus bypassing the MDS middleman supplier. In addition, other television channels ordinarily available only through cable television systems can also be provided to the room guests. These include services that are of particular interest to the business guest, such as the Cable News Network and C-SPAN. For example, the rental of daily meeting rooms to local professional associations might be increased if the AETN continuing education channel were carried. AETN offers seminars and courses in a variety of fields for professionals who must take refresher classes to maintain their legal accreditation. These include CPAs, dentists, lawyers, nurses, optometrists, pharmacists, physicians, and real-estate personnel.

The satellite also can bring in fun channels for the vacationing guest. The superstations from New York, Chicago, and Atlanta specialize in old movies and popular TV series. The children's networks can keep the youngsters happy when they are staying in the room, and the sports networks are always popular. Special programming can cater to specific hotel interests, including those innkeepers who have large numbers of black and Spanish-speaking guests.

Perhaps one of the most intriguing uses of satellite television for the hotel establishment is video teleconferencing. Such organizations as Wold Communications and Westinghouse's Vidsat produce and distribute via satellite national video teleconferences, fund-raising events, and public Fortune 500 annual meetings. In addition, special entertainment and sports events such as championship fights are also transmitted by satellite to public meeting places throughout the country.

By installing its own satellite earth station, a hotel can tap into this lucrative new market. The Reagan Presidential Committee was most successful in raising millions of dollars during the 1980 presidential campaign by tying together dozens of cities into instant satellite TV networks, using the teleconference capabilities of WETA and other PBS-affiliate stations. Thus the national politicians and the Hollywood stars who took part in these events could be in many different places simultaneously, and on a far more intimate basis than the nightly news could possibly allow.

When a national event such as a boxing match is televised through closed-circuit channels throughout the United States and Canada, in some cities the demand for tickets often exceeds the supply, and in others, no facilities exist to telecast the event. A hotel that owns its own satellite TVRO earth station could easily negotiate with the fight promoter to pick up the programming and feed it to its own auditorium and a projection TV system. Conceivably, the hotel could even feed the game to its guests' rooms on a special-fee basis. Not only would the hotel make money on the ticket sales, but the additional income derived from the patrons having dinner before the game and visiting the bar afterward could exceed the revenue from the fight itself.

The next time a charity schedules a national closed-circuit fund-raising event, the satellite-equipped hotel could become the local focal point. And as more and more corporations turn to closed-circuit television to hold national sales meetings, the hotel could capitalize on the local business. With gasoline approaching the $2 per gallon mark, the "telecommunications–transportation trade-off" is more and more apparent to corporate telecom managers. The cost of bringing a thousand people from around the country to a central location, including their air fare and housing costs, far exceeds that of a video get-together in a dozen cities via satellite. Corporations are also finding it more and more difficult to conduct annual meetings. In many cases meetings held in a public auditorium in the corporate headquarters city are overcrowded, and discriminate against the smaller stockholder who may live at a distance. By holding the annual stockholders meeting simultaneously in hundreds of locations, the corporation can maintain a closer relationship with its stockholders. A hotel, of course, is a natural place in which to hold such video stockholders meetings. Two-way participation by local stockholders can be assured by using the regular long-distance telephone network to carry their conversations back to the headquarters city.

For these and other reasons, several of the largest hotel chains in the United States have begun to install their own satellite TVRO earth stations. The first of these organizations was Holiday Inn. These folks have made a commitment to put in several hundred satellite dishes at

Holiday Inns throughout the United States, to pick up pay TV movie services directly, and to go after the expanding video teleconferencing business. The earth stations are usually 5 meters in size, and are placed in the parking lot or on the roof of the hotel building. It is fascinating to visit Fisherman's Wharf in San Francisco and to gaze across the vista from the Golden Gate Bridge to the fishing boats bobbing at the docks and to the satellite TV earth station sitting in front of the Fisherman's Holiday Inn! Other hotel chains have been quick to follow suit, including the Hilton Inns, the Hyatt Regencies, the Marriott Hotels, and the Playboy Clubs. By the end of the decade just about every major chain hotel undoubtedly will have its own satellite TVRO earth station.

The Technical Requirements

Running a hotel MATV system that offers movies in the rooms is very similar to operating a mini-CATV company. In fact, several national MDS organizations specialize in providing this service to hotels under contract. A number of problems must be solved, however, before the hotel management can switch to its own satellite terminal. First, the MATV system will have to be technically upgraded to allow for the increase in the number of channels that the satellite antenna will provide. This requires the installation of additional VHF channel modulators, and probably will require that the hotel's engineer or television company review the overall system from top to bottom. Second, some form of viewing control must be implemented to allow a room guest to obtain the movie service whenever desired, and to have the guest's account properly charged. If the hotel had been using the services of an MDS common carrier to handle the billing and collection procedures, these functions would have to be assumed by the hotel administration.

The ultimate system would allow a microprocessor-controlled headend channel switcher to send a signal automatically to the room TV's decoder box to unlock the movie channel when a guest requests that service from the headend computer. The service request could consist simply of dialing a special access code on the telephone, or depressing a button on the TV convertor box itself. Upon activating the movie service, the headend microcomputer would automatically print this status information on a small printing terminal located at the front desk of the hotel. In a somewhat simpler system, the hotel switchboard operator or "movie center operator" would feed the movie channel to

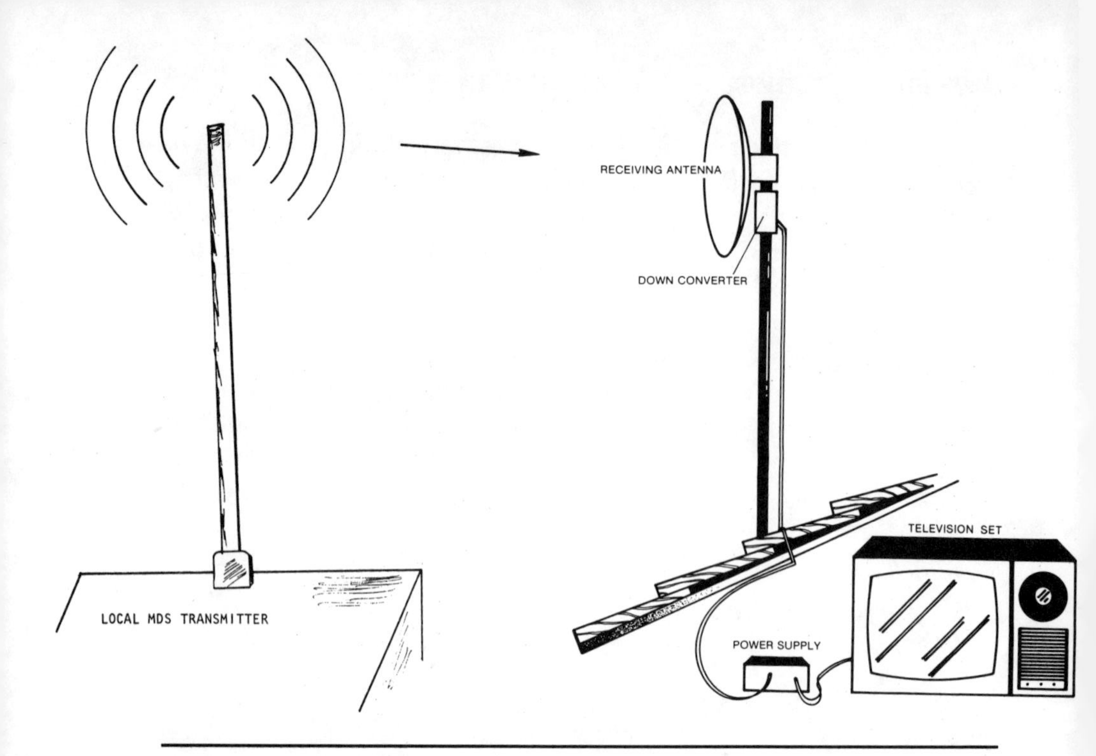

MDS system uses local microwave transmitter to carry pictures over special channels that cannot be picked up by ordinary television sets. Such a system can be used by hospitals offering joint medical educational seminars. MDS is popularly used to bring pay TV programming to big-city hotels from a central transportation point. (Courtesy of EMCEE Broadcast Products)

a particular room upon receiving a telephone request from a guest. Both of these types of systems are already in use in the United States, and a number of manufacturers sell hotel movie control systems directly (see Appendix 2).

To obtain the various services, the innkeeper will have to negotiate directly with the satellite program suppliers. Luckily, this process is quite simple. Appendix 3 provides detailed information on the types of satellite programs available, and on how to contact the satellite program suppliers. A few telephone calls should solve the programming problem very quickly.

A more difficult problem is to determine how the guest should be charged for the new satellite television service. Should it simply be given away, by including the additional costs in the daily room charge? This approach would penalize guests who are not interested in the television service by subsidizing those who do watch television. Moreover, in a highly competitive environment such as the hotel business, some of these guests might be attracted to the hotel down the street where the rates are lower—the hotel that does not provide the movie service on a "bundled" basis. Another option would be to charge each guest for each service viewed, by providing pay television movies and other entertainment channels for a per-channel charge. However, this approach is far too complicated. The third strategy would be to give away the basic television services, including channels such as the Cable News Network, WOR-TV, and ESPN, as part of the basic room charge. The movie channel could be billed on a per-evening, per-day, or even per-feature-film basis. This approach appears to be the one most prevalent today in the hotel industry.

By offering satellite TV programming, the hotel and motel can furnish not only a novel service to their guests, but one that will make the hotel more attractive to visit, and therefore more profitable, to its owners. One small eastern motel has seen its income increase more than 50 percent by promoting the fact that it offers satellite TV in its rooms. Some people actually travel several hundred miles to stay at this motel over a weekend, just to watch satellite television! And this same phenomenon is occurring throughout the country.

Taverns and Restaurants

Like motels and hotels, taverns and restaurants are profit-oriented concerns. The food and beverage industry is also totally dependent upon the service that it provides to its customers. For years even the smallest pub has known about the pulling power of the television set. Some restaurants have made a specialty of running old movies and television programs while their patrons eat. A famous Atlanta establishment has built a national reputation on the fact that it projects movies on an overhead screen above the dinner tables!

But using commercial television as a drawing card to bring people into a tavern has always involved some legal problems. One New England watering hole recently came under fire from a major television network for showing "best of" videotapes of popular daytime soap operas on certain "soap opera nights." Although the pub was jammed with people who wanted to watch, the TV service had to be stopped because the copyright laws had been violated.

By installing a satellite television earth station, a tavern or restaurant owner can legally enter into a contractual agreement with a number of sports and special events programmers, including the Entertainment and Sports Programming Network. ESPN provides 24 hours per day of original-production sports coverage. Moreover, ESPN's contract allows public meeting places such as restaurants and pubs to display its games on projection television systems. Razor-sharp feeds from the basketball court or race track can be a very strong attraction. Several other satellite programmers offer adult entertainment and adventure programming that also might be viewed over a projection television system, depending on the agreement reached with the programmer.

If the restaurant or bar is near or inside a hotel, the same satellite TVRO antenna can be used to bring both the pay TV movie service to the hotel rooms and the sports network to the pub's projection television system.

Installations can vary from a CATV-quality 15-foot parabolic dish (possibly shared by a hotel) to a consumer-quality 10-foot antenna with a low-cost satellite receiver. Even a relatively inexpensive system, costing $3000 to $4000, can provide outstanding television pictures on a projection television system, as the video output of the satellite receiver directly drives the projection TV's video input. This eliminates the need to convert the signal to a VHF television channel and the slight degradation that this process entails. Optionally, a number of organizations will rent both downlink (receive-only) and uplink satellite earth stations for special events.

A typical TR combination uplink/downlink, rented from $1500 to $7000 per day by Satellite Communications Network, New York City, can be used for special events. (Courtesy of Satellite Communications Network)

The enterprising tavern owner might also be able to negotiate directly with the major sports event promoters, such as the World Heavyweight Boxing organizers. A pub that can legally pick up and present the world heavyweight fights to its paying customers undoubtedly will be playing to "standing room only." As more and more hotels, restaurants, and taverns begin to install their own backyard satellite antennas, the major promotional organizations will begin to see these places as additional outlets through which to sell their entertainment product. Moreover, to the promoters' advantage, the satellite distribution mechanism—namely, the TVRO earth station—will already be in place.

The Hospital and Rest Home

The hospital and rest home are similar in operation to a hotel or an apartment building. Although the medical and health care institutions might not be profit oriented, nonetheless they provide services in a competitive atmosphere, and almost all medical centers have installed television sets as a standard piece of furniture in their rooms. A hospital room's TV is usually in use many hours a day, as the hospital patient often has no other form of entertainment available. By installing a satellite TVRO station, the medical care facility can provide the patient with a far greater choice of entertainment, educational, and news options. Viewing the Cable News Network, for example, is a lot like watching a video version of an all-news radio station. For the person recuperating in a hospital room, CNN can provide the link to the outside world. Likewise, for the sick child the children's networks can be a lot more diverting than the primarily adult programming found on conventional television. Finally, the patient who is convalescing over an extended period of time may appreciate the educational offerings of the Appalachian Community Service Network. Some patients or rest home guests can even obtain high school or college credits by watching the courses and seminars while resting in their rooms.

Hospitals are attempting to make their institutions more humane and friendlier for their patients. They have begun to introduce all kinds of new programs and services to allow the patient more freedom, and to help the patient feel more at home. For example, animals are now allowed in some hospitals, gift shops carry a significantly greater variety of items, food often can be brought in from nearby restaurants, and recreation facilities have been installed. In addition, most hospitals allow private telephones to be placed in patients' rooms.

One continuation of this humanizing trend is to provide the hospital patient with the option of pay TV movies. Picked up via the hospital's TVRO satellite terminal, the pay TV service can operate in a fashion similar to that in a hotel. As most hospitals readily accept national bank credit cards, the movie service could be charged to the patient's room bill, thus affording welcome entertainment for the patient, and additional income for the hospital.

Moreover, the same TVRO earth station can also be used by the hospital's professional staff to participate in the growing number of medical closed-circuit teleconferences now being beamed via satellite. The AETN continuing professional education network runs a number of seminars for the health care and hospital administration and medical personnel. Plugging in a TVRO earth station can minimize out-of-town trips to these conferences and allow the staff to keep up to date by sitting in on weekly satellite TV meetings. TVRO satellite antennas that can even handle two or more separate birds simultaneously are now available for purchase inexpensively.

Multisatellite TVRO antenna picks up 17 satellites simultaneously with 3-degree spacing between satellites. Costs $21,500 without LNAs and feedhorns. (Courtesy of Satellite Communications Network)

Convalescent and rest homes can provide satellite television services similar to those offered to tenants by apartment building managements. Many older people, especially, find the existing broadcast television programming uninteresting or not germane to their age group. Networks such as Cinemerica, which specializes in television for the over-45 population, feed their signals via satellite, and can be easily picked up by a rest home TVRO earth station. Additional pay TV movie channels can be furnished on a per-subscriber basis, as in an apartment complex. The cultural and entertainment networks such as ABC's ARTS and CBS Cable can bring symphonies, Broadway plays, the opera, the ballet, and other cultural events to older people in the comfort of their own rooms, thus affording them enjoyment that might not otherwise be available because of the limitations of transportation and the necessity for personal assistance. Like the apartment complex, the rest home facility can deal directly with the satellite television programmer by treating the installation as a mini-CATV system rather than an MATV system, and applying for service accordingly.

Chapter 7

Using Satellite TV in the Nonprofit Organization

Religious Groups, Churches, and Sunday Schools

Over the past several years religious organizations have undergone a major revitalization and change in their organization and structure. Many churches are experiencing a rebirth of faith, as witnessed by the thousands of new members joining their congregations. Attendance is up in many parts of the country, and Sunday schools have never been more popular.

Interestingly, the churches that have been most successful are the ones that have been willing to try new and unorthodox approaches to reach the people. For years one of the most successful Southern California ministries has conducted its Sunday services in a gigantic drive-in parking lot church. Recently the same organization completed a multimillion-dollar glass cathedral under the auspices of one of the United States' foremost architects (who is also designing the new American Telephone and Telegraph corporate headquarters building in New York). This church has been successful in raising hundreds of millions of dollars by expanding its ministry from Orange County, California, to the rest of the United States through the use of television and radio broadcasts.

A few years ago another widely popular television ministry was begun when a highly successful Wall Street broker received his calling from God. Pat Robertson gave up his financial career in New York City to move to Virginia Beach, Virginia, to found the Christian Broadcasting Network. In less than five years CBN has become one of the nation's most watched 24-hour-per-day television networks. More than ten million homes receive CBN programs via their cable systems and local television stations. Transmitting its programs by way of RCA's SATCOM F-3, CBN runs Christian-oriented talk shows, children's programming, family entertainment, and inspirational programming. The highly successful *700 Club* has seen hundreds of famous personalities, movie stars, and political figures make their appearance on its show.

At the same time that many religious organizations have been turning toward space to relay their inspirational programming via satellite, many local churches and interdenominational religious groups have begun to use the special private terrestrial microwave channels available for nonprofit and educational use to beam sermons and instructional programming from church to church within a local geographic area. Some of the larger religions have tied churches in various areas together by using the low-cost local video channels for Sunday school classes and social events. Thus, by communicating via large-screen projection television, the bishop of a diocese can speak to tens of thousands of parishioners, scattered over hundreds of miles, simultaneously, while they are attending local churches. The use of television as a means of reaching the people has proved so successful that by the 1990s, thousands and thousands of churches will be wired to private television networks and large-audience projection television systems. Many of these churches will be installing satellite TVRO earth stations to pick up the enormous variety of multidenominational inspirational and religious programming available on the communications satellites.

At present six major religious channels are carried by satellite. Like the Christian Broadcasting Network (now known as the CBN Satellite Network), the People That Love Television Network (PTL) and the Christian Media Network (CMN) also feed their Christian programming services via SATCOM F-3. The Trinity Broadcasting Network (TBN), headquartered in Santa Ana, California, uses the COMSTAR D-2 satellite for its 24-hour-per-day Christian programming. The National Christian Network (NCN) similarly operates on COMSTAR D-2, providing a wide range of nondenominational Christian shows and events. The National Jewish Television network (NJT) operates on SATCOM F-3 for a somewhat more limited programming day.

Although these organizations have come primarily from the nondenominational Christian movement, most of the organized religions, including the Roman Catholic Church and the American Lutheran Church, are also considering the establishment of satellite programming to reach out to the parish churches. By 1985 there may well be dozens of religious satellite channels carrying programming throughout the United States!

For a local church or Sunday school considering using the new satellite programming, the process is a straightforward and relatively simple one. Technically, a 10- to 15-foot TVRO antenna must be installed and aimed at either the SATCOM F-3 or COMSTAR D-2 satellite. Unlike a commercial mini-CATV system, the religious organization can elect to use a consumer-quality dish of slightly smaller size. Also, a single LNA/down-convertor is all that normally will be required as only one channel of programming usually will be picked up. Thus an inexpensive home TVRO terminal package such as the spherical-antenna system supplied by Downlink, Inc., or the combination of a Paraframe parabolic antenna with the Dexcel LNA/down-convertor/microreceiver configuration, for example, can be utilized. Such a system will cost between $3500 and $5000, depending on the church's geographical location, the specific supplier, and any additional options provided (such as a motorized rotatable feedhorn/LNA assembly). One option worth considering is the motorized antenna mount. For less than $1000 the normal fixed-position antenna base can be upgraded to a remotely pointable antenna, which can be easily repositioned from satellite to satellite, from a small control panel near the TVRO satellite receiver. This feature can allow the church or Sunday school to change program pickups easily and rapidly from one religious network to another.

To complete the overall installation, a consumer-grade video projection television system, such as the Advent Video Beam or the Sony Projection TV, can project the pictures to a medium-sized audience of up to 100 people or so. This will be usually sufficient for a Sunday school class, although a church congregation may require the purchase of several of these home units or a more powerful commercial projection TV system for screening in an auditorium or church hall.

In addition to the religious programming for church and Sunday school use, other satellite feeds, such as the children's networks and the educational instruction satellite programs, can be carried for Sunday school and other use. As a community service feature, the local church can make arrangements with the Appalachian Community Service Network (ACSN) to feed the continuing education and credit courses to students who attend on a daily or weekly basis, with the church auditorium or other meeting room as the satellite television classroom.

Appendix 3 presents a list of religious and other programming organizations. All of these are readily accessible to churches or Sunday schools and, in most cases, there are no charges for the programming. The typical authorization agreement allowing the religious organization to pick up the satellite television signal is usually the same one-page contract used by the cable television companies.

For a total cost of well under $10,000, a Sunday school or church can outfit itself with a deluxe TVRO earth station, high-quality projection system, and videotape recorder. This combination of equipment will enable a library of the best religious shows, Bible discussions, inspirational messages and sermons, and other programming to be built up and made available for later use.

Schools and Educational Institutions

Like the religious groups, schools and other educational institutions can also take advantage of the wide variety of high-quality television programming made possible by satellite. Depending on the requirements of the particular institution, a single-channel home-quality system such as the one described for church use can be installed to pick up a specific educational feed. As a practical matter, however, many schools may wish to receive two or more educational programs simultaneously. This will require dual LNAs and at least two satellite TV receivers, which will increase the cost of the terminal portion of the system by about 50 percent.

An enormous amount of educational programming is available by satellite, and more is being broadcast every day. The Cable Satellite Public Affairs Network feeds live coverage of the House of Representatives whenever Congress is in session. The C-SPAN channel is carried on the SATCOM F-3 satellite, and has been well received by high school civics classes and political science departments. Many people will never be able to visit Congress personally when it is in session, and C-SPAN allows the student to watch the proceedings as Congressional bills are actually being debated and voted upon. Perhaps some future president will have discovered a calling to a life of public service while a student watching the House of Representatives in session from a high school classroom.

C-SPAN provides live gavel-to-gavel coverage of House of Representatives on SATCOM F-3. (Courtesy of C-SPAN)

Formed in December 1977 as a nonprofit organization whose initial funding was provided by 25 of the largest cable companies in the United States, C-SPAN presently is carried by over eight million cable homes in the United States. In addition to its daily gavel-to-gavel coverage of the House of Representatives (from 9:30 a.m. to 6:00 p.m., Monday through Friday), C-SPAN also covers the National Press Club luncheons, conducts call-in programs and personal interviews with individual members of the House, and takes its cameras into committee meetings when the House is not in session. C-SPAN provides television coverage of the nation's political arena, and as Jeff Greenfield, television critic for CBS-TV, has said: "C-SPAN . . . may be the only way we'll find out what a campaign is all about. . . ."

From 10:25 to 10:55 a.m. (Eastern Standard Time) each day, C-SPAN, in conjunction with the Close-Up Foundation, produces a series of programs involving high school students from around the country who are brought to the nation's capital as participants in week-long seminars sponsored by Close-Up. With C-SPAN's cameras following them, the students interview representatives from the executive and judiciary branches of government, question members of Congress, and discuss the various seminar topics with lobbyists, the news media, and other individuals involved with the federal government. Recent

C-SPAN delivers TV feeds from the Capitol to its own suburban uplink antenna via its own microwave system. (Courtesy of C-SPAN)

Close-Up seminars have focused on the economy, energy, and the law —analyzing the U.S.'s energy policy for the eighties, investigating the individual's rights within the society and the criminal justice system, and looking at consumer law and the individual in the marketplace.

During the school year these C-SPAN/Close-Up seminars are telecast to hundreds of high school classrooms throughout the United States using local cable television systems. A teacher's *Issues and Answers* supplemental guide and supporting materials are also available for use by the high schools receiving the programs. The goals of the overall program are to:

1. Provide social studies teachers with timely supplementary resources, literally a living textbook on government,
2. Challenge and motivate students to examine current issues and the structure of the U.S. government, and
3. Demonstrate through peer identification how young people can participate directly in the democratic process.

In 1981 C-SPAN offered 70 half-hour programs in four separate series over its satellite channel.

Following the House of Representatives' adjournment each Friday (typically around 3:00 p.m. EST), C-SPAN televises the regular weekly luncheon speeches from the National Press Club in Washington, D.C. These speeches are heard in their entirety, and are repeated for those viewers who may have missed the first presentation. Recent speakers have included Charles Schultze, chairman of the President's Council of Economic Advisors; Ralph Nader; Luciano Pavarotti, the Metropolitan opera singer; and Jim Guy Tucker, director of the White House Conference on Families. C-SPAN is now investigating the possibility of transmitting debates live from the floor of the U.S. Senate, and may expand to a second satellite channel once the Senate authorizes such television coverage.

The Appalachian Community Service Network (ACSN) feeds its multifaceted continuing educational programming from Washington, D.C., via SATCOM F-1. Originally formed to bring educational programming to rural areas of Appalachia, ACSN has now become the nationwide satellite network for quality educational, job training, and cultural programs carried via cable television. Although it still focusses its energies on the 13-state Appalachia region, which reaches from New York to Mississippi, ACSN's programming is viewed nationwide.

ACSN courses include *Humanities Through the Arts,* an introductory humanities and art appreciation course examining seven art forms: film, drama, music, literature, painting, sculpture, and architecture. *It's Everybody's Business* is a program that provides an overall picture of business operations in the United States, including analyses of specialized fields within business organizations and the role of business in modern society. *Earth, Sea, and Sky* is an introductory earth science course. *Designing Home Interiors* introduces the topic of interior decoration and design. *Applied Sketching Techniques* examines the basics of free-hand drawing and then goes on to more advanced drawing techniques, covering some of the "tricks of the trade." *Introducing Biology* provides a strong introduction to the concepts of biology, covering the earth and the universe, plant biology, human anatomy, physiology, behavior and development, genetics, evolution, and ecology. Among other courses presented are photography, speed learning, dealing with children and family problems, auto maintenance, and American government.

Many of these courses can be taken for credit at the undergraduate college level, and to receive such credit an interested student enrolls in an ACSN telecourse through a local college or university. Although students must meet at the college for an orientation session and for course examinations, the seminar work itself can be done at home by watching the ACSN television channel.

Other special programming includes the broadcast of major teleconferences such as a recent U.S. Department of Agriculture Extension Services program and the annual conference of the American Council on Education.

Community colleges and other educational institutions that wish to offer ACSN telecourses either can negotiate with their local cable company to carry the ACSN programming in their territory, or can install their own TVRO satellite terminal to pick up the ACSN programming for distribution on campus or in regional community centers, etc.

The American Educational Television Network (AETN) distributes its nondegree continuing educational programming to cable companies throughout the nation via the SATCOM F-3 satellite. AETN works with professional associations to offer courses required by various regulatory bodies to provide professionals with ongoing training. Architects, CPAs, dentists, engineers, attorneys, nurses, pharmacists, doctors, and real-estate professionals all must complete a minimum number of educational credits each year in order to maintain their licenses to practice or to obtain a particular certification within their profession. There are more than 30 million such practicing licensed professionals in the United States. To participate in an AETN course, an individual would typically pick up the seminar through his or her local cable television company, and officially enroll in the curriculum through the national society or association operating in that field. Participants pay a fee ranging from $50 to $150 per course, and their successful completion of the educational materials, as indicated by a final test, is relayed to the applicable licensing agency of the state in which the participant resides.

As AETN's educational programming is primarily distributed by CATV systems, there are many locations to which the service is not carried. In these cases an individual could install a stand-alone TVRO earth station, or a local professional association or organization could install a TVRO earth station to enable its professional members to take these required continuing education courses on its premises. Hospitals, professional societies, and clubs that have meeting room facilities could make this service available to their membership and staff by purchasing the least expensive home-version satellite TVRO terminal.

During the summer AETN carries supporting course programming of a more general nature, which can be viewed by the general populace, as well as applied to the continuing education requirement of a number of professional fields.

Although most of these satellite TV educational programmers deliver their courses to cable companies for distribution, any motivated

1981

Summer Program Schedule

August

Time	Date	Time	Date	Time	Date	Time	Date
							1
						5:00pm	**LIVING ENVIRONMENT-10** *Engery Alternatives*
						5:30pm	**CASE STUDIES SM. BUSINESS-5** *Running the Show*
	3		**5**		**7**		**8**
6:00pm	**LIVING ENVIRONMENT-11** *Conservation of Vital Resources*	6:00pm	**LIVING ENVIRONMENT-12** *Economic Geology*	6:00pm	**LIVING ENVIRONMENT-13** *Solid Waste*	5:00pm	**LIVING ENVIRONMENT-14** *Wildlife Management*
6:30pm	**CASE STUDIES SM. BUSINESS-6** *The Balancing Act*	6:30pm	**CASE STUDIES SM. BUSINESS-7** *The Breaking Point*	6:30pm	**CASE STUDIES SM. BUSINESS-8** *Their Own Brand*	5:30pm	**CASE STUDIES SM. BUSINESS-9** *Dealing and Wheeling*
	10		**12**		**14**		**15**
6:00pm	**LIVING ENVIRONMENT-15** *Forest and Man*	6:00pm	**LIVING ENVIRONMENT-16** *Land Use in the City*	6:00pm	**LIVING ENVIRONMENT-17** *Water Resources*	5:00pm	**LIVING ENVIRONMENT-19** *Air Pollution*
6:30pm	**CASE STUDIES SM. BUSINESS-10** *Taking Off*	6:30pm	**GREAT PLAINS EXPERIENCE-1** *The Land*	6:30pm	**GREAT PLAINS EXPERIENCE-2** *Lakota: One Nation*	5:30pm	**GREAT PLAINS EXPERIENCE-3** *Clash of Cultures*
	17		**19**		**21**		**22**
6:00pm	**LIVING ENVIRONMENT-20** *Impact of Political Science*	6:00pm	**LIVING ENVIRONMENT-21** *Impact of Economic Systems*	6:00pm	**LIVING ENVIRONMENT-22** *Myths of Technology*	5:00pm	**LIVING ENVIRONMENT-23** *Individual Involvement*
6:30pm	**GREAT PLAINS EXPERIENCE-4** *Settling of the Plains*	6:30pm	**GREAT PLAINS EXPERIENCE-5** *The Heirs to No Mans Land*	6:30pm	**GREAT PLAINS EXPERIENCE-6** *Four Portraits*	5:30pm	**GOING METRIC-201** *Measurement of Length*
	24		**26**		**28**		**29**
	OFF		OFF		OFF	5:00pm	**LIVING ENVIRONMENT-23** *Individual Involvement*
						5:30pm	**LOOSENING THE GRIP-4** *Signs & Symptoms*
	31						
6:00pm	**LIVING ENVIRONMENT-24** *Solutions and Projections*						
6:30pm	**LOOSENING THE GRIP-11** *An Ounce of Prevention*						

Time Zone Reference Tables

Pacific DST	Mountain ST	Central DST	Eastern DST
2:00 PM	2:00 PM	4:00 PM	5:00 PM
2:30 PM	2:30 PM	4:30 PM	5:30 PM
3:00 PM	3:00 PM	5:00 PM	6:00 PM
3:30 PM	3:30 PM	5:30 PM	6:30 PM

American Educational Television Network schedule. (Courtesy of AETN)

individual or educational institution should be able to deal directly with the satellite programmers themselves, especially if the local CATV system does not carry the particular channel desired. Such a direct-reception arrangement would benefit both the programmer, who achieves a greater viewership, and the national society or organization associated with the particular professional discipline. With permission, a local educational institution or affiliated branch of a national professional association could videotape the programs, and build up an outstanding library of continuing education tapes for use by its local membership at any time.

Charity and Fund-raising Events

Satellite television can be used by many charitable organizations to raise funds. Such major national charities as the Muscular Dystrophy Association have employed the television satellites for years to tie together dozens of cities in multilocation telethons and teleconferences. Many events of this type are carried via satellite annually, and affiliated chapters of the national associations as well as other interested organizations can easily plug into a national teleconference or telethon event by installing a local TVRO satellite terminal. For a teleconference, the satellite terminal might feed its signal into several 25-inch color television monitors located in a classroom or meeting room reserved for the event. In the case of a major semipublic or public telethon, the satellite video signal might be fed, instead, into a large auditorium where the picture would be projected via a professional wide-screen color television video projector, rented for the occasion.

In lieu of purchasing a satellite TVRO earth station, the nonprofit charity or educational organization might elect to rent this facility. Both two-way TR (transmit-receive) and downlink-only TVRO satellite terminals are available for this purpose. As satellite TV installing dealerships spring up nationwide, more and more of these organizations will have portable 5-meter parabolic antennas mounted on trailers for use in their sales demonstrations. Such a facility could easily be rented over a weekend by an interested organization to pick up a one-time satellite television event.

Even if a fund-raising or teleconference event is held two or three times a year, it may be more cost effective for nonprofit organizations to rent a satellite terminal on an occasional-use basis than to purchase the equipment outright. Appendix 2 provides a list of installing dealers and manufacturers who could refer one to a local dealership that might be able to furnish this service.

The local charity or nonprofit group need not confine its ideas to its national parent organization's television efforts. Many other independent charities and organizations may not have the luxury of affiliation with a major national, highly recognized institution. In these cases the local nonprofit group can still utilize satellite TV for its own fund-raising events by picking up special programming such as a major sporting event or world boxing match totally by itself. By obtaining the video feeds at wholesale prices from the satellite programmer, a charity can sell tickets for the attraction, and realize as a profit the difference between the admission price or donation and the cost of the satellite-delivered programming negotiated with the satellite programmer. Although this approach may require a bit of imagination on the part of the management of the local nonprofit group, many of the satellite programmers will be happy to deal with any and all commercial and nonprofit organizations to ensure that its event is seen by as many people as possible. The Muhammad Ali fights have been carried by cable television systems, subscription (pay) TV stations, MDS carriers, and via closed circuit to auditoriums worldwide. Most promoters of world-class boxing events will enter into arrangements with any legitimate organization, providing that the attendance can be monitored, admission fees collected, and an accounting performed after the event is over. During several of the recent boxing matches, for example, enterprising theater owners installed temporary TVRO satellite terminals to carry the events live in their auditoriums. A nonprofit organization could employ satellite television in the same way for a charity fund-raiser.

Chapter 8

Setting Up a Satellite TV Dealership

Like any consumer electronics business dealing in high-technology sales, the successful satellite TV dealership must be able to solve the five puzzles that apply to the management and operation of the business—people, technology, money, market demand, and competition.

Properly trained and motivated people are the main key to a profitable retail business. The attitude that the salesforce and installation crew bring to their jobs will make the difference between a marginally successful operation and a viable business. This means that the salespeople have to be adequately trained in the concepts of satellite TV technology, and should have a real enthusiasm about the subject. Individuals who have worked in the consumer electronics field, especially in the premium hi-fi and video markets, should be good candidates.

The technology itself is inherently uncomplicated. A TVRO earth station has only four or five major components; a sophisticated stereo system may have over a dozen complicated parts with many knobs and dials to adjust. Many satellite television owners will have already outfitted themselves with projection television sets, video-cassette recorders, color cameras, cable television selectors, subscription television decoders, and complicated switching devices. In addition, many sophisticated video hobbyists—and there are over two million in the United States—are already familiar with the satellite television concept, and the installation of a satellite TVRO terminal will be a simple process. For the dealer, however, the selection of the appropriate mix of earth stations, receivers, LNAs, and related components may be a complicated process, given the large number of manufacturing organizations already in the field and the necessity of evaluating the various products individually. Later in this chapter, a few of the leading suppliers are reviewed briefly and their products and capabilities discussed. Appendix 2 presents a fairly complete listing of those major organizations in the TVRO terminal business today.

Financing the start-up business is an important consideration, and the amount of money required will vary from dealership to dealership, depending on its size and its marketing strategy. Individuals who wish to enter the mail-order satellite earth station sales business, delivering the product on a "drop ship" basis, may find that several thousand dollars of initial seed money will be sufficient to place classified ads, print sales documents and brochures, and cover the initial overhead, assuming that the manufacturers will be willing to ship their products in a single-unit quantity, and that the direct-mail customer is willing to pay up front for the product before delivery is made. The major stereo retailer or video store will require a different level of financing, probably between $50,000 and $100,000, to purchase the necessary display and demonstration equipment, and to stock the TVRO earth station components so that installations can be made rapidly, without relying on back-ordering the equipment from the factory.

Then there is the question of financing the customer's purchase. Although a number of banks will now extend personal loans to the buyer to cover the equipment cost, the valuation of a backyard TVRO earth station cannot be performed along the lines of, say, an automobile purchase. The industry is just too new—and the number of present TVRO satellite terminals can be counted in the tens of thousands, not the tens of millions.

The market demand is another piece of the dealership puzzle that must be solved. Individual buyers in a specific city or town must be identified and reached. One way to accomplish this is through selected mailings to subscribers to the leading video enthusiast magazines, such as *Videoplay Magazine* and *Home Video.* Also, mailing lists of subscribers may be rented from most publications for $50 to $100 per thousand names; they represent a good source for leads because these individuals are already paying money for information concerning the home video field. Displaying and operating TVRO terminals at local shopping centers and other public locations can create buyer interest, and demonstrations of the equipment on local TV talk shows and interviews with the local press and radio call-in shows can also increase the market awareness of the specific dealership, and market demand for the TVRO earth station product.

Although many individuals and organizations will be entering the marketplace to set up dealerships over the next few years, the competition for a well-thought-out installing distributorship should continue to be quite small, as the demand for home satellite TV terminals has been growing at near exponential proportions.

The market for a satellite earth station dealership is multifaceted, and at the retail "installing dealer" level the firm may elect to sell equipment to individual home consumers, install complete equipment for mini-CATV systems in apartment buildings and condominium complexes, provide the satellite earth stations for MATV systems in hotels, motels, and hospitals, and rent the equipment to various organizations for special events. Thus, an installing dealership can focus its energies by opening a retail store in a shopping center to cater to the local and walk-in traffic for that area, or it can concentrate on the commercial marketplace, possibly by operating in an industrial park or office building.

The TVRO Marketplace Distribution Mechanism

There are six major components in the distribution chain for the TVRO earth station market (see below). The electronics parts manufacturer, such as Hewlett-Packard, fabricates the actual transistors and other circuit components that make up the equipment components or modules. These people sell their products directly to the electronic component manufacturers.

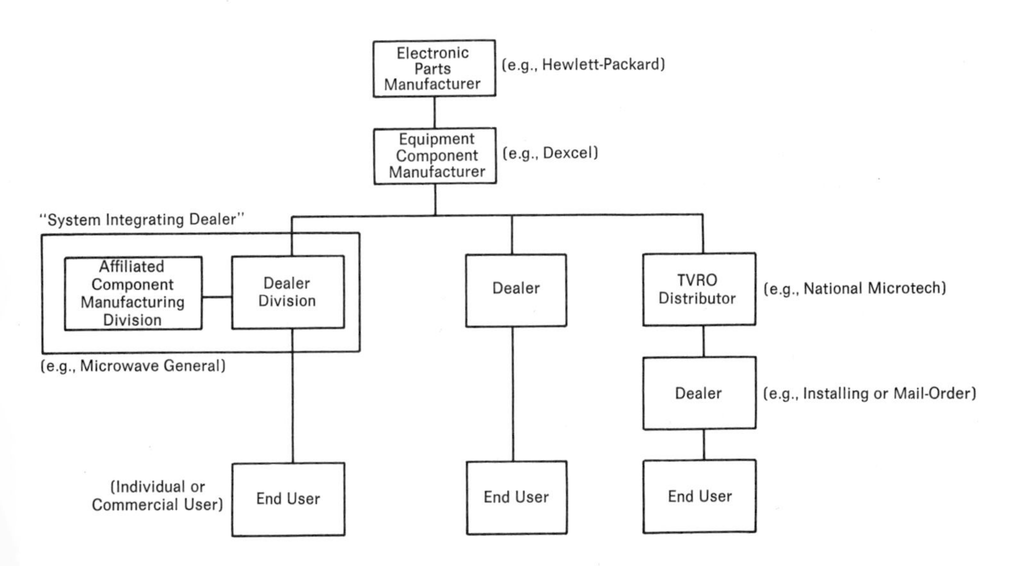

The electronic component manufacturers, such as Avantek, Inc., producers of low-noise amplifiers, purchase the basic electronic components from the parts manufacturers, and assemble them into complete modules, or system components such as LNAs, antennas, and receivers. These organizations usually prefer to sell their products directly to distributors or installing dealers, as their ability to provide detailed service and installation support at the consumer level is limited. Occasionally, however, the manufacturers will deal directly with the end consumer, especially if that user is a sophisticated video hobbyist or commercial buyer such as a mini-CATV system.

The component distributors, such as National Microtech, Inc., sell their products at retail to the end user, and at distributor wholesale cost to independent dealerships. As the marketplace becomes more sophisticated and the number of installing dealerships increases to several thousand or more, we can expect to see the component distributors establishing their own independent dealerships, dealing directly with the end consumer, and only selling to dealers at the distributorship level.

At present two kinds of dealers provide TVRO earth stations: the mail-order supplier, and the local installing dealer. The mail-order organization does not usually have the sales or service support capabilities that a local installing dealership maintains, but may offer somewhat lower prices for equipment that is purchased by the knowledgeable user. The installing dealership, on the other hand, will usually sell its product for a higher overall cost, but provides a turnkey installation, including ongoing maintenance support. Similar relationships have existed in the consumer electronics, stereo, and video marketplaces for years.

Marketing Strategies

Distributor

To be a stocking distributor in this business, long-term commitments need to be made with a dozen or so component manufacturers, to afford "second sourcing" on the receivers, dishes, and LNAs. Distributorships normally are negotiated with the factory at the highest levels, and ordering commitments range from 100 units per year and up. A distributorship is essentially a stocking and reinvoicing operation, providing an interface between the manufacturer and the dealers. Distributors may supply somewhat limited technical support in training and servicing their dealerships, but the distributor is primarily a purchasing vehicle through which the buying dealer must go to obtain the product for immediate delivery. Although some manufacturers elect to retain the distributor function (and profits) at the factory level, most will enter into agreements with independent wholesale distributing organizations with exclusive and nonexclusive territory options. As exclusive sales territories are difficult to enforce legally, many distributors have found themselves competing both for product shipments from their suppliers and in sales to their customers, the dealers. Thus a distributorship organization needs to establish its own unique group of dealers who will retain their loyalty to the specific distributor. This is often accomplished by providing the dealer with immediate delivery on product orders, personnel training assistance, and preferential financing terms on established dealer purchases. For the average person, acting as a distributor of TVRO earth station equipment will probably not be the best way to enter the market.

Dealer

A mail-order dealership, however, is a relatively low-cost operation that can be run by an individual at home part-time during the evenings and weekends. Mail-order dealers should focus their attention on reaching those potential buyers who are knowledgeable video enthusiasts, people who are already familiar with the hobby electronics field in general, and home satellite television in particular. These buyers may have already visited a local satellite TVRO dealership, have attended a national convention on the subject, and have read a number of recent articles in video magazines. The best publications in which to run classified advertisements for a mail-order dealership are those video and electronics magazines with large national audiences. These include *Home Video, Videoplay Magazine, Video Magazine, Video Review, Popular Electronics, Radio Electronics,* and *Coop's Satellite Digest.* The amateur radio publications, such as *QST,* might also be good places in which to try out a classified ad or two. Study some examples of classified ads that have appeared over the past few months in consumer electronics publications. Writing a good ad is, in itself, an art form of the higher type. And we all think we are experts at it! Sadly, many people who enter the mail-order business do not follow the few basic rules that are covered in any of a number of best-selling books on the subject. For those people who follow the rules, however, and understand both their market and the business they are in, mail order can indeed be the best way to enter the TVRO dealership business.

One of the most useful tricks is to look at a competitor's ad—one that has appeared again and again over a period of months. The reappearance of an ad usually means that it is pulling well enough to warrant its reinsertion in the magazine, which means the dealer is making money!

The installing dealership has higher initial costs than its mail-order cousin, but an installing dealer can often realize a greater net profit on gross sales because the customer receives a more complete package, including the value-added installation, assistance in proper equipment selection, and ongoing maintenance support. Some installing dealers have met with success by running small ads in the weekly TV magazines published by their local newspapers. The installing dealership can also be started on a part-time basis, with the sales and installation of the TVRO terminals handled during the weekends. A very effective marketing tool that can be used to the installing dealer's advantage is a trailer-mounted 10- to 15-foot portable TVRO terminal. The terminal can be hitched to the back of an automobile and pulled onto the customer's site for an "off-the-air" demonstration. And this same selling tool can be used by the dealer for occasional rental to groups and organizations that might require a TVRO earth station for a special event or periodic function. (See Chapter 7 for examples.)

Although many dealers offer only a single system configuration, a more sophisticated dealership will probably want to carry three types of TVRO terminal. The "do-it-yourself" package is a bare-bones system supplied off the shelf. The equipment is already boxed, and ready

Portable 4-meter earth station on a trailer. Used by dealer to demonstrate satellite television system at the customer's home.

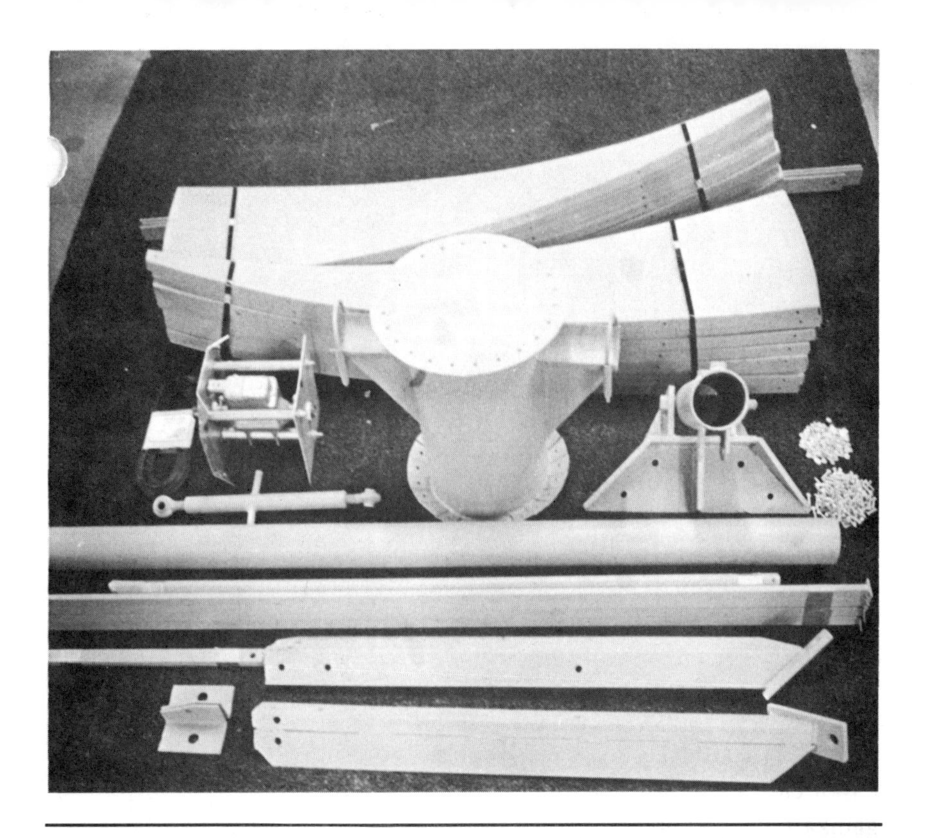

Some antennas are shipped "knocked down" for easy assembly. (Courtesy of Antenna Development & Manufacturing Inc.)

to be taken away by the customer. An 8- to 16-foot spherical antenna can be assembled by the customer directly. Components of the system can be sold separately, and can often be charged against a customer's credit card with an appropriate call to the authorization center to confirm the credit limit. Such a system should be sold only to the handyman who is not afraid to tinker when putting the antenna together. The dealer should provide an hourly, expert consulting service to assist the customer, if necessary, after the antenna has been erected and put in place, but the dealer should not actually assemble the system.

The middle-price standard model dealer-installed system consists of a 10- to 13-foot parabolic dish, with manually adjustable antenna mount, single LNA, and 24-channel, home-quality tunable receiver. This system will probably account for 50–60 percent of the dealer's sales. The deluxe dealer-installed TVRO system will probably include two LNAs, a motorized antenna mount, and a deluxe receiver with remote-control capabilities and multiple room feeds. This top-of-the-line system should always be on demonstration in the dealership to allow customers a "hands on" feel of what the best satellite TV system can do.

The fully-assembled ADM antenna. (Courtesy of Antenna Development & Manufacturing Inc.)

In addition to selling TVRO terminals, the installing dealer can carry add-ons and support products, including videotape recorders, television sets, books and magazines covering the video field, and home computer-controlled motorized antenna mounts. Computer stores already serving the home market may wish to expand into the TVRO earth station product line, as there appears to be a tremendous overlap in the community of interests between these two consumer electronics products. Indeed, one of the major satellite TVRO organizations, Downlink, Inc., was founded as a spinoff of Black and White Enterprises, Inc., a leading home computer store in Connecticut.

One option a number of installing dealers have elected to adopt is that of the combination dealer and partial system component manufacturer. In this arrangement the dealer purchases some, but not all, of the TVRO components from other manufacturers. Typically, the receiver

Side view of popular ADM antenna. (Courtesy of Antenna Development & Manufacturing Inc.)

and LNA are purchased from manufacturers who specialize in producing these components. The antenna surface (dish) itself or the antenna mount is often built by the dealer and integrated into a package that bears the dealer's unique brand. Many of these "integrating dealers" have entered the business from other fields from which antenna-building skills can be readily transferred. For example, shipbuilders who are good at producing accurate fiberglass hulls can also manufacture fiberglass parabolic dish surfaces. Many people who have worked with sheet metal feel comfortable producing spun-aluminum antennas. Other people who have worked in the building trades have found that the construction of wood and metal screen spherical antennas is a relatively straightforward task. There is quite a bit of detailed information in the marketplace concerning the specific requirements for designing a parabolic or spherical antenna of any size. Satellite Television Technology

Some antennas, like the Paraframe, require a crane for installation. (Courtesy of Paraframe Inc.)

of Arcadia, Oklahoma (see Appendix 2), has published a number of excellent manuals on this subject.

By acting as both a dealer and a specialty manufacturer, these installing dealers have been able to integrate the product line vertically, increasing their profits on the TVRO earth station package by including the profit of producing the antenna at the manufacturing level in the overall profit of selling the complete system. Many audio high-fidelity stores have done similar limited manufacturing/integrating for years by having local furniture and cabinet makers build finished speaker enclosures in which the dealer installs a tweeter and woofer speaker set, thereby creating a unique speaker brand.

Assuming that a 15-foot parabolic antenna sells for $2500 at dealer cost, and that this antenna is used as the basis for a $10,000 TVRO home satellite system, the dealer's profit can be increased significantly, rising from perhaps 20 to 35 percent.

How to Become a Dealer

It is almost a truism in the consumer electronics business that before you sell anything you should know the product inside out. This means that a dealer's first installation should be the dealer's own. If completed correctly, this first TVRO terminal is likely to become your heavily used demonstration system, and so it is important that the equipment involved be of the highest quality affordable. It is easy to sell a lesser system to someone who has been shown the penultimate; it is far more difficult to inspire a potential customer to purchase a more expensive system after having seen a marginal-quality TVRO satellite terminal. Therefore, a good demonstration system is absolutely essential.

Assuming that the first system is installed at home, there may be some initial problems which, from the outset, make entering the dealership business more difficult. For example, you might be living in Vermont. At this latitude the problem of "look angle" must be taken into consideration. The farther north one travels, the lower the antenna must be pointed to the horizon, and an antenna in northern Vermont is tilted upward only about 8 degrees. At this very low look angle, not only will a building across the street block the view of the antenna, but even an automobile parked in its way could affect the signal. By selecting the proper site, possibly by placing the antenna on the roof, this problem can be minimized or eliminated. The antenna, of course, must have a clear and unobstructed view by line of sight to the satellite 22,300 miles away.

The second problem—for people who live in metropolitan areas—is that of terrestrial microwave interference. Whereas interference from telephone company-owned microwave communications towers will occur very rarely in suburban or rural areas, urban areas—typically one in ten, or more, locations—will have a problem with interference. Terrestrial microwave interference varies significantly in its effects on satellite television reception. In some instances all of the channels on all of the satellites will be affected, with strong crosshatch interference pat-

RANE

Installation of a roof-mounted parabolic antenna. (Courtesy of Helfer's Antenna Service)

terns occurring in the picture and a gnawing, buzzing sound affecting the audio. In other instances the telephone company interference will occur only on a specific transponder or two, and produce only a slightly fuzzier picture. Often, moving the satellite TVRO antenna by only a few feet will eliminate the terrestrial interference completely. In severe cases the satellite antenna may have to be shielded with additional metal panels, or even placed in a hole, using the earth as a natural barrier to the unwanted microwave interference. Anyone who has installed a television antenna on a roof, and has had to twiddle with it to minimize ghosts in the picture, can appreciate the importance of proper antenna placement. Luckily, even a 16-foot antenna is not overly difficult to move a few feet before it has been permanently attached to a concrete or other fixed base.

One of the most effective ways of determining if a specific customer's site will be free from interference is to use a portable collapsible TVRO antenna, such as the Luly 12-foot parabolic antenna. It folds up like a big beach umbrella, and can be easily transported to the site and mounted on a sturdy, 35-mm tripod for testing. Such an antenna can also be used as an effective demonstration tool, with the warning that it is extremely lightweight and thus prone to acting like a sail in winds of any strength.

Once the decision has been made to focus on a quality installation for the first demonstration TVRO satellite antenna, the dealer then needs to investigate both the various types of systems that can be sold, and the problems and pitfalls of financing the customer's purchase.

Financing Equipment

Let's assume that you wish to sell a deluxe turnkey TVRO terminal for $10,000 (see Table 8-1). Such a system would include a 15-foot parabolic antenna, a 120-degree LNA, a 24-channel receiver with built-in down-convertor, a motorized antenna mount, miscellaneous cabling and parts, an antenna pad, and installation.

There are four financing options available to the dealer to help in assisting a customer to purchase the equipment.

First, the dealer can work with a local bank to establish a mechanism whereby the bank lends money to the customer directly for purchase of the equipment on an installment note, typically secured by the equipment itself. Such signature notes, while common in the $2000 to $5000

TABLE 8–1.
Financing the Customer's Purchase

Example #1: TVRO turnkey deluxe system:
Cost of $10,000 includes:
(1) 15-foot parabolic antenna
(2) 120-foot LNA
(3) 24-channel receiver w/down-convertor
(4) Motorized antenna mount
(5) Cabling and miscellaneous parts, antenna pad, etc.
(6) Installation
(7) 90-day guarantee

Financing options:	
(1) Bank loan	Consumer (personal) 70% (when available)
(2) Credit card	Typically 30% ($3000) maximum
(3) Dealer financing	Dealer carries note with 25% down
(4) Dealer financing	Dealer sells note to a commercial factor

Dealer Financing

$10,000	Cash price
−2500	Down payment to dealer
$7500	Dealer finances at 48 months at 24% interest: Monthly payment is $244.51
	Total repayment to dealer is: $11,736.66 (interest plus principal)

range, are much more difficult to arrange at the $10,000 level. When possible, a consumer will typically have to make a down payment of at least 30 percent, and commercial rates of 24 percent would not be unreasonable to expect. While the availability of bank money will improve as more and more TVRO systems are installed, this line of financing may be limited in many areas.

The second option is to establish a merchant's account with the Visa and MasterCard organizations to allow the customer to charge the purchase to the customer's Visa or MasterCard account. Most bank cards, however, carry a limit of well under $3000. Recently, a number of savings and loan organizations have begun to offer Visa cards to their customers who deposit funds into their savings accounts or money market funds. To attract these funds, the savings and loan association offers a charge card account with a line of credit equal to 80 percent of the money on deposit. Thus, a customer who deposited $10,000 into a savings and loan account would receive a Visa or MasterCard credit limit of $8000. This strategy could be used to finance some purchases, but would limit the possibility for sales.

A third financing technique requires that the dealer carry the loan,

Installation of large parabolic antenna mounted on cement base. (Courtesy of Channel One Inc.)

with perhaps a 25 percent down payment. On a $10,000 system price, assuming the customer pays the dealer a deposit of $2500, dealer financing for 48 months at a nominal 24 percent annual interest would permit the customer manageable monthly payments of $244.51. The total repayment to the dealer of principal and interest over this four-year period would amount to $11,736.66, giving the dealer an additional income of more than $4000, if the dealer is able to finance the purchase internally. This approach would be quite useful to the part-time dealer who has a full-time job and can afford to underwrite the customer's purchases.

The fourth option would allow the dealer to finance the customer's purchase directly and then sell the note to a financing organization or commercial factor group, which would purchase it at a discount. A number of organizations exist to provide a similar function in the hi-fi/stereo marketplace, and as more and more exotic and expensive video systems, including videotape recorders, color TV cameras, and projection television sets, are introduced to the marketplace, these financing groups will expand throughout the country.

Of course the fifth option—the customer pays cash—is always available!

Pricing the Typical System

If the dealer decides to offer three types of TVRO terminal—the basic turnkey system, the deluxe package, and the do-it-yourself handyman special—it appears that three different sets of components must be stocked. However, certain savings can be realized by stocking specific components for use in all three systems. The low-noise amplifiers (120°K model) will be used by all system types. Unless a combination LNA/down-convertor (such as the Dexcel combo unit) is used in a specific system configuration, the LNAs will be interchangeable among all systems. Even when the Dexcel combo LNA/down-convertor is selected, the manufacturer will allow the dealer to place quantity order purchases combining the combo units with straight LNAs to arrive at a total package purchase price.

In the case of our original $10,000 system (to be sold to the customer for, say, $9995), the dealer costs amount to approximately $8050 (see Table 8-2). At a selling price of $10,000 the net profit to the dealer will be $1950, a 19.5 percent markup. However, if the equipment is sold for

$10,995, the dealer's profit will be $2950, or a 37 percent profit margin, an acceptable return for a system of this type.

Let's consider a basic installed turnkey system (see Table 8-3).

TABLE 8–2.
Example 1—Dealer Costs for $10,000 TVRO System

(1)	15-foot antenna (parabolic)	$2500
(2)	Deluxe receiver (home quality) w/ down-convertor	2000
(3)	120° LNA	800
(4)	Motorized mount w/ remote control	1500
(5)	RF channel modulator w/ splitter	200
(6)	Cabling and miscellaneous components	150
(7)	Installation (25 hours at $20/hour)	500
(8)	Shipping and insurance	400
	Total Costs to Dealer	$8050
	Net to Dealer	$1950 (19.5% profit)

Such a system could use a 12-foot parabolic antenna with mount, circular feed, and remotely controlled polarization rotator. H&R Communications' Model 12-K wholesales to the dealer for $1500. The Dexcel Dex-1000 satellite receiver and combo LNA/down-convertor cost the dealer roughly $2200. The Microverter II RF TV channel modulator sells for $60, and coaxial cable concrete pads, shipping, and insurance are an additional $500. Installation at $20 per hour should take 12 hours, costing the dealer $240. The total price to the dealer for this basic turnkey installed system will run about $4600.

Such a system could be sold to the consumer for from $5795 to $6995. The dealer profits would range from 25 to 52 percent. Perhaps the "best" selling price (one that maximizes income to the dealer) would be $5995, which would give the dealer a profit of $1495 per system sold, on a dealer markup of 30 percent.

TABLE 8–3.
Example 2—Turnkey Basic System

Dealer cost:	(1)	H&R Communications Model 12 K 12-foot parabolic antenna w/mount, feed, and polarization rotator	$1500
	(2)	Dexcel DRX1000 satellite receiver with LNA down-convertor	2200
	(3)	Micro-Verter II RF TV channel modulator	60
	(4)	Coax cable and hardware	100
	(5)	Concrete pad and miscellaneous parts	200
	(6)	Shipping and insurance	300
	(7)	Installation (12 hours at $20/hour)	240
		Total cost to dealer	$4600

At $5995, selling just two such systems a month will generate income to the dealership of $33,480 per year. A small dealership with sales at this level could be handled on a part-time basis from one's home, with the dealer spending one weekend a month demonstrating and selling the equipment and another two weekends installing the newly sold systems.

Selling Price to Customer	Dealer Markup	Dealer Profit
$5795	25%	$1195
6995	52%	2395
5995*	30%	1395

*At $5995, sales of two such systems per month generate income to the business of $2790 per month, or $33,480 per year. Such a business could be run on a part-time basis.

The mail-order dealer or installing dealer who wishes to provide a do-it-yourself hobbyist system may elect to carry a bare-bones package consisting of an 8-foot spherical antenna and lower cost receiver. McCullough Satellite Systems, Inc., in Arkansas produces such an antenna for kit sales (see Table 8-4). Although the per-system profit on a do-it-yourself package might amount to only a few hundred dollars, the attraction of a complete satellite terminal selling for less than $3995 is significant, and the total number of these systems sold may be substantially higher than the turnkey installed systems, depending on the market strategy employed and the location of the dealership. However, the dealer must continue to emphasize to customers that such a system is for the advanced enthusiast and electronics hobbyist only, and the dealer should be able to provide a well-documented training aid such

TABLE 8–4.
Example 3—Hobbyist "Do-It-Yourself" System

Dealer cost:	(1)	McCullough "8-ball" spherical antenna	$500
	(2)	SAT-TEC R2A receiver	750
	(3)	Avantek 120° LNA	650
	(4)	Micro-Verter II RF modulator	50
	(5)	Coax and connectors	80
	(6)	Shipping and insurance	100
		Total cost to dealer (24+ quantity)	$2130

Note: This system must be put together by the user. Therefore, no installation charge is included. A detailed dealer-provided instruction booklet and preferably a complete videotape showing actual installation by the dealer of such a system must be provided the customer. It should be emphasized that such a system be sold to only an experienced video enthusiast on a "do-it-yourself" basis.

as a videotape to show in detail how the kit should be put together. Also, sales of the do-it-yourself kit may remove some of the buyers from the marketplace, people who ordinarily would have purchased the basic dealer-installed turnkey system at a slightly higher cost. These are decisions, of course, that each dealer will have to make after analyzing the dealership's own unique marketplace.

Selling Price to Customer	Dealer Markup	Dealer Profit
$2499	17%	$369
2799	31%	669
2995*	41%	865
3495	64%	1365

*Recommended selling price

Assuming that the dealer will be purchasing equipment either directly from the manufacturers or on a discount basis from distributors, lot purchases of 24 or more units must be made before the dealer can obtain the necessary discounts to purchase the equipment for the prices that have been outlined above (see Table 8-5).

As an example, consider another basic turnkey system using an ADM 11-foot parabolic antenna (ICM Model 4300 remotely controllable receiver and 120-degree Avantek LNA combination). In mid-1981, for quantities of 24 or more, the LNAs could be purchased for approximately $600 each. The antenna systems were selling to dealers for $1800, and the receivers wholesaled at $800 apiece. Including the addi-

TABLE 8–5.

Example 4—Dealer Financing for a Typical System

Total cost to dealer for equipment*			
Initial delivery (first month)	6 LNAs	$3,600	120° LNA (Avantek)
	2 Antennas	3,600	11/12-foot parabolic with
	2 Receivers	1,600	polarization rotator
	2 Modulators	120	and cyl. feedhorn;
	2 Cable sets	200	e.g., ADM II 11-foot—ICM Model 4300
	2 Miscellaneous parts	500	Includes antenna base
		$9,620	(poured concrete)
Following 11 months (total):	18 LNAs	$10,800	
	22 Antennas	39,600	
	22 Receivers	17,600	
	22 Modulators	1,320	
	22 Cable sets	2,200	
	22 Miscellaneous parts	5,500	
		$77,020	

Or an average of two systems delivered per month for a monthly dealer cost of $7001.28.

*Assuming 24+ lot purchases spaced out over one-year delivery period, based on mid-1981 prices for a "standard-system" installed TVRO terminal of $3610 equipment, $240 labor.

tional cable sets, miscellaneous parts, and hardware and TV modulator, a total TVRO terminal employing this system configuration would have cost the dealer $3610. Twelve hours of labor at $20 per hour to install this system would increase the cost by $240.

If the initial delivery included two of each component and six LNAs, the dealer would need to write a check for $9620 for the first month of operation. (Note that LNA manufacturers do not like to ship piecemeal.) It is assumed, however, that as the demand for low-noise amplifiers and subsequent backlogs increase, the manufacturers will be willing to partial-ship a 24-unit order in several lots, with the first delivery consisting of six units (minimum). Over the following 11 months an additional 18 LNAs and 22 other system components will be delivered, which will cost the dealer approximately $77,020. This means that an average of two systems per month will be delivered for a dealer cost of approximately $7001 per month.

Looking at the cash flow projections for such a dealership, let's assume that of the two units delivered during the first month, one is retained as a demonstration and the other unit is sold for $4995, providing a 38 percent markup. Assume thereafter that two units are sold per month. This dealership thus will require about $5000 in investment, up front, to cover the negative cash flow during the first two months of operation. By the end of the third month the dealership will be marginally profitable. By the end of 12 months, the dealer will have netted $22,734 in this example. Now, if system sales are slowly increased to two per week, which could be accomplished by two people working on a part-time basis, such a dealership would realize a net profit (before business expense overhead) of about $100,000 per year, providing new satellite television service to about 100 people per annum! (See Table 8-6.)

Reviewing the Suppliers

The home satellite TVRO earth station business is characterized by both old-line reputable manufacturers and newer, less experienced organizations. Because the industry is so new, many backyard operations have sprung up to manufacture satellite dishes and electronics, and a number have already disappeared. Therefore, for the beginning dealer it is important to find out as much about a vendor's operation and business strategy as possible. When writing or calling a manufacturer,

TABLE 8–6.
Cash-Flow Projections
TVRO "Standard System"—Equipment Plus Labor Costs
for ADM/ICM Combination Package of Example 4

	Month											
	1	2	3	4	5	6	7	8	9	10	11	12
Equipment	$9,620	$7,001	$7,001	$7,001	$7,001	$7,001	$7,001	$7,001	$7,001	$7,001	$7,001	$7,001
Installation[1]	240	480	480	480	480	480	480	480	480	480	480	480
Total cost	9,860	7,481	7,481	7,481	7,481	7,481	7,481	7,481	7,481	7,481	7,481	7,481
Sales[2]	4,995	9,990	9,990	9,990	9,990	9,990	9,990	9,990	9,990	9,990	9,990	9,990
Net profit	(−4,865)	2,509	2,509	2,509	2,509	2,509	2,509	2,509	2,509	2,509	2,509	2,509
Cumulative cash flow	(−4,865)	(−2,356)	153	2,662	5,171	7,680	10,189	12,698	15,207	17,716	20,225	22,734

Notes:
[1]Average 12 hours at $20/hour labor.
[2]Assuming one unit sold in month one; thereafter two units sold per month.
First unit retained for personal demo system.
Selling price $4995, providing a 38% markup.

a set of specific questions should be asked to help clarify the product offerings and corporate capabilities and strategies.

For *antenna manufacturers,* basic questions include the following:

What antenna sizes do they make?
What type (parabolic or spherical) is offered?
How much do they cost?
What is the antenna gain (in dBw) for each antenna size?
Are motorized mounts available for the antennas?
What types of feedhorns are used?
Are dual vertical/horizontal feeds available?
Is a motorized polarity feedhorn rotator available?
Does it have remote control?
What is its cost?
What is the shipping weight of the dish?
Is it petalized (that is, does it break down into pieces)?
What material is the dish made of (fiberglass, wire mesh, spun aluminum, etc.)?
Will the dish require a crane or other heavy-duty equipment for installation?
How many hours will it take to assemble the dish?
What payment terms and delivery schedules are available?

For *manufacturers of receivers,* another set of questions is applicable:

Is the receiver a frequency-agile fully tunable 24-channel model?
How much does it cost in quantity?
What are the price breaks?
Can the receiver be remotely controlled and tuned?
What is the FM threshhold of the receiver (should be 7–8 dB, at least)?
How many audio subcarriers can be tuned in by the receiver? (RCA satellites use one subcarrier frequency, Western Union WESTAR satellites use another).
Are the audio features switch-selectable from the front panel?
If not, how easily can they be changed?
Is this an optional feature? If so, what is the additional cost?
Is a built-in RF modulator available?
If not, how does the receiver manufacturer recommend the receiver be attached to an ordinary TV set?
Are remote controls for the feedhorn polarity rotater built into the receiver?
Is an LNA power supply built into the receiver? (LNAs typically require 15–28 volts of direct current to be supplied by a separate power supply if they are not ordered with a built-in AC power supply option.)
What other features are available on the receiver?
Does it come in a wood cabinet or rack mount?
What is its size?
What payment terms and delivery schedules are available?

Although the *LNA manufacturers* all offer mostly similar products, a number of questions can still be asked these firms:

What types of LNAs are available?
Which LNA is best for use in your location (80°K, 100°K, 120°K, 150°K)?
What do they cost?
Is a built-in down-convertor option available to provide a combination LNA/down-convertor package?
How much does it cost?
Is a built-in power supply provided?
What is its cost?
If not, what are the DC power supply requirements of the LNA (typically 15–28 volts at a hundred milliamps)?
Are the industry standard mounting flanges (WR229 type), probe, and isolator components provided?
What payment and delivery terms are available?

A similar set of questions can be asked of the *suppliers of the necessary innerconnecting cables:*

What diameters and lengths are available?
What are the losses in decibels per linear foot?
On what types of connectors are the cables terminated?
What are the costs, payment and delivery terms, etc.?

The *modulator manufacturers* should be queried on the following:

Does the modulator come in a separate box or can it be mounted inside a receiver?
What kinds of connectors are required?
Does it have a built-in power supply?
What are its costs, payment terms, delivery schedules, etc.?

Finally, the *automated motorized antenna mount manufacturers* should be asked:

Which antennas can be easily coupled to the mounts?
Can multiple remote-control panels be connected?
What are the costs, payment and delivery terms, etc.?

When the foregoing questions have been carefully and adequately answered, the prospective dealer should have a pretty good feeling for the capabilities and professionalism of a specific vendor. It is always important, of course, that the dealer "stay on top of" the vendor once payment has been made for a specific product. Factory shipping delays are always possible and it will often be necessary to follow up on delivery times to make sure that the equipment is actually shipped on schedule.

The following list summarizes a few of the well-known TVRO equipment manufacturers.

Antenna Manufacturers

Several dozen manufacturers presently build parabolic and spherical TVRO antennas available for the home and commercial marketplaces. Detailed information concerning the manufacturers and their addresses may be found in Appendix 2. Not all of these manufacturers are listed here but those most commonly found in the home satellite and mini-CATV markets are included.

A-B Electronics. A-B Electronics has been in the marketplace for several years, offering complete "high-ticket" TVRO terminal packages. A-B purchases Paraframe 5- and 6-meter antennas and marries them to their motorized remote-controlled polar mounts. AB has shipped products throughout the Caribbean area and South America, and also acts as an installing dealer for mini-CATV systems in Florida.

ADM (Antenna Development and Manufacturing, Inc.). The popular Model AD-11 parabolic 11-foot antenna is relatively lightweight (200 pounds), and employs a 24-petal aluminum dish surface that is shipped knocked-down in a small box. The antenna uses a standard polar mount (weighing an additional 265 pounds) and the package can be assembled on site without the use of a crane. ADM is a well-respected and popular manufacturer.

Andrew Corporation. Andrew sells primarily to the CATV marketplace, offering a series of parabolic antennas, of which hundreds have been installed nationwide. Heavily used by the telephone industry, these are rated as extremely high-quality but expensive products.

Anixter-Pruzan. Oriented to the CATV market, Anixter-Pruzan manufactures its own 5-meter parabolic dish and also acts as a general distributor of CATV products and equipment. The firm is well respected in the telephone industry and, like Andrew, makes a quality but expensive product.

H&R Communications, Inc. A highly successful company primarily selling to the home satellite TV industry, H&R Communications manufactures a series of parabolic antennas ranging in size from 13 to 20 feet. The organization has become the industry's largest volume distributor of TVRO terminal systems and equipment, including receivers, LNAs, trailer-mounted antennas, and other products. Multiple mounting bases are available, and the fiberglass antennas are relatively lightweight.

Harris Corporation. Some of the largest earth stations in the world are manufactured by the Harris Corporation, and the organization specializes in commercial CATV and INTELSAT tracking and transmit/receive stations. The Harris Model 6200A, a 6-meter parabolic antenna with Cassegrain feed and azimuth-elevation mount, is popular with CATV companies. The antenna mount can be motorized and remotely controlled (Harris also manufactures a full line of satellite TV receivers for the CATV receivers). The company's products are extremely well respected and quite expensive.

HERO Communications. A range of 16- to 24-foot parabolic antennas of the company's own design has enabled HERO to capture sales throughout the Caribbean and South America as well as the United States. Weighing well under 300 pounds, the 6-meter antennas can be assembled without a crane and can be coupled to a heavy-duty motorized base assembly that will allow the antenna to be steered from horizon to horizon in less than a minute. This is a well-built product for the consumer TVRO marketplace, often used to view the more exotic INTELSAT and Russian international satellite systems.

Home Satellite Television Systems. Presently manufacturing a fully motorized complete 10-foot parabolic antenna system, HSTS provides the easiest unit to operate on the market today. To switch from one satellite to another, the user simply pushes the desired button on a ten-digit TouchTone keypad; the antenna automatically turns to position itself on target for the new satellite! HSTS is a well-made reasonably priced consumer TVRO system. A 13-foot antenna is also available.

Hughes Microwave Communications Products. Hughes is the industry leader in satellite communications, having built most of the communications satellites for the common carriers. The organization primarily specializes in military and commercial satellite communications, and its 5-meter dish is particularly popular with CATV companies. Hughes also manufactures receivers and other CATV equipment, including the Model IDC-472 down-convertor, which can handle up to 12 channels simultaneously for each polarization type for down-conversion feeding to 12 separate individual receivers. As an industry standard, Hughes is outstanding but expensive.

Lindsay Specialty Products Ltd. Lindsay has been in the Canadian CATV antenna business for several decades. Its 10-, 12-, and 15-foot parabolic antennas are heavy-duty, well-made units, available for both the mini-CATV and home consumer marketplaces. The products sell in the medium-price range.

Robert A. Luly and Associates, Inc. Bob Luly manufactures a unique, low-cost home-user parabolic antenna, designed to fold up like a beach umbrella. A collapsible metalized Mylar antenna that weighs just 15 pounds for the 15-foot version and 12 pounds for the 10-foot version, it is a brilliant design that sells for around $1000. The Luly antennas can easily be shipped in the collapsed state, and can be mounted on an inexpensive heavy-duty 35-mm tripod for immediate installation. Because of their exceptional flexibility and light weight, however, the Luly antennas are subject to the slightest of breezes, and must be enclosed in a geodesic dome or other housing. Luckily, large recreational tents are available from Sears, Roebuck and camping supply stores that can be used to house the antenna in the backyard, or even on a roof. The Luly antenna makes an excellent demonstration device for the installing dealer who wishes to set up an actual working system at a customer's home. It could be coupled to a hobby-type motorized telescope base for remote-control pointing of the antenna.

McCullough Satellite Systems, Inc. The leading manufacturer of consumer backyard spherical antennas, McCullough (previously known as Vidiark Electronics) makes the popular 8-ball 8-foot TVRO antenna and other spherical antennas ranging in size from 6 to 15 feet. The antenna is shipped unassembled in kit form and consists of wooden support framing with a wire mesh reflective surface. The prices are low and the quality excellent.

Microdyne Corporation. A major manufacturer of commercial and CATV antenna systems, Microdyne has not officially entered the home satellite TV marketplace, but its products, including receivers, are of high quality and medium price, and available for the top-of-the-line home TVRO terminal.

Microwave Associates Communications. A major CATV and commercial supplier, Microwave Associates manufactures a number of antennas, and other microwave-related products. The products are expensive but of high quality.

Microwave General. This company provides a top-notch, home parabolic-dish TVRO system that includes a fully motorized remote-control antenna mount and pointing system. For the complete top-of-the-line deluxe home satellite terminal, the Microwave General antenna with remote-control motorized mount and integrated receiver system is the way to go.

Mini-Casat. This company manufactures a 10-foot parabolic dish, sans the antenna mounting base and feedhorn. One of the least expensive parabolic surfaces available, the reflector sells for less than $700 in single quantity and below $600 each for ten or more.

Paraframe Inc. One of the granddaddies of the home satellite TV industry. Paraframe entered the marketplace in 1977 to produce a stressed parabolic wood-framed structure that offers tremendous strength and highly accurate surface tolerances at very low cost. The Paraframe parabolic antennas are shipped knocked-down in component parts and can be easily assembled on site without using a crane. It is a good product for kit sales at reasonable prices.

Scientific-Atlanta Inc. The Scientific–Atlanta corporation is the primary commercial supplier of CATV antennas. Its parabolic line is exceptionally complete, ranging from 3- to 10-meter sizes, and coupled to a complete range of receivers, modulators, and other CATV equipment. The Scientific–Atlanta name is recognized throughout the industry and hundreds of hotels, motels, and apartment buildings are presently using Scientific–Atlanta's extremely expensive systems.

United States Tower Company (USTC). A commercial organization popular in the mini-CATV business, USTC manufactures a series of parabolic and spherical antennas at reasonable prices for the high-quality, home TVRO terminal. The company prefers to continue specializing in CATV equipment sales.

Receiver Manufacturers

In 1977 a typical CATV-quality TVRO earth station receiver sold for $4000 to $6000 and up. Today, over a dozen new companies manufacture receivers whose prices have dropped to as low as $800, and are continuing to fall. Receiver manufacturers range from the very smallest start-up companies, which often are producing the leading state-of-the-art product, to the largest old-line commercial suppliers.

AVCOM of Virginia Inc. An excellent receiver, the AVCOM unit is priced somewhere between the typical home-quality consumer receiver and the commercial CATV receiver, with performance equal to that of the best commercial equipment. The AVCOM model includes a remote-control option, a vertical/horizontal polarization rotation switch, and other useful features.

Comm-Plus (Communications Plus). This Canadian firm produces a good-looking, high-quality receiver in the medium-price range for the home TVRO marketplace.

H&R Communications, Inc. Two different receiver systems are available. The first is a 24-channel, tunable unit with top-of-the-line options, including automatic scan tuning, built-in meters, and automatic selection of audio subcarriers. The second receiver is packaged with a 10-foot parabolic antenna, providing most of the electronics at the antenna itself, allowing an inexpensive coaxial cable to bring the picture into the house and directly to the television set. H&R Communications is a major manufacturer and distributor, and is well respected in the industry.

International Crystal Manufacturing Company (ICM). One of the major suppliers of home satellite TV receivers, the company's popular Model 4200 and the later Model 4300 have sold in the thousands of units and provide a top-quality remotely controllable system with multiple options at extremely low costs. ICM has been in business since 1946 manufacturing high-quality quartz crystals for the commercial and amateur radio marketplaces and is well respected in the field.

KLM Electronics Inc. A well-made receiver offered by a relatively new company, the Sky Eye 1 provides multiple-transponder tuning and continuous audio subcarrier tuning to pick up the various hidden audio channels on the different birds. Utilizing a split-receiver arrangement with the majority of the receiver electronics and down-convertor system designed to be mounted at the antenna dish, regular inexpensive coaxial cable can be used to bring the video signal into the final satellite tuning selector located near the television set.

Ramsey Electronics Inc. Manufacturers of the Sat-TEC R series receivers, Ramsey Electronics has been well received, and has shipped hundreds of receivers that retail for under $1000. Theirs is a well-made, excellent product, available at low cost.

Other receiver manufacturers not mentioned include *Hughes Microwave* and the *Harris Corporation,* which were covered under antenna manufacturers, as well as a number of CATV and commercial suppliers who are also in the marketplace. (See Appendix 2 for further information on these organizations.)

Low-Noise Amplifier (LNA) Manufacturers

There are basically three major low-noise amplifier manufacturers in business selling products to both the CATV industry and the home TVRO marketplace. In addition, a fourth manufacturer provides LNA kits and complete low-cost LNA packages for the home user.

Amplica Inc. This company manufactures both 120-degree LNAs and 100-degree devices at competitive prices in lots of 25, 50, 100, and up. Amplica is a well-respected microwave component manufacturer.

Avantek. Another major commercial manufacturer, Avantek has been instrumental in reducing the price of 120-degree LNAs to under $500 each in dealer quantities of 100 or more.

Dexcel Inc. A CATV-microwave component supplier, Dexcel was first to see the outstanding potential of home TVRO satellite terminal systems, and offers inexpensive LNAs in ranges from 85°K to 150°K. Dexcel also supplies a popular combination LNA and down-convertor, and with the development of its new Drex 100 microminiature satellite receiver has become one of the major suppliers of home TVRO home terminal components.

Simcomm Labs. This Kersey, Colorado, organization specializes in producing low-cost microwave components and receiver assemblies for the consumer marketplace. The 120-degree LNA retails for under $600, and dealer costs in quantity can push this figure to below $400. Simcomm does not have the frame or production capacity, however, of the "big three" LNA-25 manufacturers.

Other major manufacturers marketing TVRO products include *Chaparral Communications Company,* manufacturer of the outstanding Superfeed circular feedhorn. The Chaparral feedhorn is inexpensive, costing less than several hundred dollars, and can add up to 1.5 dB valuable gain to a home TVRO satellite antenna system. This can often make the difference between perfect TV reception and a picture full of noise.

ATV Research manufactures the popular MicroVerter II RF modulator, which converts the baseband video output of the satellite receiver to a standard television channel that can be tuned in by an ordinary television set.

A B Electronics, Microwave General, Home Satellite Television Systems, and *HERO Communications* all manufacture deluxe remotely controlled motorized antenna mounts, which enable the home viewer to reposition the antenna from one satellite to another from the comfort of the livingroom.

Chapter 9

Putting Together a Typical Satellite TV System: Three Choices

For the satellite TV enthusiast there are essentially three possible ways to put a satellite in the backyard: build one yourself, buy one in component form (antenna, receiver, LNA) and assemble it over a weekend, or have a local dealer install a complete TVRO terminal for you. The purpose of this chapter is to look at the pros and cons of each of these choices, to help you determine which arrangement would be best for your own needs.

Buying from Scratch for the Electronics Expert: The "Cheapie" System

Back in the mid-1970s Taylor Howard, a professor of electrical engineering at Stanford University, having worked with satellite microwave technology for a number of years, decided to put together his own backyard TV earth station to pick up the then brand-new cable television feeds. Ty, or Tay as he is known to his friends, proceeded to modify an old military surplus parabolic antenna to track the new television satellites. To cut down on the expense of purchasing a professional satellite video receiver (they were then selling for upward of $10,000), he designed his own circuit boards for a home-brew receiver,

Taylor Howard's original Navy-surplus backyard antenna. (Courtesy of Howard Engineering)

which he built in his basement. After a number of months of tinkering with his new toy, Ty was able to pick up excellent television signals direct from his backyard with an overall investment of less than several thousand dollars. About the same time, another enterprising engineer —Robert Coleman in South Carolina—purchased an old telephone company surplus parabolic microwave antenna and went about building a similar backyard system using surplus military electronic parts and inexpensive hobbyist equipment. Bob was able to bring his total costs down to well under $1000, and later experimental development work done by two other TVRO experts, Clyde Washburn, Jr., and Nelson Ethier, proved that it was possible for amateur radio operators and electronic hobbyists to put together their own receivers and parabolic antennas using readily available inexpensive components.

Another of Ty Howard's backyard military service antennas. (Courtesy of Howard Engineering)

The inside of a typical "do-it-yourself" receiver kit.

By 1980 all of these electronics innovators had been contacted by Bob Cooper, Jr., of Satellite Television Technology in Oklahoma. Bob, the original editor of *CATJ,* the journal of the Community Antenna Television Association, persuaded them to document their developments and publish them in the form of receiver and antenna manuals, which STT is currently selling.

The response was immediate and overwhelming. Several thousand copies of these experimenters' "how-to" instruction guides have been sold for $30 to $40, and the low-cost antenna and receiver designs described in detail in the manuals have been developed by other people who now produce commercial, fully packaged equipment for sale to individuals and dealers, and individual circuit boards and specialty components that are sold at cut-rate prices to the "do-it-yourself" electronic hobbyist.

The printed circuit cards used in the Howard receiver consist of five functional components. They are available from Robert M. Coleman, Route 3, Box 58-A, Travelers Rest, SC 29690. They sell, less components, for $99.

Similar basic receiver modules are available from Sigma Interna-

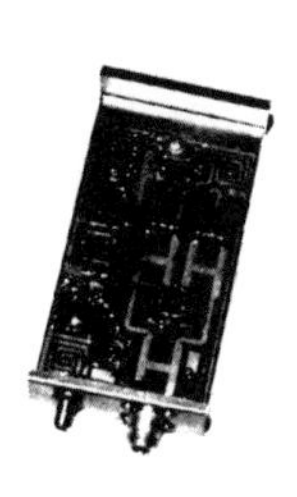

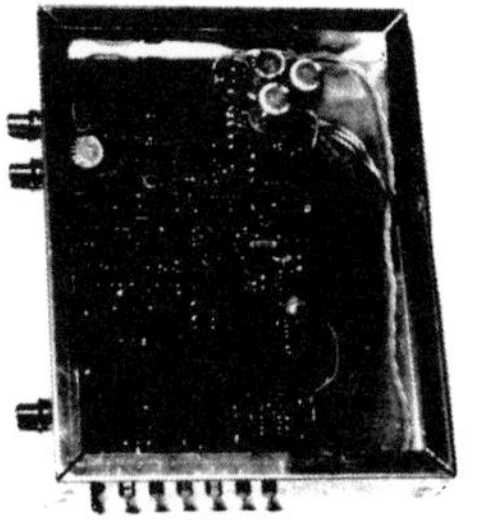

INDRA TVRO VI

BASIC RECEIVER MODULES

STATE OF THE ART OPERATIONAL P.C.B. MODULES

Enter the fast growing TVRO Industry with INDRA's readily marketable modules.

INDRA offers to the manufacturer a basic receiver package with a wealth of options easily adaptable to your private packaging and labeling requirements.

With INDRA's uncompromised design, your end user will experience: Sharper pictures with purer colors, less noise and interference and undistorted sound.

O.E.M. and DISTRIBUTOR INQUIRIES ARE WELCOME
HIGH VOLUME DELIVERY AVAILABLE

ADVANTAGES

- Simplifies your package design
- Uses your private label
- Continued R&D support
- Warranty service
- Low cost installation
- No development costs
- Small package design
- Virtually unlimited options
- High reliability
- Superior performance
- Prepackaged down converter

STANDARD FEATURES

- Theoretically ideal for greatest FM advantages
- 150 Mhz IF frequency to reduce cable tilt.
- Sharpskirt bandpass filter
- Theoretical 1.8db carrier to noise ratio
- Built in D.C. power block for LNA
- No alignment, nothing to become detuned.
- LNA power, down converter power and tuning using RG-59 Coax

INDRA SATELLITE COMMUNICATIONS, INC.
MARKETED NATIONALLY AND INTERNATIONALLY BY:
SIGMA INTERNATIONAL, INC.
MAILING ADDRESS: P.O. Box 1118 Scottsdale, Arizona 85252 U.S.A.
TELEPHONE: (602) 994-3435 TELEX: 165 745 Sigma CABLE: SIGMAS
OFFICE ADDRESS: 617 North Scottsdale Road Scottsdale, Arizona 85257 USA

Magazine advertisement for receiver kits available from Sigma International, Inc. (Courtesy of Sigma International, Inc.)

70 MHz DEMODULATOR CARD

The Sat-tec D-1 demodulator is the last block in a TVRO system, it is where the 70 MHz IF signal is converted to video and audio. The D-1 contains a PLL demodulator, video processor (CCIR de-emphasis, 4 MHz low pass filtering and 30 Hz clamp), dual sound sub-carrier demod and AFC circuitry. The power requirement is small, 15VDC @ 200ma., signal input is -20dbm @ 70 MHz. AFC will enable the user to lock most any VTO L.O. with no problem whatsoever. Video and audio outputs are a standard 1 volt p-p suitable for driving any monitor, VTR, or modulator.

SPECIFICATIONS:

Signal input 70 MHz at -20dbm (22mv)
AFC lock range: greater than 5 MHz
Sound subcarriers: 6.2 MHz and 6.8 MHz fully independent
Video level out: std. 1 volt p-p
Audio level out: 1 volt p-p
Power requirements: 15VDC @ 200 ma
Demodulator NE564 PLL IC
Tuning voltage out: 2 to 13.5 volts
Tuning voltage in: 0 to 15 volts max.

D-1 Demodulator Kit $99.95
D-1 Demodulator PC board only $49.95

Part Number	Description	Price Each
Avantek GPD-1002	1GHz, 12 db gain TO-8 can amplifier, 15VDC	$45.00
Watkins-Johnson V802	2.5-3.7GHz VTO, lower noise than Avantek types	120.00
Watkins-Johnson V705	600-1000MHz VTO, lower noise than Avantek	120.00
Signetics NE564	PLL selected to operate at 70MHz	7.50
Vari-L DBM-500	4GHz mixer, SMA connectors	85.00
Amperex ATF-417	1GHz, 25db gain hybrid amplifier, 20-24VDC	19.00
Motorola MWA-110	400MHz, 14db gain, -2.5dbm	9.00
Motorola MWA-120	400MHz, 14db gain, +8dbm	9.75
Motorola MWA-220	600MHz, 10db gain, +10.5dbm	12.40
Motorola MWA-230	600MHz, 10db gain, +18.5dbm	13.50
Motorola MWA-310	1GHz, 8db gain, +3.5dbm	12.40
Motorola MWA-320	1GHz, 8db gain, +11.5dbm	13.50
Motorola BFR-90	3GHz F_t NPN transistor, 15db gain @ 1.2GHz	2.50
Motorola MRF-901	3GHz F_t NPN like BFR-90 but 2 emiter leads	2.75
Regulators: 7800 Series	5V, 8V, 12V, 15V, 1A TO-220	1.50
Regulators: 7900 Series	-5V, -8V, -12V, -15V, 1A TO-220	1.75
IF Transformer	10.7MHz IF can be padded to 6.2 or 6.8MHz	1.25
Tuning capacitor	10pf multi-turn for filters, PLL, etc.	.95
Coil form + can set	Nice coil form set for filters, good to 120MHz	2.00

Sat-tec Systems; Box 10101
Rochester, NY 14610; (716)381-7265

Recent ad for receiver modules by SAT-TEC. (Courtesy of SAT-TEC)

tional, Inc., Box 1118, Scottsdale, AZ 85252, and a number of other Canadian and U.S. organizations (see Appendix 2).

About the time that Ty Howard was at work on his new backyard satellite terminal, an enterprising manufacturer by the name of Hayden McCullough formed McCullough Satellite Systems, Inc., in Salem, Arkansas, to build the inexpensive, spherical antennas originally developed by Oliver Swan. The 8-ball comes in three sizes: 8-foot, 10-foot, and 12-foot models. The antennas list for $650, $710, and $750, respectively, and can be put together by an individual over a weekend. The McCullough spherical antennas are about the cheapest on the market, and unlike a parabolic antenna, don't resemble a satellite terminal. They look more like the wood and screen latticework on which one might grow ivy! The antennas are built from 1½- by 1½-inch angle iron and 1- by 2-inch redwood strips running horizontally and spaced 8 inches center to center, attached to 1- by 3-inch strips running vertically. The redwood horizontal and vertical strips are fastened to the angle iron frame with adjustable bolts by which the exact curvature of the reflector surface is set. The reflector surface itself consists of inexpensive aluminum screening attached to the horizontal strips with staples. Sitting out about 15 feet in front of the antenna is a focal horn that picks up the signals that the antenna has reflected and concentrated from the distant satellite.

Unlike the parabolic antennas, the spherical antenna can receive several different satellites simultaneously, and these satellites can be picked up by placing two or more focal horns in the proper positions about 15 feet in front of the antennas.

Utilizing the combination of a home brew receiver constructed from the circuit boards developed by Ty Howard and others and the inexpensive Swan spherical antenna kit sold by McCullough, a do-it-yourself hobbyist can put together two-thirds of a home TVRO system for well under $1000. The last component, the low-noise amplifier, can be purchased "off the shelf" from one of the commercial LNA suppliers, or can be built using a basic LNA component kit available from several other manufacturers. One of these organizations, Simcomm Labs (Box 60, Kersey, CO 80644), now makes a low-cost LNA available in assembled form for $495 in single quantity. The ultralow-noise GaAs FET LNA transistors sell for less than $50, and LNA kits can be acquired for under a couple of hundred dollars.

By purchasing a spherical antenna kit (or building an antenna from scratch using instructions available through STT in Oklahoma), the do-it-yourself electronics hobbyist can assemble a backyard home TV satellite system for well under $1500, and a real pro could build a system from component parts for less than $800!

However, delving into the world of microwaves is an exotic venture indeed, and unless one has a number of years of experience in building receiver equipment that operates at these extremely high frequencies, and has access to the tens of thousands of dollars of the specialized test equipment required to align the sensitive circuits, it is not recommended that the novice, even if knowledgeable in electronics, tackle the construction project of a home backyard satellite terminal from the ground up.

This does not mean, however, that a do-it-yourself system cannot be built by an electronics hobbyist. For just a little more money, completely aligned and basic receiver modules can be obtained, and low-cost 24-channel completely assembled receivers are now available for under $800 at the home satellite TV shows held twice yearly by Satellite Television Technology.

NOTICE!! — Vidiark Electronics, pioneer of the famous "8-Ball" satellite antenna has developed and is now in production of an all new **high performance** satellite receiver. Also, from now on, will be operating under the name of **McCullough Satellite Systems, Inc.**

The 8-BALL

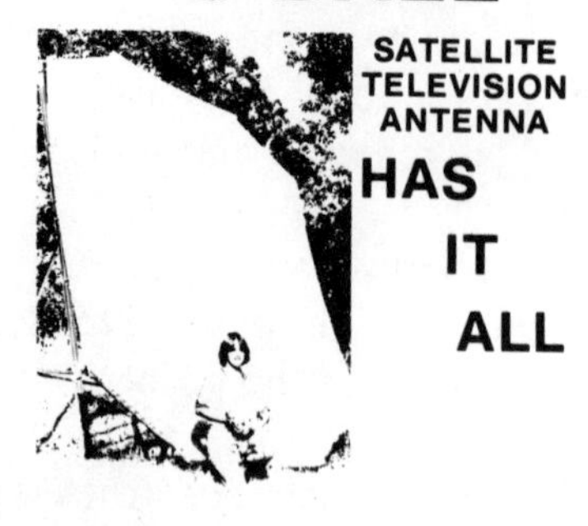

PRICE: Less than half the cost of other antennas.
PERFORMING: The "8-Ball" consistently out performs all other antennas of equal size.
DURABILITY: With its open mesh surface and wide stance frame. The "8-Ball" will survive winds fatal to other antennas.
APPEARANCE: Blends in pleasantly with the environment.

Size	1-3	4-Up	Heavy Mesh	Extra Bracing	Galvanized Frame
8-ft.	$495	$395	30	25	65
10-ft.	550	445	45	40	80
12-ft.	595	475	60	50	100

Avantek LNA (120° In Stock) $650.00
Microdyne Commercial Receivers.... $2600.00

The McCullough Receivers

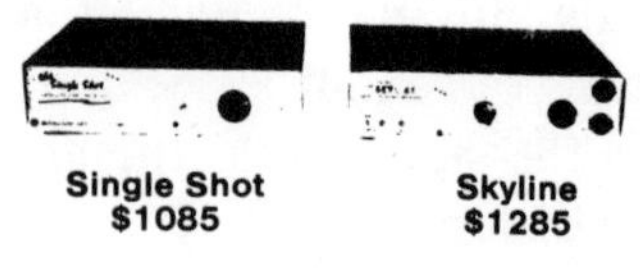

Single Shot $1085 — Skyline $1285

— FEATURES —

- Removable downconverter - can be left in receiver **or** placed at antenna
- Single conversion with image reject mixer
- 6.2 & 6.8 Audio - others on request
- Discriminator type video detector
- Polarity rotator switch
- LNA Power Jack (+18vdc)
- Modulator power jack (+12vdc)

Additional features on skyline model
- Smooth action 10-turn tuning control with LED bar channel indicator
- Remote control included
- LNA circuit monitor

Crystal controlled modulators (type used in home video recorders) - Ch 3 & 4, wired to plug in rear of receiver - Cost $75.00.

— Feed Horns —
Galvanized $25
Aluminum $40

McCullough Satellite Systems, Inc.
P. O. Box 57 — Salem, Arkansas 72576
Phone: 501-895-3167
501-895-3318

The McCullough 8-ball spherical antenna sells for $495! It has been joined by low-cost receivers. (Courtesy of McCullough Satellite Systems, Inc.)

Assuming that the backyard satellite TV builder buys a McCullough spherical antenna, which can be easily assembled by following the instructions, and purchases an already assembled LNA from Simcomm Labs, the total cost, including a professionally built receiver, should be under $1900. The whole project can be completed in two days, thus enabling the user to watch a first-run Sunday night movie if construction is begun on Saturday morning.

So, the "cheapie" system will cost from $800 to $1900, depending on the electronics sophistication of the builder. Anyone who has put together a color TV kit or built a stereo system from construction plans should have no problem at all in assembling a television satellite terminal using a prebuilt and prealigned receiver and LNA. The entire system should take less than 15 hours to complete, and the spherical antennas can be easily assembled by a single person. The professional

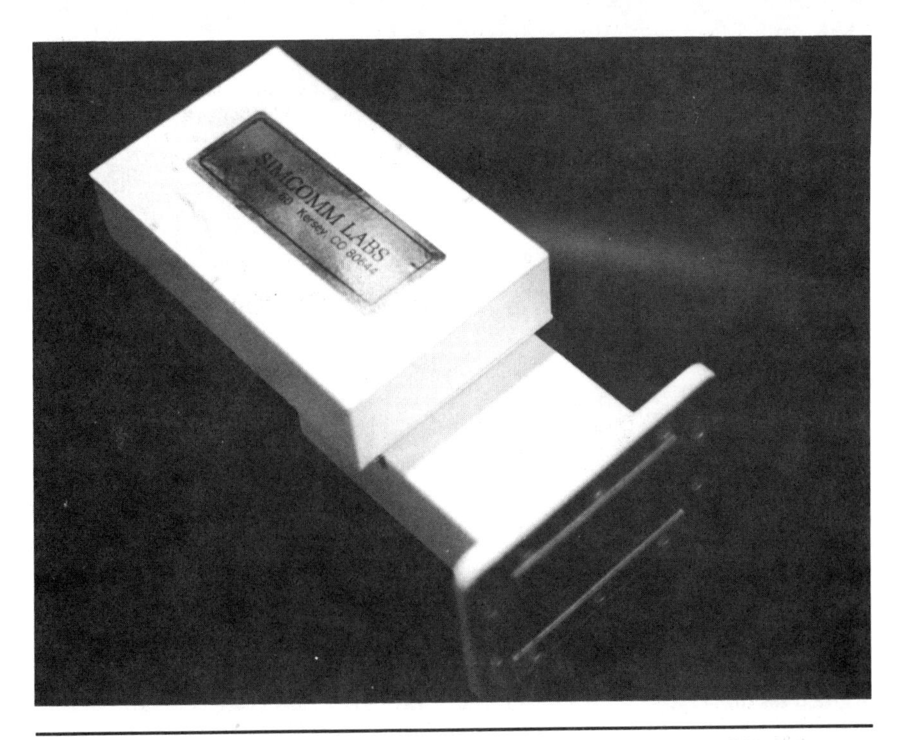

Simcomm Labs' low-cost LNA sells for under $500 and less in kit form! (Courtesy of Simcomm Labs)

electronics engineer who has access to the necessary test equipment could come up with a home backyard satellite terminal for well under $800, using the component printed circuit boards designed by Ty Howard. The system would take about 30 to 50 hours to construct and install.

Buying Assembled Components: A Medium-Priced System

For most people the best route to follow might be to buy completely assembled components from mail-order or local dealerships. A number of distributor organizations, including Interstar Satellite Systems, Downlink Inc., Starview Systems (H&R Communications, Inc.), and National Microtech Inc. will supply complete satellite TVRO terminal packages, less installation, which can be shipped directly to the site and put together in less than a day by the home satellite TV user. For example, the National Microtech P-1100 satellite receiving system consists of an ADM 11-foot parabolic antenna and feed assembly, an Avantek 120-degree LNA Model 4215, a Sky Eye-I receiver, and down-convertor, complete cables and connectors, and a video cassette that provides step-by-step pictorial installation and operating instructions. At the dealer level (quantity of five or more) Starview will ship its Model 12-K system via UPS. This TVRO earth station package consists of H&R's own 12-foot antenna with AZ/EL mount, prime focus and button-hook feed, 120-degree LNA, ICM's Model 4300 receiver, cables and connectors, and complete system installation instructions. The total system package price is $2295.

Likewise, for $3995 Downlink Inc. will provide a complete package of antenna, LNA, receiver, and cables using a 12-foot spherical antenna and the wood-cased D-2X receiver.

By shopping carefully among the various suppliers (the back pages of the electronics magazines and the *SAT Guide* are good resources here), the home satellite terminal buyer should be able to find the right system for between $3000 and $4000. This includes antenna, receiver, LNA, cabling and connectors, and RF modulator to convert the video signal to a normal television channel that can be tuned in by an ordinary television set.

The first time through, a typical person should take about 10 hours

An inexpensive high-quality home satellite receiver (top unit). Can be combined with hi-fi equipment. (Courtesy of Dexcel Inc.)

to assemble a small, petalized parabolic dish antenna, and align it so that it points correctly at the desired bird. Specific information on how to find the birds and also how to calculate just how large an antenna might be required at your location appears in Appendices 6 and 7. One advantage of putting together a parabolic antenna is that the surface of the dish is usually precisely formed at the factory, and after assembly on site the dish can be pointed relatively quickly to the proper location in space with only minor tweaking necessary. Dish surfaces must be precisely shaped, with tolerances that can vary only by one-quarter inch or so over the entire surface of the antenna. Variations in shape greater than this amount will significantly limit the effectiveness of the antenna, and may prevent any signal from being received at all! For this reason, the spherical antenna, which is usually assembled piece by piece, requires far more care in aligning its surface shape to the spherical template or proofing instructions supplied. Therefore, for the home satellite TV enthusiast who simply wants to open up the shipping box and "plug 'er in," the parabolic antenna is probably preferable.

A number of videotapes and instructions for installing and properly aiming a satellite TV antenna are available from the suppliers, including Satellite Television Technology in Oklahoma. After repointing an antenna from bird to bird a couple of times, the aiming process becomes very simple. The old pros who have installed dozens of antenna systems throughout the country have been known to be able to assemble and aim a TVRO in a couple of hours, and for those antennas that come out of the shipping container in one piece, a seasoned installer can have the dish pointed to the SATCOM F-1 satellite within a matter of moments.

With proper selection and preparation of the installation site (determining beforehand where the antenna will go, verifying that nothing will block the view of the satellite, clearing the ground for the necessary mount support, etc.), a "do-it-yourselfer" who has put up a regular television antenna on a chimney should have little trouble in installing his or her own backyard TVRO satellite terminal.

Ordering a Turnkey System from an Installing Dealer

Perhaps the easiest way of putting in a home TVRO satellite or terminal is simply to pick up the phone and place an order with an installing dealer. In this arrangement the dealer will make a preliminary analysis of the location to determine where the antenna should be located, and then make a specific recommendation as to particular antenna size and receiver combination.

Turnkey systems vary considerably in price, ranging from a low of perhaps $4000 to a high of $20,000 or more, depending on the size of the antenna installed and the features selected. The typical price for a fixed-mount single-satellite antenna with 24-channel tunable receiver and motorized vertical/horizontal polarity rotater control mounted on a small concrete pad will typically be between $5000 and $7000. A fully remote-controlled motorized antenna mount, allowing the dish to be easily switched from satellite to satellite, might raise the price another $2000 or more. If two people wish to watch different television shows in different rooms, then a dual LNA/dual feed will likely be required, with a signal power splitter allowing two or more satellite receivers to be connected simultaneously to the same antenna. This would enable two, three, or even four or more satellite receivers to be located throughout the home, each one connected to a television set in a different room. A multireceiver system of this kind can easily increase the overall costs by $3000 to $5000 or more, as the entire configuration approaches the quality and complexity of a commercial mini-CATV system.

All dealers are not equal. Experience varies from company to company, and quality of workmanship and understanding of importance of service in the consumer electronics business vary from dealer to dealer. Therefore, it is important for the consumer to check out the dealership and its customers carefully before taking delivery on any equipment. One way of doing this is to visit an installation in the area done by the dealer. By talking with previous customers, the potential buyer should get a pretty good feeling as to the level of quality and service the dealer is able to provide. Was the satellite TVRO terminal carefully assembled? Was the base built sturdily? Was the incoming coax cable buried underground and properly insulated in its entrance through the outside wall of the house? If follow-up service was required, was the dealership able to provide it quickly, competently, and at reasonable cost? What kind of warranty was offered?

Another way of investigating a local dealership is to contact the suppliers who sell to the dealer and ask for their honest opinions. Has the dealership been late in making payments? Does the dealer have competent and adequately trained installation and service personnel on staff? Does the dealer sell a sufficient number of satellite terminals to be considered a major dealer in the area?

Of course, by reading this book, the home TVRO satellite terminal buyer will have obtained a pretty good handle on the overall industry, including the various manufacturers in the business. One thing that can be easily checked is the size of the antenna recommended for purchase by the dealer. Will it be adequate, or should it be larger? A dish that is too small may produce picture quality that is unacceptable to you. Conversely, if, for example, a 12-foot antenna is sufficient to bring in top-notch pictures, then the expense of purchasing a 15-foot antenna may be unnecessary. One way of verifying the proper approximate antenna size is to perform the calculations for your geographic location as described in Appendix 7.

There are literally tens of thousands of home satellite terminals now in operation in the United States and Canada. Many of these satellite TV owners installed the systems themselves. However, the majority of installations were done by installing dealers, a number of whom, like the Helfer's Antenna Service in Pleasantville, New York, travel throughout the United States to make an installation. As the number of home satellite TV terminals reaches the hundreds of thousands, many dealerships offering competitively priced equipment will be operating in every area of the United States and Canada.

Chapter 10
The "Secret" Signals on the Birds

All About Subcarriers

Up until this point we have discussed satellite transponders in terms of their ability to carry television video and audio signals, but satellite transponders are also capable of carrying wide-range, high-fidelity radio channels, audio newswire feeds, special news teletypewriter channels, high-speed stock market and commodity exchange data feeds, private telephone and other audio channels, and new teletext data services that bring "electronic newspapers of the air" to ordinary television sets nationwide.

The basic NTSC (National Television Standards Committee) color television system requires a communications channel with a bandwidth of 6 MHz. Considering that an ordinary telephone conversation takes up a communications channel whose bandwidth is only 3000 Hz (or 3 kHz), it would be possible to squeeze 2000 such voice conversations onto a single video circuit! This in fact is done regularly through a technique known as multiplexing.

Because the communications satellites share the same sets of microwave frequencies as do the terrestrial telephone company microwave communications relay circuits, an unacceptable amount of interference would occur if some additional technique were not used effectively to separate the two conflicting microwave signals. Although the communications satellite's transmitter is 22,300 miles away, the telephone company's nearest microwave system may only be three blocks down the street, and an antenna pointed out to space would still pick up the undesired telephone signal coming from half a mile away.

The satellite carriers solve this problem by dispersing the 6-MHz video signal over a 36-MHz bandwidth, spreading out the 6-MHz TV signal so that it takes up a much greater bandwidth than ordinarily required. By using this technique, any terrestrial telephone company microwave interference present in a 6-MHz piece of the total 36-MHz spectrum will be effectively ignored in the signal detection process. So, although satellite transponder channels appear to be 36 MHz wide and are spaced at 40-MHz intervals (including a 4-MHz guard band of unused space separating each transponder channel from its upper and lower neighbors), the effective width of the satellite transponder is limited to far less than that. Typically, no more than 8–10 MHz of bandwidth is ever used, and an ordinary U.S. color television signal takes up no more than 6 MHz of space. Since satellite transponders are not limited to carrying just a 6-MHz television channel, additional "secret" signals are often inserted between 6 MHz and 10 MHz.

Audio and data signals that take up less than a full video channel's spectrum are typically known as subcarriers; these are simply frequency-modulated carriers of narrow bandwidth that either take the place of the video signal (when the transponder is not used for carrying television pictures) or are combined with a video carrier to provide for simultaneous video and audio data feeds.

As the subcarrier chart on the next page shows, the normal audio portion of a television program is placed onto a subcarrier ranging from 5.41 MHz to 7.5 MHz, depending on the satellite. In the North American NTSC color format, a special 3.5-MHz color burst subcarrier containing the color video information is also present in the overall television signal's spectrum, and in the European PAL and SECAM color television systems, this color information is found instead on a 4.33-MHz subcarrier. Ordinary television sets have special color decoder circuits that look for their respective color burst signals to reconstruct the necessary coloring information of the transmitted picture.

Since the television signal itself only requires a portion of the full baseband frequency spectrum to transmit and receive (or recover) a color video picture, any frequency above approximately 4.2 MHz may be used to carry further unrelated audio or data channels. In the RCA SATCOM satellite systems, four subcarriers at 5.8, 6.2, 6.8, and 7.4 MHz are available for use, one of which transmits the program audio channel. Thus, on SATCOM F-3, transponder 3 carries not only the video and audio portions of the WGN-TV signal from Chicago, but also

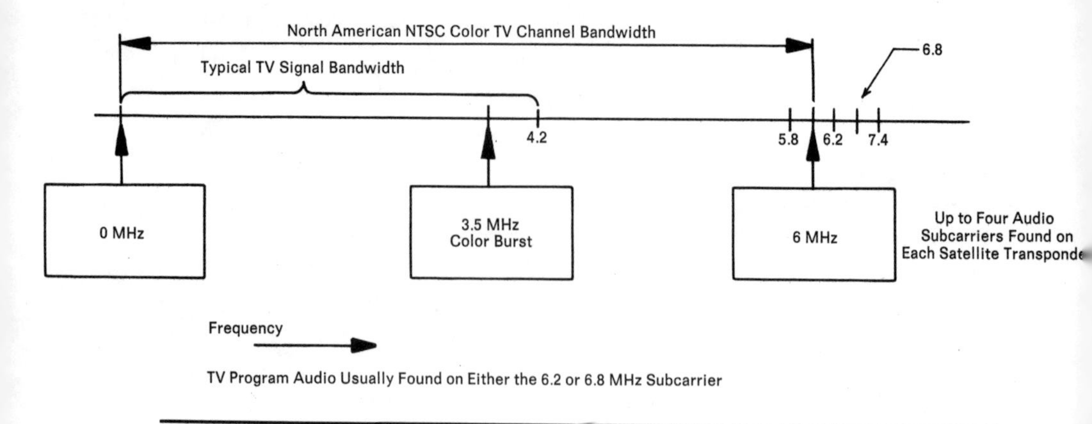

Chart of popular subcarriers used on TV satellites. (Courtesy of James Lentz)

the fine arts WFMT stereo-FM radio station from Chicago on a 5.8-MHz subcarrier, and a Seaburg music channel on a 7.6-MHz subcarrier.

Most of the home satellite video receivers allow for the front-panel switch selection of the program audio, which is placed on a subcarrier at either 6.2 or 6.8 MHz, depending on the satellite used (RCA SATCOM satellites use 6.8 MHz for the program audio subcarrier). A number of receivers provide a variable tuning audio knob to allow any subcarrier to be decoded from 5.8 to 7.4 MHz. This type of receiver, such as the KLM Sky Eye-I, can pick up not only the audio of a television program, but also other "secret" audio signals transmitted over the same transponder.

When a transponder is not used to rely a television signal, then the portion of the spectrum ordinarily reserved for the TV video can be used to carry other audio and data signals instead.

In normal telephone company parlance, this technique of squeezing multiple signals onto a single transponder, or carrier, is known as multiplexing, and the most common type employs a technique called single-sideband modulation. In this process unrelated audio or data signals are simply stacked up in ascending frequency, one above the other. The next chart illustrates this concept. Normal telephone channels have a bandwidth of 4000 Hz (4 kHz), the very highest frequency that a human vocal cord can generate. A single telephone circuit (known to them as a carrier) is combined with 11 others to form a group of 12 voice conversations. Five groups with a total bandwidth of 240 kHz when combined together form a supergroup. A supergroup of these 60 voice circuits can be delivered directly to the satellite-coupling earth station transmitter, where they can be inserted anywhere from 0

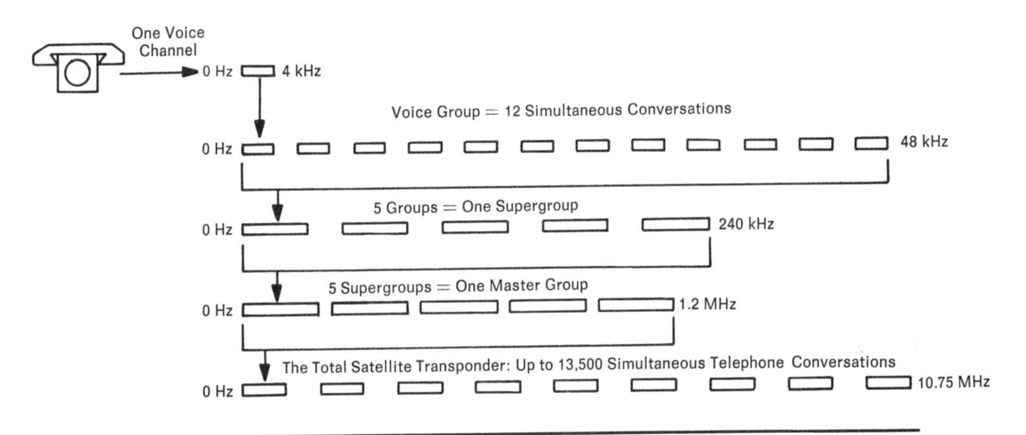

Multiplexing multiple audio channels onto a satellite transponder. (Courtesy of James Lentz)

MHz up to 10.75 MHz—the maximum upward usable limit of the satellite transponder's frequency spectrum. A master group consists of five supergroups combined together, or 300 separate audiotelephone channels. A group occupies 48 kHz of bandwidth, a supergroup takes up 240 kHz of spectrum, and a master group requires 1200 kHz of space. Many master groups can be stacked one on top of the other on a typical satellite transponder, with each supergroup in a master group being separated by an 8-kHz guard band.

Using this clever technique of multiplexing many individual carriers together, a satellite common carrier can literally squeeze thousands of audio-telephone channels onto a single satellite transponder.

This process of creating carrier systems of groups, supergroups, master groups, and jumbo groups (three master groups!) gives the telephone company tremendous flexibility in arranging and routing telephone circuits throughout the country. By careful administrative design, specific supergroups or master groups can be assigned for the exclusive use of particular geographic regions or cities. Thus, a given supergroup of 60 voice channels might be arranged for exclusive use between New York and San Francisco, while the supergroup next in frequency might be dedicated for use between Washington, D.C., and Los Angeles, and the supergroup above it used between Los Angeles and San Francisco. In this way the complexity of arranging thousands of independent satellite-provided telephone circuits between any two points in the United States is reduced to routing a series of easily manageable 12- and 60-channel collections of conversations. Significant equipment savings are also possible as the cost to multiplex, modulate, and demodulate 12 or 60 channels simultaneously is only slightly higher than the cost of multiplexing, modulating, and demodulating a single-voice circuit.

However, in some cases it is clearly advantageous to enable a single audio-telephone channel to be uplinked to a bird from a specific location and manipulated independently rather than as part of a group or supergroup collection of signals. Another technique known as single channel per carrier (SCPC) is offered by the satellite common carriers to accomplish this task. SCPC transponders do not use the frequency spectrum as efficiently when operating in the single-carrier mode, however, and not nearly as many of these types of audio channels are carried on the satellites. But SCPC signals do have the ability to be transmitted up to the bird with more power, and the satellites relay these audio signals to earth at greater power levels, allowing smaller receive-only antennas to be utilized. Because of this important feature, a number of the news wire services use these special-tariff SCPC channels to deliver high-quality audio news feeds to radio stations nationwide, employing small 3- to 5-foot receive-only parabolic antennas mounted on the radio station roofs. In addition, these same SCPC channels can carry high-speed teletypewriter and data signals also used by the news wire agencies and newspaper chains to relay news stories instantly throughout the United States. Both the Commodity News Service (CNS) and Bonneville Satellite Corporation use this technique to carry commodity data from the exchange floors to points in various parts of the country.

Picking Up the Signals

Now that we know there are hundreds or thousands of private audio, data, and telephone channels squeezed on the nonvideo satellite transponders (see Appendix 3 to determine which transponders carry these nonvideo signals), how can the TVRO satellite television terminal be used to pick up these interesting signals? The answer is that it can't, directly. But with the addition of one of the popular short-wave communications receivers that will tune the 100-kHz to 30-MHz range, the thousands of hidden channels can be picked up as well. Companies such as Radio Shack sell these communications receivers, which are popularly used to listen into short-wave radio signals. One of the more popular manufacturers is the Kenwood Corporation of Japan, whose R-1000 receiver provides an outstanding demodulation system to enable the individual audio channels to be demultiplexed from the transponder's multichannel nonvideo signal.

To detect these narrow-band voice and data signals, the antenna input of the communications receiver is connected to the video baseband output of the TVRO satellite receiver. (See below.)

In operation, the TVRO satellite receiver is first tuned to the desired transponder, such as transponder 15 on the SATCOM F-3 satellite. The communications receiver is then switched into the single-sideband mode and set to either the lower sideband or upper sideband position. Beginning with the lower sideband selection, set the receiver to its lowest frequency (100–200 kHz) and simply begin tuning upward across the frequency spectrum. Every 4 kHz or so, a carrier will appear. The carriers will either contain telephone conversations, national radio network audio feeds, private voice communications circuits, or teletypewriter data that will be heard as a series of high-speed varying audio tones.

To convert these audio tones to teletypewriter signals that can be printed out on a small printer or home computer terminal, a special

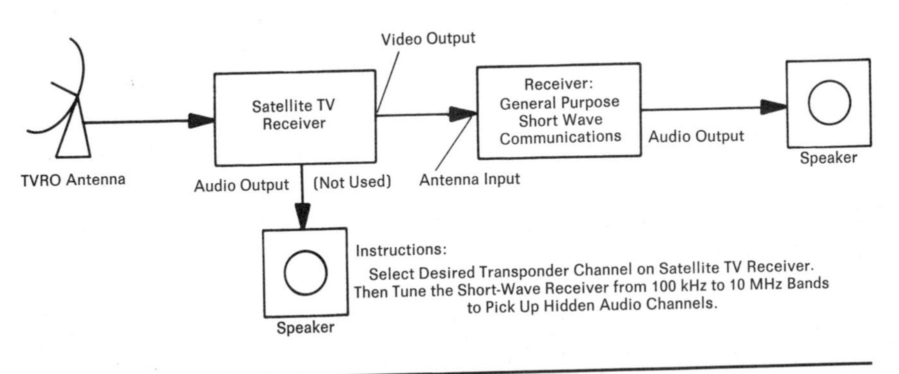

Configuration to pick up the "secret" satellite audio signals. (Courtesy of James Lentz)

teletypewriter decoder modem, known as an RTTY demodulator, must be connected to the audio output of the communications receiver (see next page). A number of organizations sell these RTTY demodulators, popularly used by amateur radio operators, and the ham radio magazines such as *QST, SQ, 73,* or *Ham Radio* advertise these devices, which retail from $200 in kit form to over $1000 for a commercial version.

If, when tuning the short-wave communications receiver upward in frequency past 4.2 MHz, the signals suddenly disappear, this probably means that the manufacturer of your satellite TV receiver has installed a special filtering circuit that allows only the typical 4.2-MHz video channel to be passed. Many manufacturers build in these circuits to

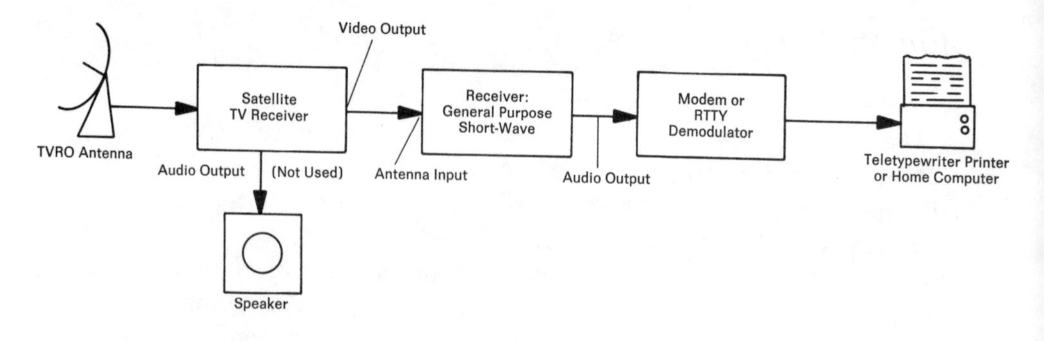

Configuration to pick up the "secret" satellite news wire and teletypewriter signals. (Courtesy of James Lentz)

minimize the possible noise interference to an ordinary television set created by unwanted signals outside the television spectrum. These filters, internal to the TVRO satellite receiver, are considered a positive feature in the satellite television receiver's design. Unfortunately, a nonvideo transponder carrying many channels may extend up in frequency to well beyond 10 MHz, and a number of the more interesting audio signals and data feeds will be lost. However, many manufacturers are now providing a second, separate video output jack arranged in front of the filter to allow for a short-wave communications receiver to be used in picking up these nonvideo signals. It is also easy to have the installing dealer modify the receiver by adding a new output jack, effectively bypassing the internal filter. Once this is done, the full spectrum of "secret" signals will be available to the listener.

A note is in order concerning the Communications Act of 1934, specifically Section 605, the "secrecy clause," which establishes guidelines concerning the interception of private nonbroadcast communications. Chapter 12 discusses the Communications Act in detail, and specific sections that apply to the viewing of satellite TV signals, including the listening in on audio conversations and news wire teletypewriter traffic. Tuning into a satellite is conceptually the same as listening into a police or fire department channel or mobile telephone circuit using a scanning receiver. Although millions of these receivers are now in use, common good sense as well as FCC regulations prohibit the divulging of any information received without approval of the person or organization providing the conversation or audio program.

Audio Programs

The Radio Station Feeds

Although WFMT, the fine arts radio station in Chicago, is transmitted as a 5.8-MHz subcarrier on the WGN-TV signal (transponder 3, SATCOM F-3), most audio programming is delivered using the single channel per carrier or FDM multiple-audio-signal nonvideo multiplexing technique. Still, a number of audio programming services are available on the video transponders, and can be tuned in by the TVRO satellite television receiver without requiring the purchase of a short-wave communications receiver.

The Seeburg Music Service provides Lifestyle, a trendy, upbeat 24-hour-per-day uninterrupted background music service on a 7.695-MHz subcarrier of WGN-TV (transponder 3, SATCOM F-3). The Satellite Music Network furnishes two different stereo-audio services, also on subcarriers of the WGN-TV signal. At 5.8 MHz a popular-music adult-format audio feed is provided, and at 5.94 MHz the format is country and western music. JISAL, a Christian-oriented easy-listening music service, is located on a 7.4-MHz subcarrier on WTBS (transponder 6, SATCOM F-3). The JISAL easy-listening music offers an assortment of commercial-free songs, ranging from oldies to contemporary and country to jazz, along with Christian music and commentaries. Contiinental Broadcasting (previously Christian Broadcasting Network) includes three stereo, adult contemporary-format audio music networks and a fourth audio news network on a SATCOM F-3 transponder (transponder 8) at 5.58 MHz, 5.94 MHz, 6.30 MHz, and 7.56 MHz.

The Satellite Radio Network, a similar project of the PTL Network, is also carried on its own transponder as a 6.2-MHz subcarrier (transponder 2, SATCOM F-3).

With regard to the nonvideo transponders, a veritable cornucopia of selections is available to the network audio listener. These channels require the use of a short-wave communications receiver and range from 92 kHz (as tuned in by the short-wave communications receiver) to 8000 kHz, and higher.

One of the most interesting satellites is SATCOM F-2, located at 118.9 degrees west longitude. On transponder 7, the Alaskan Forces Satellite Network, playing middle-of-the-road and Top-Forty music 24 hours per day for the U.S. Armed Forces personnel in Alaska, may be found at 294 kHz (lower sideband mode). Anchorage radio station KBYR is located on SATCOM F-2 (transponder 7) at 333 kHz (LSB). Radio station KYAK from Anchorage, a country and western station, can be located at 393 kHz (USB), same transponder. At 958 kHz (LSB) a black soul station feeds the Alaskan Forces Satellite Network, and at 1190 kHz (NSB) the Alaskan Forces station provides another audio channel.

The Mutual Radio Network operates on 1702 kHz (NSB) on SATCOM 2 (transponder 7), and ABC Radio Network news feeds may be found at 2358 kHz (LSB).

National Public Radio (NPR) may be found at 134 kHz (LSB) on SATCOM-2 (transponder 11) with a second NPR audio channel at 602 kHz (LSB). At 2802 kHz on transponder 11, Anchorage radio station KBYR duplicates its signal during the day, and in the evening a Christian-oriented programming channel may be found. Anchorage radio station KFQZ is located at 2854 kHz (LSB).

Switching to the even SATCOM-2 transponders, CBS Radio sports feeds may be found at 421 kHz on transponder 4 (USB), and NBC Radio news feeds at 469 kHz (USB). On transponder 10, Mutual Radio Network feeds carriers at 2454 kHz and at 2082 kHz (LSB), but the most interesting transponder is 16, chock full of narrow-band audio and data services.

The Alaskan Forces Satellite Network (AFSN) appears again at 166 kHz on transponder 16 (LSB), and additional CBS Radio news feeds may be found at 238 kHz (LSB). Sports feeds appear at 310 kHz (LSB), and Philadelphia radio station KYW is located at 552 kHz (LSB). NBC Radio News has an additional feed at 694 kHz (LSB). Additional feeds for Voice of America and other independent radio networks are also located on this transponder.

Transponder 20 of SATCOM F-2 carries another NBC sports feed at 190 kHz (LBS), and at 7382 kHz (USB), NBC-TV uses a "talk-back" coordination channel to connect TV crews in the field with the New York City Broadcast Operations Control Center during special sports events carried on television by transponder 8 (the NBC–TV contract video channel).

News Wire Services and Special Circuits

Although there are literally dozens and dozens of special audio circuits, some of the more interesting ones include the AP Radio news feeds at 950 kHz (LSB) on SATCOM F-2 transponder 10, and the UPI Audio news service at 682 kHz (LSB) on SATCOM F-2 transponder 16. The AP Radio Network Newsbrief service can be found at 700 kHz (LSB) on transponder 16, and the National Black Radio Network contains several audio feeds on this transponder as well.

The Anchorage Airport Control Tower is at 1938 kHz (LSB) on

SATCOM F-2, transponder 7, as well as at 1234 kHz (LSB) on SATCOM F-2, transponder 11. The Kenai Flight Service Weather provides an extensive weather summary and forecast for Alaska at 1946 kHz (LSB) on SATCOM F-2, transponder 7, and the Valdez Weather Advisory Service for Alaska, including the Bering Sea and the U.S.S.R., may be found at 2366 kHz (LSB) on SATCOM F-2, transponder 11. Over on WESTAR I, the National Public Radio Service feeds 12 separate audio channels, using the single-channel-per-carrier technique, on transponder 6. These can be located easily because far fewer audio channels are present that use the SCPC transmission technique.

The News Wire Teletypewriter Feeds

All of the major news services provide dozens of multiple teletypewriter channels covering every conceivable type of news and sports information, including stock market reports, gold and silver bullion prices, weather reports, and regional news stories. The most popular transponders are SATCOM F-2 transponder 7 and SATCOM F-2 transponder 16; SATCOM F-2 transponders 4 and 10 are less heavily used.

The decoding of these teletypewriter channels requires the special

Dow Jones Cable News service is carried on SATCOM F-3, transponder 3—WGN. (Courtesy of Dow Jones Cable News)

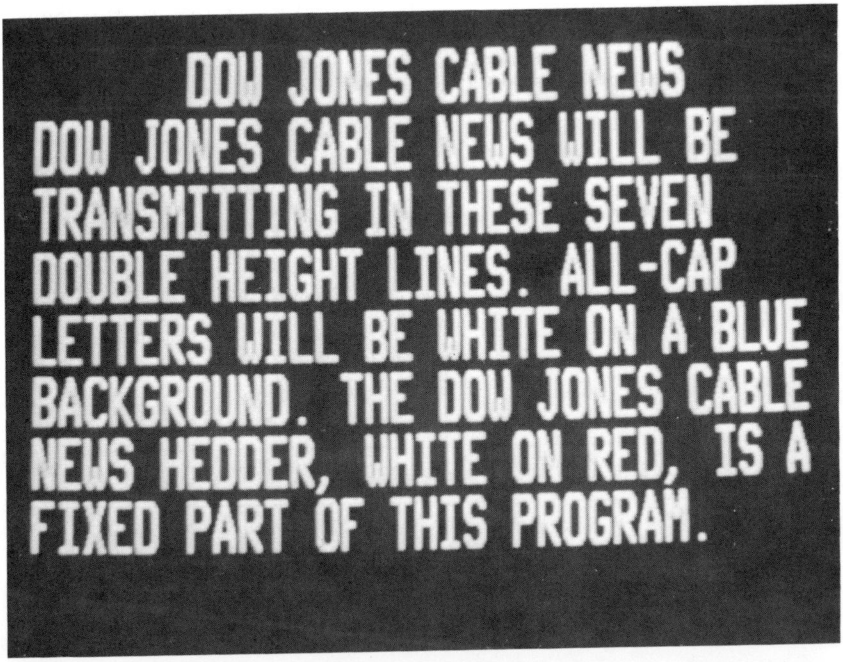

Informative

WASHINGTON: PRESIDENT CARTER TOLD ORGANIZERS OF AN ANTI-NUCLEAR RALLY IN WASHINGTON THAT THE NATION'S NUCLEAR PLANTS MUST CONTINUE TO OPERATE. BUT HE PROMISED TO PUSH THE COUNTRY TOWARD ALTERNATE ENERGY SOURCES AND CONSERVATION MEASURES.

General News
...Thirty pages of up-to-the-minute national and world news gathered by hundreds of Reuter correspondents world-wide. Around the clock, around the world, Reuters is the leading source of breaking news for newspapers and broadcasters. NEWS-VIEW brings the same news into the homes of cable television subscribers *as it happens.*

NYSE-A MOST ACTIVES PAGE 1 1456

FAIR CAMR	$54\frac{7}{8}$	$+\frac{5}{8}$
RAMADA IN	$12\frac{1}{4}$	$-\frac{7}{8}$
SEARS ROE	$19\frac{1}{4}$	$-\frac{1}{4}$
BANKAMER	25	$-\frac{1}{8}$
CHARTER WTS	$10\frac{3}{4}$	$-1\frac{7}{8}$

Financial News
...PLUS, business and financial news as the brokers see it. Reuters is the world's largest business news organization. Brokers and bankers around the globe depend on us to make their financial decisions. In addition to market-moving news, NEWS-VIEW presents *the only available real-time market indices and prices of the most active stocks,* on both the New York and American stock exchanges, available to home viewers. Operates Monday-Friday, 9 a.m.—8 p.m. Eastern time.

Entertaining

—SPORTS HEADLINES—

...U.S. WINS WOMEN'S FEDERATION TENNIS CUP

...RENALDO NEHEMIAH BREAKS OWN WORLD MARK IN 110-METER HURDLES

Sports News
...Inning-by-inning or quarter-by-quarter, NEWS-VIEW sports keeps fans in touch with what's happening in every professional sport. Scores, league standings, league results—it's all on NEWS-VIEW Sports.

AQUARIUS
JAN. 20—FEB. 18

YOU ARE TEMPTED BY A NEW IDEA BUT ARE UNCERTAIN ABOUT PROCEEDING. YOU SHOULD TAKE A CHANCE AND IMPLEMENT SOMETHING NEW. IT WILL ADD EXCITEMENT TO YOUR LIFE AND COULD LEAD TO MAJOR SUCCESS LATER.

Features
...Supplied by one of the nation's leading astrologers, NEWS-VIEW Horoscopes appear every two hours and include the day's outlook for every zodiacal sign.

...Traveling? NEWS-VIEW offers vacationers and business travelers weather forecasts in more than 45 cities across the country. In winter, skiers turn to NEWS-VIEW to see ski slope conditions at an equal number of resorts.

Horoscopes and travel weather/ski resorts are just two exciting features on NEWS-VIEW. Also included are items on Pet Care, Sports Quizzes and general interest features.

The Reuters News-view service. (Courtesy of Reuters, Ltd.)

RTTY demodulator, whose signal must be fed into a teletypewriter or computer terminal.

Teletext Services and the Vertical Blanking Interval

Perhaps one of the most exciting hidden services is teletext, an electronic "newspaper of the air," which is carried along with a number of television signals in an unused portion of the television picture itself.

Invented in Great Britain in the early 1970s, teletext services have sprung up worldwide, and several different standards are now being developed in the United States. Basically, teletext allows the home viewer equipped with the proper decoder box to pick up 200 pages or more of alphanumeric text and colorful graphics designs that are displayed on the television set. In operation, a small keypad is used with the decoder to select among the desired news, weather, and other pages of information available on the systems.

Videotext, a CATV teletext service offered by Oak Communications Inc. (Courtesy of Oak Communications Inc.)

The teletext systems accomplish this by sending a high-speed digital data stream over the television channel during that period of time between television picture frames in which the electron gun in the television set is momentarily switched off. This period of time, known as the vertical blanking interval, occurs every thirtieth of a second when the electron gun has scanned to the bottom of a television frame and must be brought back up to the top left of the picture tube to begin scanning the next frame. During this retrace interval the teletext signal can be sent safely without causing the data stream to be visible on the television picture tube itself.

One of the leading organizations presently experimenting with teletext is the CBS Television Network. Using their KNXT television station in Los Angeles, and in conjunction with the Public Broadcasting Station KCET and the Boston-based WGBH Closed Captioning Center for the Deaf, CBS Television is presenting a Southern California teletext service that includes weather maps, news briefs, sports summaries, children's puzzles, airline schedules, and other pages of information of interest to the greater Los Angeles viewer. This service probably will be extended to satellite as CBS, along with NBC and ABC, begins to use the COMSTAR D-3 satellite heavily in 1983.

A number of cable television satellite programmers also use the vertical blanking interval to send out nationwide teletext and special

Teletext signals can be delivered via satellite to MDS, CATV, or broadcasting stations. (Courtesy of Antiope Systems)

Source-to-user data routing

Instead of taking a picture of a page of text, the teletext performs a sequentiel digital processing of the characters contained on the page and transmits the related information over the TV-network.

The user then receives it through his antenna (or through a cable distribution network) and requests the desired pages by composing an identification number on his control keyboard.

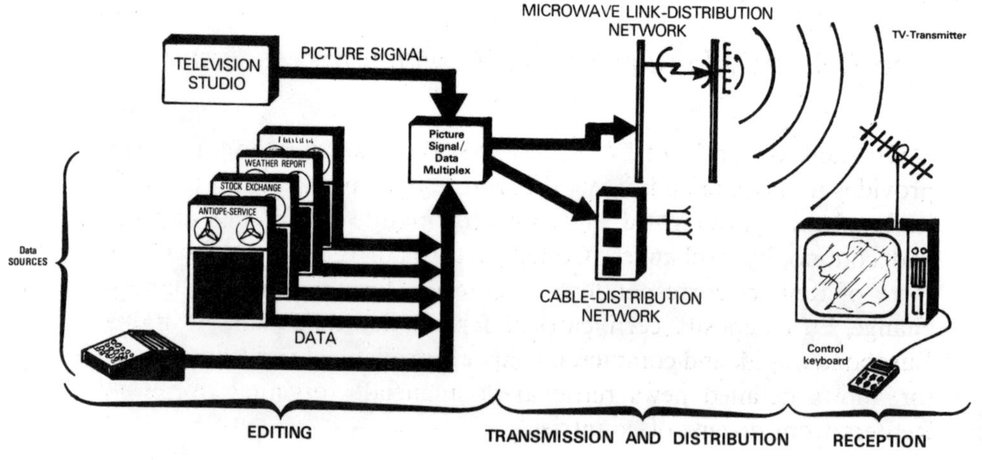

services data signals. In the WTBS vertical interval (transponder 6, SATCOM F-3), both United Press International and Reuters transmit their cable news and news view services consisting of continually changing news text. Dow Jones Cable News, a similar service, is carried over the vertical interval of television station WGN (transponder 3, SATCOM F-3). In addition, several combination text, audio, and slow-scan television data feeds are produced by North American Newstime, carried over the 6.2-MHz subcarrier of WTBS, and The Woman's Channel, a specialty service oriented to modern working women and homemakers, is carried on the 7.4-MHz subcarrier of WTBS.

As more and more organizations enter the teletext field, including the Bell Telephone System, which has announced its intention to develop a compatible national interactive videotext service, dozens of satellite programmers will begin to utilize the vertical interval space to squeeze these additional teletext signals onto the transponder channel.

A number of firms manufacture subcarrier decoders for use by the cable television companies. One such supplier is Wegener Communications, Inc., of Atlanta, Georgia. Wegener's decoders have been installed in hundreds of CATV systems. Southern Satellite Systems, the resale common carrier delivering WTBS nationwide via satellite, has developed a special cable text decoder for use with its teletext service, and is making these decoders available to cable television systems. For those satellite enthusiasts who would like to build their own teletext decoders, a number of articles have been published over the years in *Wireless World,* a British magazine, and teletext decoder kits based on the British teletext standard are available for order from companies who regularly advertise in the back pages of that publication.

The Reuters Monitor Service Global Market System

On transponder 18 of SATCOM F-3, the Reuters Monitor Service provides immediate and direct access to money market rates from the leading banks of the world, and up-to-the-minute information on stock and commodity exchanges. Collected through the international network of Reuters correspondents, financial information on foreign exchange, Euro deposits, certificates of deposit, domestic money markets, Eurobonds, gold, and commercial paper is available. In addition, Monitor allows detailed news retrieval of financially oriented stories by Reuters from dozens of countries.

Hand-held keypad can be used to retrieve teletext data as well as select cable channels. (Courtesy of Oak Communications Inc.)

INDAX is the Cox Cable Company's CATV teletext service. (Courtesy of Cox Cable Co.)

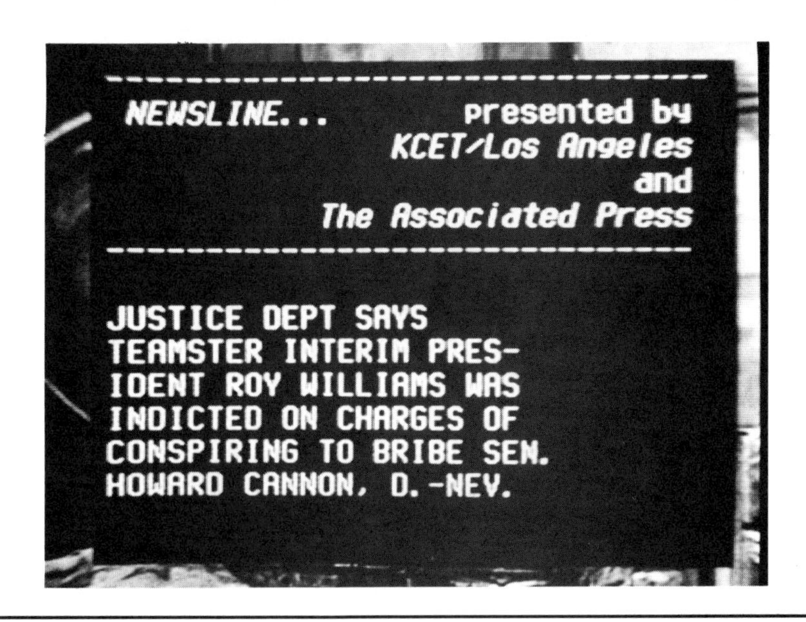

In a significant advancement of television's usefulness as an information service, KCET/Los Angeles has, through a cooperative venture with the Associated Press, become the nation's first television station to provide a continuous feed of the AP news wire service during broadcast hours. Dubbed Newsline, the service can be received in the approximately 3000 Southern California homes already equipped with decoders used to allow the hearing-impaired access to closed-captioned programs. (Courtesy of M. Trumbo/KCET)

KNXT teletext pages from this CBS-owned Los Angeles station. (Courtesy of KNXT)

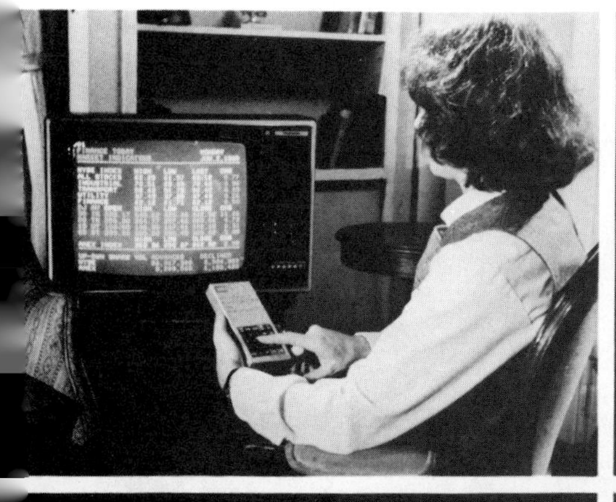

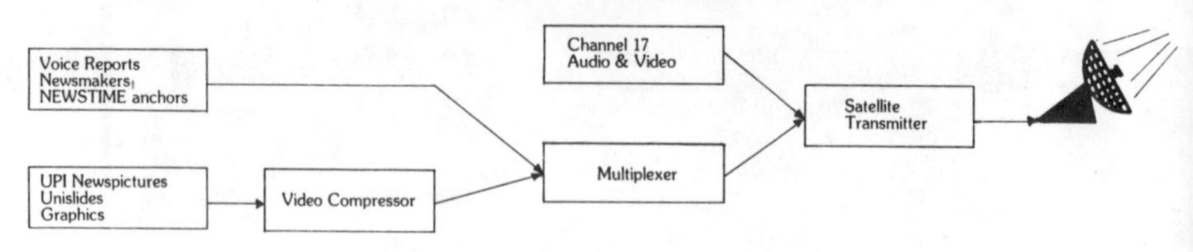

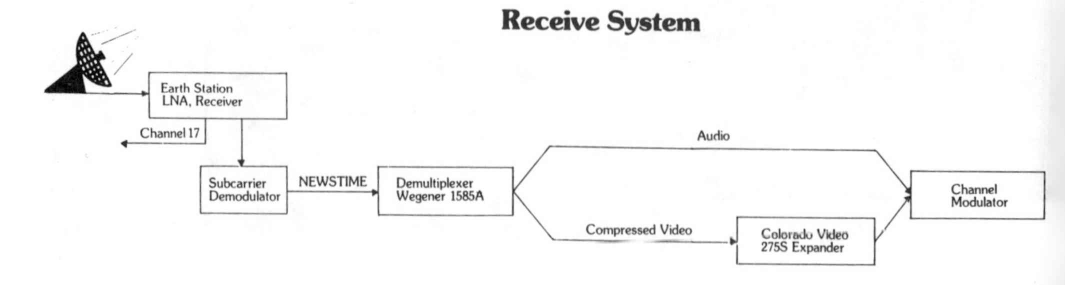

Newstime—a slow scan wire photo service carried on a subcarrier of SATCOM F-3's transponder 17. (Courtesy of Newstime)

Example of Newstime photos being transmitted, each picture taking about a minute to be received. (Courtesy of Newstime)

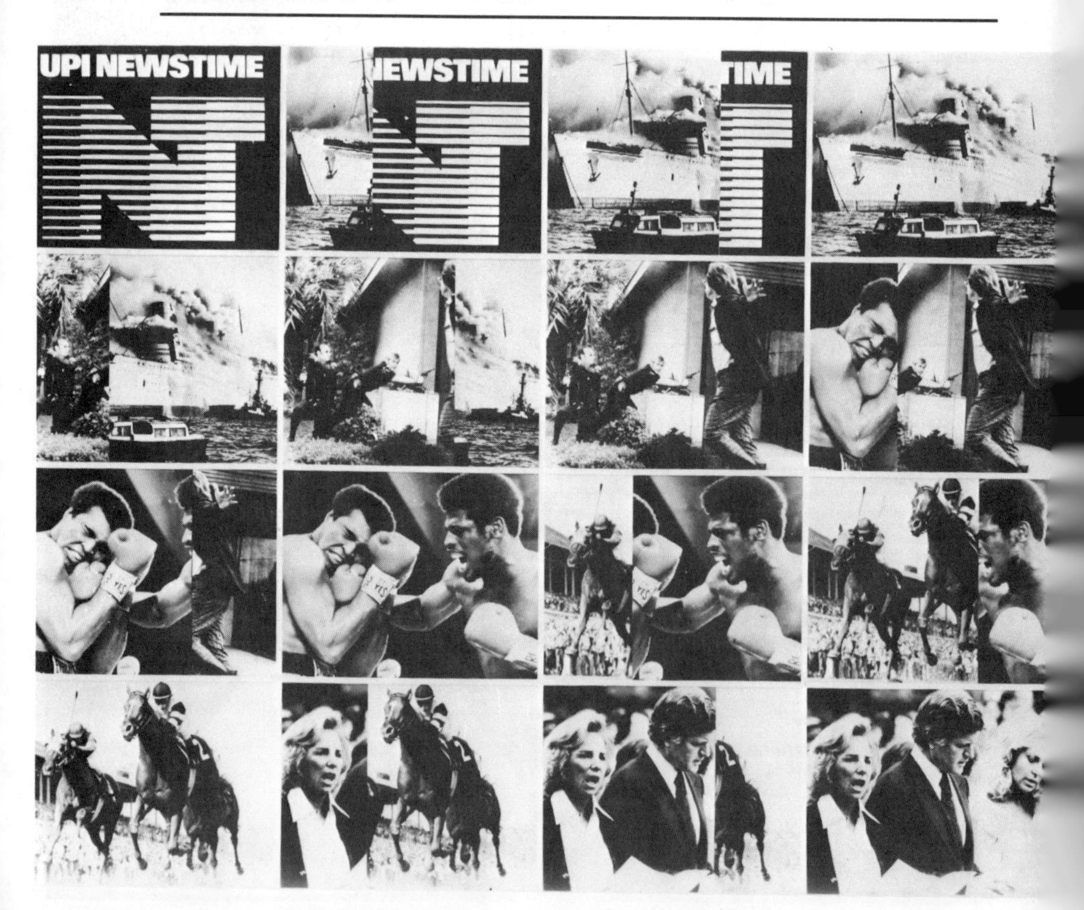

The Canadian Telidon system combines one-way teletext with two-way videotext capabilities to allow video shopping from one's easy chair. (Courtesy of Dept. of Communications, Ottawa, Canada)

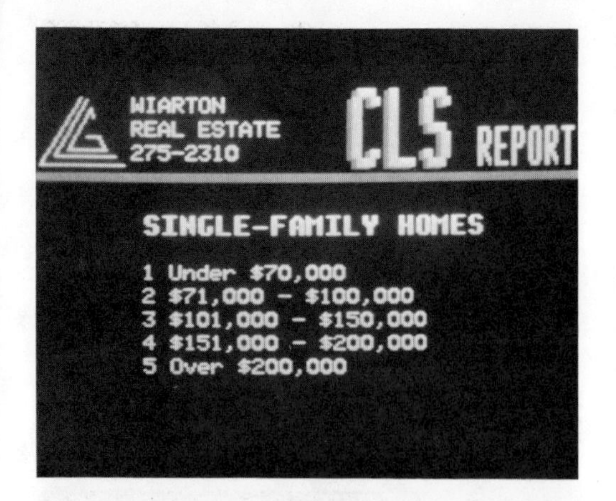

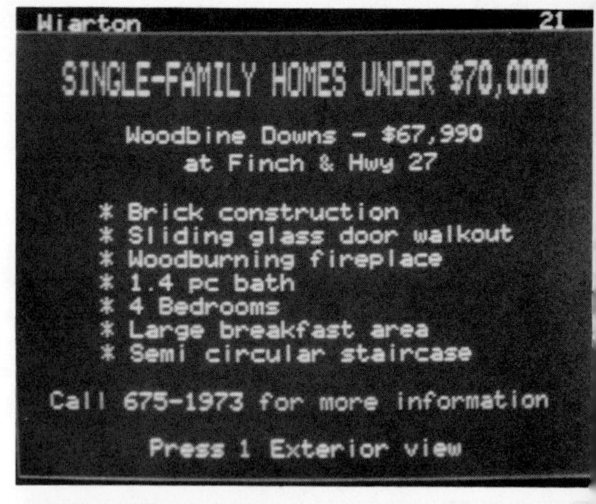

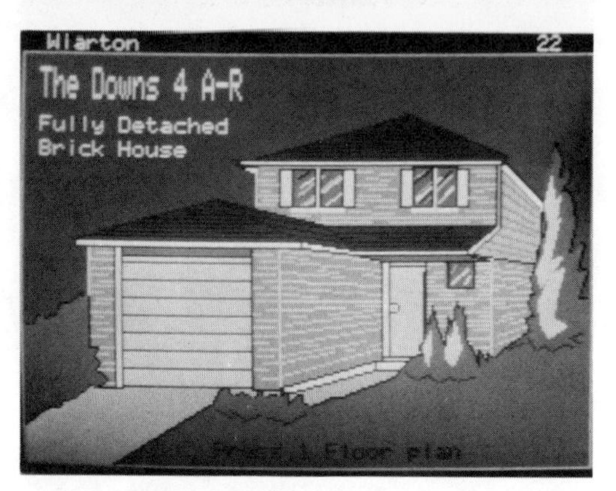

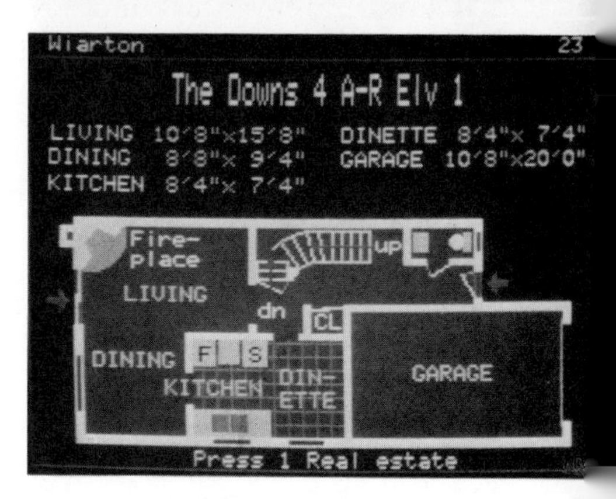

Top left: The shop-at-home house hunter presses buttons on the Telidon videotext controller associated with the home television receiver to look at single-family homes, and is provided with a list of five different price categories.
Top right: The shopper is given the features of several homes in the $70,000 price category, one of which is shown.
Bottom left: If more information on that house is desired, the viewer is then shown a graphic picture of the house.
Bottom right: This can be followed by the house's floor plan. (Courtesy of Dept. of Communications, Ottawa, Canada)

The Reuters Monitor Service takes an entire transponder to transmit its high-speed data signal to special decoder systems and terminals located throughout the United States. A number of CATV systems, such as Manhattan Cable, have entered into agreement with Reuters to carry the Monitor data signal over one of their spare cable television channels. In various offices in lower Manhattan, the incoming cables provided by the CATV company are connected to desk-top microcomputer-based decoders, video terminals, and keyboards in-

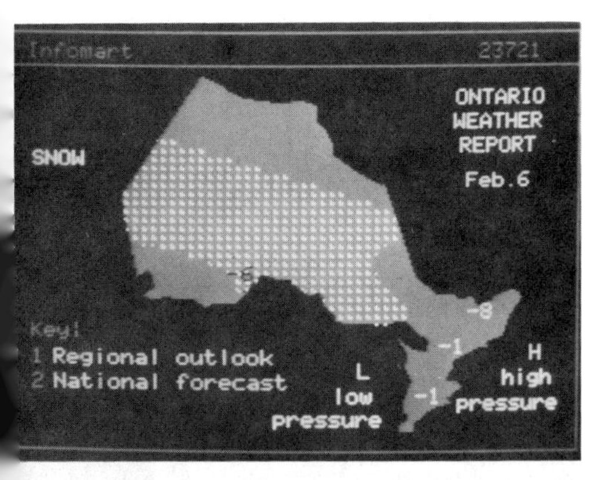

Up-to-the-second news information is constantly available to subscribers of the Telidon videotext system.
Top left: a weather map.
Top right: news picture of Albert Einstein.
Bottom left: news picture of President Ronald Reagan.
Bottom right: cover of *Time* magazine.
Graphics quality can include superior alphageometric displays as well as actual photographic representation. (Courtesy of Dept. of Communications, Ottawa, Canada)

stalled by Reuters personnel. With the aid of the Reuters Monitor directory, financial customers have immediate access to the entire market information and news retrieval of the Reuters Monitor Service from the central data bank. An optional teleprinter allows a hard copy to be produced to save information.

Through negotiation with Reuters directly, home and business owners of satellite TVRO terminals can obtain the proprietary Monitor decoder system on a monthly rental basis. This allows individuals and

At the studio, the pages of text and graphics can be created by a simple editing terminal. (Courtesy of Dept. of Communications, Ottawa, Canada)

financial organizations located in even the most isolated communities to plug into the world's number one financial news network.

One of the disadvantages of the old system of person-to-person inquiry for money rates was that the inquiry itself could have the effect of driving the market one way or another. With the Reuters Monitor Service, a subscriber can interrogate any financial market, including the stock market and commodity exchanges of North America and Europe, in complete anonymity. Several thousand terminals are now plugged into the Reuters system in North America, to receive the world's financial information directly from outer space!

Chapter 11

The Other TV Stations: MDS, LPTV, and STV

The decade of the eighties may well be known as the years of the television revolution. In addition to the TV satellite that can deliver many channels to one's backyard TVRO terminal, new and different television stations are springing up. They include multipoint distribution service (MDS) channels, new low-power TV (LPTV) minibroadcast stations, and subscription television (STV) broadcasting stations.

Each of these services is a unique concept, employing different technologies and television frequencies to deliver its video signals. As a group, they are totally unlike the television satellite signals coming from space: MDS, LPTV, and STV stations all use local transmission facilities whose signals typically travel less than 40 miles to reach the viewer's antenna. They also operate using power signals that are significantly higher than the typical 5-watt output that a satellite transponder produces 22,300 miles away.

As a consequence, the cost of the TV receiving equipment, a direct function of the antenna size and complexity, is much lower for these new terrestrial television services. Unfortunately, in many states it is illegal to pick up these scrambled and pay TV signals and to watch the television pictures using one's privately owned receiving equipment. Therefore, it is important to understand the difference between watching satellite television pictures coming directly from space and clandestinely viewing these "private channels" on an unauthorized basis. Pulling in the world from your backyard is OK—as long as the antenna is pointed out to space. A more thorough discussion of the legal and regulatory issues concerning this subject can be found in the next chapter and Appendix 5. In addition, SPACE, the Society of Private and Commercial Earth Station Users, is an active and powerful lobby group whose more than a thousand members are owners of home backyard TVRO stations. Membership in SPACE plugs one not only into a world of television enjoyment, but also into a network of friends and enthusiasts the world over.

MDS: The Local Microwave TV Service

Multipoint distribution service, or MDS, is a unique television communication service created by the Federal Communications Commission in 1970, which utilizes a specially reserved set of microwave frequencies to carry television and data communications from a central transmission point to the surrounding community. MDS is a common carrier service, not a broadcasting service. This means that MDS channels, of which there are presently two per city, are not considered public frequencies, but rather come under the same regulations that govern the operation of the telephone company microwave channels that carry the bulk of North American telephone traffic.

MDS (multipoint distribution system) uses a special private microwave TV channel. (Courtesy of Conifer Corporation)

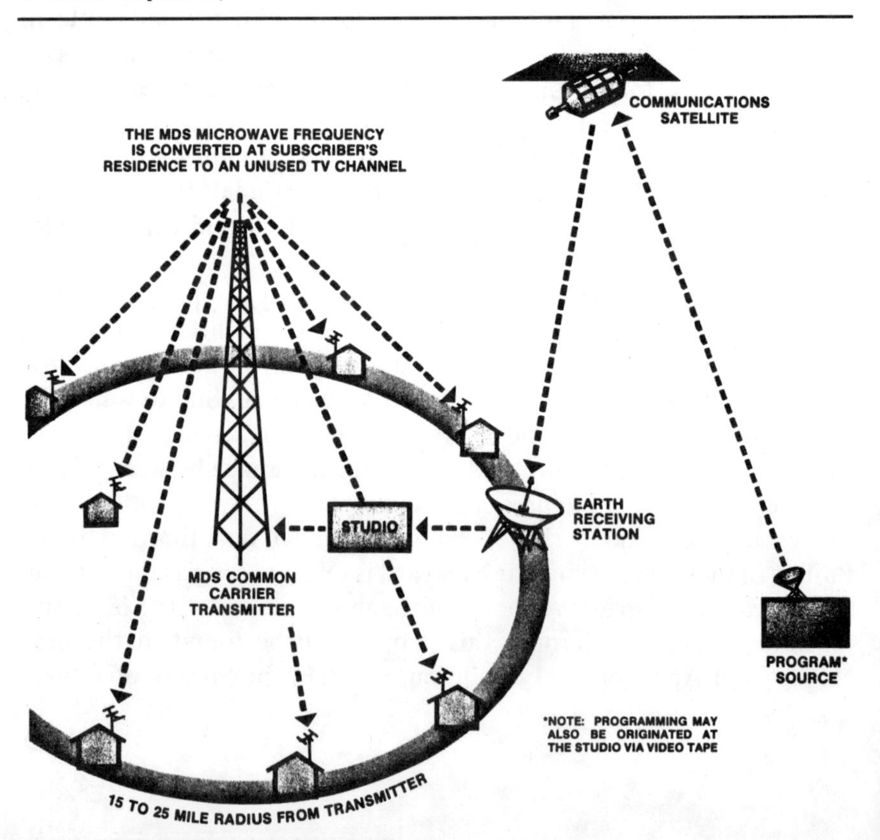

Although MDS is similar to broadcasting in that it is an over-the-air omnidirectional multireceiver system, it differs in two important ways. First, while broadcast stations aim at mass audiences within a given locality, MDS is a narrowcasting service designed to reach a selected target audience. In the late seventies MDS channels were used primarily by subscription television programmers (such as Home Box Office) to deliver pay movie services efficiently and inexpensively from a central transmitter (usually located on top of a tall building or mountain) to hotels, motels, and apartment complexes in the surrounding communities. These programmers now use MDS as well to deliver pay TV programs directly to tiny roof-top antennas for home reception. MDS, as a microwave service, operates on a line-of-sight basis only, and can reach out only to those selected subscribers who wish to tune into its channel.

Second, whereas commercial advertisers shoulder the exorbitant cost of broadcast television, the MDS transmission costs are borne by the subscription television programmer, at prices that are extremely low. A typical MDS transmitter produces a power output of from 10 to 100 watts, and costs between $25,000 and $100,000, a tiny fraction of the

How MDS works. (Courtesy of Standard Communications)

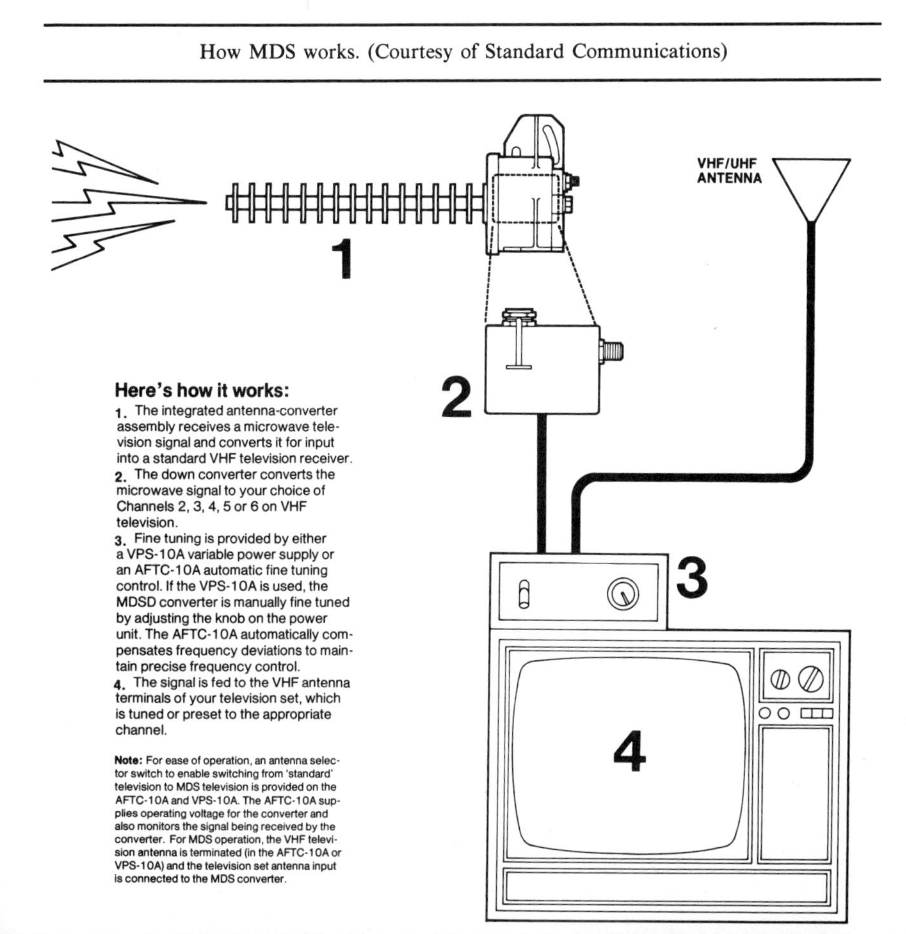

expense that a typical television broadcasting station must pay for its television transmitter facilities.

It is the cost analysis that has made MDS such an attractive service for the specialized pay movie programmer. MDS transmission charges bear no relationship to charges normally extracted from commercial advertisers by broadcasting stations. Indeed, air time purchased from an MDS operator (the owner of the local MDS transmitter facility) would amount, in comparative terms, to only a tiny fraction of that which would be charged for identical air time by a TV broadcasting station.

In fact, MDS transmission charges more nearly approximate the telephone-line charges normally tariffed by telephone companies. But again, because MDS is an over-the-air system, it can in most instances provide a cheaper and more efficient way to distribute specialized pay television programs than that offered by telephone or cable companies. MDS signals can penetrate a metropolitan area, reaching a selected audience on their own television sets, without the expense or burden of stringing unsightly cables. Simultaneous transmission to multiple points separated by large distances can be offered at costs significantly lower than those charged by common carriers operating existing land-line facilities. In contrast to other systems, omnidirectional wireless transmission charges do not increase in direct proportion to the distance between points of communication. Rather, MDS costs relate primarily to the purchase of air time and to the expenses associated with equipping selected receive locations.

To pick up an MDS channel and television signal, a special microwave receive antenna and low-cost microwave receiver must be installed on the subscriber's premises. Originally, costs for these uniquely designed systems ran into thousands of dollars. While these expenses were acceptable for an MDS operator or pay television programmer using an MDS channel to feed a hotel, for example, they would clearly be unacceptable to the average home subscriber. Luckily, just as the price of microwave equipment has collapsed in the satellite television market, so too the price of MDS receivers has fallen dramatically. Today, an MDS pay television programmer can purchase a complete antenna/receiver package to be installed in a subscriber's house for well under $50.

Because of this, MDS growth has been explosive over the past several years. Dozens of MDS operators have licensed their transmission facilities to pay TV programmers throughout the country, and the typical MDS system has between 10,000 and 30,000 subscribers who pay monthly fees ranging from $15 to $25 to have the small microwave

An example of an inexpensive roof-mounted MDS receiving antenna located at the subscriber's home. (Courtesy of Lindsay Specialty Products Ltd.)

antennas installed on their roofs for reception of the first-run evening movie programs.

Although the MDS channels are considered common carrier services, and therefore are not treated as public air-wave facilities in the same way that the TV broadcasting stations are, a number of electronics entrepreneurs have sprung up to manufacture and sell illegal MDS receiving equipment, often at outrageously high prices running from $150 to $500. These outfits boast such colorful names as Pirate TV and Bootleg TV, as well as more obscure names such as The Home Television Reception Company. It is estimated that tens of thousands of Americans have purchased these unauthorized MDS receiving units and are now using them to watch the first-run movie programs that are carried over MDS systems on a nonscrambled basis in the top 50 U.S. cities. Possession of the receiving equipment is not in itself illegal, but the manufacture and sale of bootleg equipment have been made unlawful in a number of states. Thus, picking up the single MDS movie channel locally is not at all the same as tuning in TV channels being beamed down to earth from the television satellites in outer space.

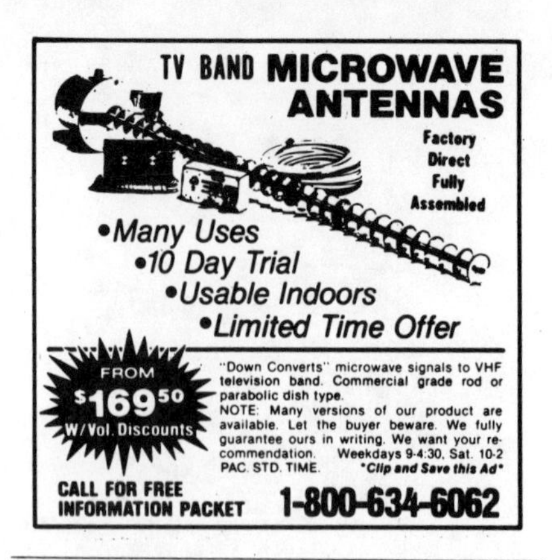

SUBSCRIPTION TV DECODER KIT $39.00. Includes Parts, Manual, and Etched Board. Manual Only $4.60. MICROWAVE TV DOWNCONVERTER KIT $169.00. Includes Everything. Catalog $2.00. J&W Electronics, P.O. Box 61, Cumberland, R.I. 02864.

SUBSCRIPTION TV: We are offering UHF converters for Dallas, Fort Worth, Fort Lauderdale, Miami, Los Angeles, Phoenix, and Chicago. Send today $179.95 for kit, $298.95 for assembled unit, $14.95 for manual, or $1.00 for details, plus list of other cities we offer to: United Electronics, 206 Delmar Ave., #4, Cincinnati, Ohio 45217.

SUBSCRIPTION TV. Complete 2300 MHZ Microwave Antenna System. 18 inch dish, probe type unit. Six month warranty, $279.95. Detailed plans, $9.95, refundable towards antenna purchase. Program guide & parts, also available. DynaComp Electronics, Dept. AMA, POB 4181, Scottsdale, AZ 85258. (602) 941-9395.

TV ads for bootleg MDS and subscription television (STV) station decoders and receiving equipment are popping up in the back pages of electronics magazines.

Subscription Television (STV) Broadcasting Stations

Over-the-air television broadcasting was introduced to the United States in the 1940s. By the early 1950s Blonder-Tongue and other equipment manufacturers had begun to experiment with transmission of scrambled over-the-air television signals that could be used to carry pay TV programming supported directly by subscription fees paid by the TV viewer. By the mid-1950s several experimental subscription television stations, as they were called, were in operation, furnishing commercial-free television shows and programming to narrowly targeted audiences in New York City and elsewhere. The project was a disaster.

Difficulties in transmission techniques and unreliable technology coupled with poor programming and "pay per view" billing assured the premature demise of the infant STV operations. For almost two decades the pay TV concept lay dormant.

By the end of the 1970s, however, the introduction of satellite television technology allowed the newly formed Home Box Office subsidiary of Time-Life, Inc., to deliver its first-run movies to the burgeoning cable TV market. STV operators were soon to follow, and two particular organizations—ON-TV (a subsidiary of Oak Industries, a leading CATV decoder manufacturer) and WOMETCO TV—began over-the-air STV operations in Los Angeles and several other U.S. cities. Today almost 100 STV stations are either in operation or have been applied for nationwide, and the original Los Angeles ON-TV UHF channel now boasts almost half a million subscribers who pay over $20 per month each to watch commercial-free first-run movies on their television sets.

In operation, the STV organization buys first-run movies from the major motion picture studios. These films are then transmitted over conventional UHF television broadcasting stations, which have had their video and audio signals scrambled to prevent ordinary television sets from receiving the pay TV signals successfully. The scrambling techniques popularly in use and approved by the FCC consist of synchronization suppression systems that prevent the ordinary television set from successfully locking onto the TV's video signal. Many of the systems also use "barker channels," which provide an audio announcement inviting the ordinary TV viewer to pick up the phone and subscribe to the movie service. When a regular television set is tuned to the STV channel during the scrambled-mode transmission time, only the barker announcement can be heard. The program audio is hidden away on a special subcarrier that can be received only by the proper STV decoder equipment.

For subscription fees ranging from $15 to $25 per month, the subscriber to an STV channel's programming receives a special descrambler unit that converts the jumbled STV signal back into an ordinary television picture. Some of these STV decoders are quite sophisticated, and by using microprocessor technology can allow multiple levels, or tiers, of programming to be presented at various times throughout the day. For example, the basic subscriber might pay $20 per month to receive a first-run G-rated movie service running from 7:00 p.m. to midnight. For an additional $10, the subscriber might obtain the second tier of programming, perhaps an adult-film R-rated service programmed between midnight and 3:00 a.m. Additional tiers may support one-time-only special events programming such as the world heavyweight boxing championship fights.

Since STV stations are broadcasting systems and not common carrier services, they come under the rules and regulations of the FCC's Broadcasting Bureau. Thus, although the signals are sent over the air in a secure format, a number of organizations provide construction plans and information on how to build one's personal decoder box. In addition, several electronics hobbyist magazines have run articles describing how the systems work and how to put together an STV decoder. Clearly, though, most STV systems technology is proprietary and patented, and any massive purchase of bootleg STV equipment will undermine the whole STV market and may, in some cases, actually drive an STV operator out of business. STV stations presently rent the subscriber decoder boxes on a monthly basis as a part of the overall service, but new regulations pending at the FCC will allow home users to buy their own decoder systems outright as long as they agree to continue to pay an ongoing monthly programming fee to the STV operator. In the meanwhile, STV, like MDS, is a rapidly expanding new locally broadcast service that is significantly different from the television satellite signals coming directly from space.

Low-Power Television (LPTV)

The third innovative video service is the low-power television station, 3000 of which will ultimately be in operation.

LPTV is a spinoff of the television translator service, a service that has been in existence since the early 1950s. A television translator is an extremely low-power TV transmitter and receiver combination that picks up a distant city's VHF television channel and rebroadcasts it to a local community on a different UHF television channel. TV translators are often used in rural areas and in those pockets of the United States and Canada in which the normal television reception of a broadcasting station is interfered with by hilly terrain or other natural phenomena that block the signals' paths. There are more than 5000 such translators operating in the United States today, typically transmitting with a power output of from 1 to 10 watts.

On October 17, 1980, the FCC announced a new low-power TV service, which would allow these micro TV stations to increase their power output significantly to a maximum of a thousand watts. This increases the range of an LPTV station from a few miles to perhaps 15 to 20 miles or more in radius. In addition, new applicants for these low-power TV stations could provide for origination of programming from the stations' own studios, or reception of programming from television satellites, as well as conventional retranslation of a distant TV station's signal. The response was instantaneous and overwhelming.

By April 9, 1981, when the FCC placed a freeze on new LPTV applications, over 5000 requests to construct and operate low-power television stations had been received by the Commission in Washington. Nonprofit, educational, and commercial programming groups sprang up throughout the country to file for these new licenses. One organization, KUSK-TV of Prescott, Arizona, filed to create more than 140 new stations nationwide. These stations would pick up KUSK's country and western television programming via satellite for local distribution throughout the country. The C&W programming would be commercially sponsored, instantaneously creating a new, fourth minitelevision network. KUSK is owned by Allstate Venture Capital Corporation, a subsidiary of Sears, Roebuck and Company.

Other innovative programming networks have formed to provide first-run pay TV movies, educational programming, and commercial entertainment on a nationwide basis to local LPTV affiliates, which will also pick up the video signals through the use of 15-foot TVRO satellite antennas. One such network could boast of over 73 affiliates as of July 1, 1981. The new network system allows a central computer facility to address and command any subscriber's decoder box. In this way, a viewer who wishes to watch a special event or change subscription from one service type to another can simply call a national subscriber center on a toll-free (800) number. The order desk personnel enter a command into the central computer, which transmits its data signal to the subscriber's decoder unit as a subcarrier on the national television channel.

Low-power TV opens up a new vista of opportunities for satellite-based television networks and programming. By the mid-1980s dozens of specialized television networks will be distributing their signals locally via LPTV transmitters, with an average range of 15 miles. Some of these networks will be commercially sponsored, others will be educationally oriented, and still others will be subscriber supported. In any event, the television programs will all be delivered by new TV satellites launched beginning in 1981. By 1985 the combination of MDS, STV, and LPTV stations operating nationwide will create 50 new television networks beaming their signals directly to one's backyard, providing a new wealth of television programs to the home satellite TVRO terminal owner.

Chapter 12

The Government and You: The Rules That Govern Satellite TV

Various rules and regulations govern the reception of satellite television signals. Which ones are applicable to you depends upon whether this viewing is done by you in your private home, or is provided by an organization in the business of reselling the satellite TV programming, such as a minicable television system.

In general there are three aspects to the legal and regulatory question: Copyright issues involve picking up and reselling, at a profit, pay TV and other programming. Cable television regulation issues concern technical and other standards as established by the FCC. The "Secrecy Act" issues concern the unauthorized publication of communications as defined by Section 605 of the Federal Communications Act of 1934, which established the Federal Communications Commission.

If that all sounds confusing, that's because it probably is! Dozens of lawyers in Washington, D.C., make good incomes arguing these points of law daily. The purpose of this chapter is to review the various rules and regulations as they might affect both the individual viewer and the commercial mini-CATV organization.

The Mini-CATV System

In establishing a CATV system that operates by using cables strung across city property, the operator of the system comes under the jurisdiction of the federal government as well as of local, state, and city authorities. In most cases, obtaining permission to operate a cable television company is a time-consuming process that involves submission of competitive bids from a number of different organizations to the city or local government regulatory body. In the bigger cities, obtaining a CATV franchise is a multimillion-dollar project requiring years of planning and negotiations. In the smaller, rural communities, however, it is still possible to establish a profitable satellite-based cable television system for which permission is readily granted by the township or city council. Thousands of such public CATV systems have appeared nationwide, with dozens more going into business monthly.

However, the fact that a cable television system is located completely on private property will usually eliminate the necessity for obtaining detailed CATV construction and operating permits from city and state agencies. These mini-CATV systems are, in fact, not considered cable television companies at all, but master antenna television systems (or MATVs). A mini-CATV system can be constructed in a high-rise apartment building, in a housing development community, or in any other jointly or individually owned private-property multiple-dwelling project. In these cases the primary regulatory agencies governing the operation of the mini-CATV systems will be the Copyright Tribunal and the FCC.

The FCC considers a mini-CATV system to be a cable television operation only in certain narrowly defined instances. First, any mini-CATV system that is jointly owned by the users (such as a condominium or housing association) does not fall under the cable television system definition. The CATV rules were originally developed as an outgrowth of the common carrier regulations for organizations that rent their facilities to the general public. In this case it is clear that a jointly owned system that makes its services available only to its owners is not a common carrier service, and is not providing its signal-carrying service to nonrelated entities. Thus, no matter what television signals are carried on a condo association's mini-CATV system, the organization will be exempt from FCC rules. If the mini-CATV system is owned, however, by an independent party, such as an apartment building management company that provides the mini-CATV service to its tenants on a rental basis, the system may be considered a cable operation under FCC rules and regulations, depending on the types of signals the mini-CATV system carries.

In all cases, mini-CATV companies that have fewer than 50 subscribers are not defined by the FCC as cable television systems. Thus a mini-CATV company with 48 subscribers need not file any forms with the FCC, or even reveal its existence to that government agency.

Mini-CATV systems with more than 50 subscribers may, however, have to file with the FCC, and will have to abide by FCC regulations, including equal employment opportunity, carriage of local television signals, and record keeping as required by the Commission. On the other hand, there are certain instances when a mini-CATV system with more than 50 subscribers is exempt from FCC rules and regulations. The FCC rules were established to regulate broadcasting, and the orderly and equitable reception and retransmission of these over-the-air television channels via cable companies is of primary concern to them. These broadcasting-oriented rules do not cover the reception and carriage of satellite television programming signals that do not originate at broadcast TV stations. Therefore, if a mini-CATV system picks up a local TV station's signal from an "off the air" TV receive antenna and carries it on its cable, the CATV system will fall under FCC jurisdiction (provided it has more than 50 subscribers). Likewise, if a mini-CATV system picks up Ted Turner's WTBS Channel 17 television station from Atlanta by way of the RCA F-3 satellite, the mini-CATV system is still carrying a television station's signal and so is answerable to FCC regulations. However, if the mini-CATV system picks up, via satellite, Ted Turner's Cable News Network, a programming channel that is not originated by a television station (Cable News Network has its own, totally independent and separate nationwide studios), then this television source of programming does not come under FCC jurisdiction.

Thus, if a mini-CATV carries only nonbroadcasting services that it picks up via satellite or other sources (but not from television station broadcasting feeds), then the CATV system will be totally exempt from FCC rules and regulations. Such a mini-CATV system could carry several dozen channels of programming, including the Cable News Network, C-SPAN, the Continental Broadcasting Network (originally the Christian Broadcasting Network), a movie service such as Showtime, a children's channel such as Nickelodeon, and one or more of the commercially sponsored programming channels such as the USA Network or the Modern Satellite Network. Operating this kind of CATV system would allow the owner to provide for complete CATV reception of all the cable programming content via a TVRO satellite terminal arranged with multiple receivers tuned to the various transponders.

Although the FCC jurisdiction would not extend to such a system, the regulations governing the use of copyrighted materials would certainly apply. Therefore, permission would be required by the mini-CATV system to carry some or all of the programming through contractual agreement.

Some satellite programmers, such as C-SPAN, produce television programs that are in the public domain. This means that no copyright has been asserted. It thus is possible to pick up C-SPAN and distribute the gavel-to-gavel coverage of the House of Representatives without signing any copyright permission agreements. However, C-SPAN does require that an authorization agreement be signed with them, which they will readily grant, to enable them to obtain statistical information on who is watching their channel. The religious programmers such the Continental Broadcasting Network provide their services at no charge to any interested mini-CATV company. Although many of these programs are copyrighted by the religious organizations, they will readily issue single-page authorization agreements permitting the mini-CATV system to carry the programming free of charge. Again, these organizations require the formality of a contractual agreement to enable them to track the viewership, and therefore success, of their satellite television programming.

The pay TV movie programmers require perhaps the most all-encompassing usage and copyright protection agreements to be signed with the CATV companies. The movie suppliers are themselves middlemen who buy the first-run motion pictures from the major Hollywood studios on a limited-use basis. Their agreements with the studios allow them to deliver these movies to nonbroadcast, pay TV distribution systems (such as cable television and MDS companies) for several showings over a two- to four-week period. The pay TV satellite programmers do not actually own the films, but rather rent the right to distribute the product via satellite to their customers.

Since the pay TV programmers are themselves limited by their contractual agreements with the movie studios as to when, where, and how the first-run movies can be presented, the pay TV programmers pass these restrictions on to their cable television customers in their user contracts. These agreements prevent the CATV company from tape-recording or manipulating the content of the movies, and authorize the CATV companies to carry the movies only on an immediate, undelayed basis upon their reception from the satellite.

Some of the pay TV suppliers argue that their contracts with the motion picture studios prohibit them from providing their movie service to any organizations other than duly recognized cable television companies or MDS carriers officially licensed by some branch of government. In this case it may be necessary for the mini-CATV company to register with the FCC, and possibly the Copyright Office, as a way of proving to the pay TV supplier that the mini-CATV system is in fact eligible to pick up and distribute the first-run movies from the satellites.

Some pay TV programmers will readily sell their movie services to satellite-based MATV systems in apartment complexes and housing communities, but have other restrictions concerning the use of their programs by individuals or non-CATV entities.

In any event, to receive and resell legally the movie service provided by the pay TV satellite-based suppliers, the CATV organization must have signed a valid authorization letter and agreement with the pay TV programmer. Most of these agreements call for the payment of monthly fees on the basis of the total number of CATV viewers who are subscribing to or receiving the particular pay TV movie service. Typical fees range from $2 to $5 per subscriber per month, averaging about 50 percent of the retail selling price for which the average CATV company markets the service. Without a properly signed agreement, the CATV organization that carries a pay TV movie channel and distributes the service to its subscribers, either free or for a fee, is in violation of the Copyright Act of 1976. As the Motion Picture Association has a diligent and active enforcement division and violations carry both criminal and civil penalties, this process is not to be recommended!

If the mini-CATV company wishes to pick up off-the-air television broadcasts from local television stations, it can do so without permission from the television programmer or broadcast station. The same copyright act that significantly increased the penalties for theft of copyrighted materials also gave CATV companies a big break in the form of a compulsory license that automatically authorizes a CATV company to pick up and carry such broadcasting programming. To obtain this license, a CATV company must register with the Copyright Office in Washington, D.C. Depending on the size of the CATV system and also on the number of the off-the-air broadcasting signals the system carries, a semiannual copyright royalty payment must be made by the CATV company to the U.S. government. These fees range from $30 a year upward, and the money collected is eventually distributed by the federal government among the copyright owners and producers.

The Home Satellite TV Viewer

The home satellite TV viewer is in a world apart from the commercial satellite company. Unlike the mini-CATV organization, the home viewer is not watching the satellite to make a profit or resell the service, but rather is viewing the TV pictures for personal edification. Consequently, the FCC regulations concerning the operation of a CATV system do not apply to the home satellite TV viewer. In addition, it can easily be argued that the reception of the superstations such as WTBS in Atlanta, WOR in New York, and WGN in Chicago is not covered by FCC cable television regulations per se as these are ordinary broadcasting stations that a viewer could watch who was physically located within the over-the-air television signal coverage area of the TV station.

Questions concerning the reception of copyrighted protected materials pose a thornier problem. Does the home viewer who puts an earth station in the backyard and picks up first-run movies direct from space have a right to view these movies without further permission? Or should these movies only be watched if the home user has paid a fee to the pay TV programmer? The answer varies, depending on whom you go to with your question. The Motion Picture Association of America and most of the pay TV programmers argue, of course, that these movies from space are private property, and should not be seen without permission. Consumer advocates have honored this argument, however, by pointing out that the satellite signals are received on one's private property and that the home owner's rights are being "violated" by the pay TV programmer who is beaming a movie service to earth and in effect "trespassing" on private property without permission! The Society of Private and Commercial Earth Station Users (SPACE) has taken the logical position that their members, who represent thousands of private TVRO owners nationwide, should be able to obtain authorization from the pay TV companies to tune into the movie channels by paying a nominal monthly subscription fee. Such a fee would, of course, be far less than the monthly rates charged to home viewers by commercial CATV companies who are acting, in effect, as retail distributors of the pay TV product.

A number of the movie programmers have accepted this argument, and, primarily through the efforts of SPACE, it is possible to obtain permission to watch first-run movies via one's backyard TVRO home satellite terminal by signing a simple one-page agreement. For further information on this straightforward process, a home TV satellite viewer should contact SPACE in Washington, D.C. (see Appendix 4 for the address).

The viewing of nonpay TV services on the birds is a much simpler process. All of the religious programmers will give permission to pick up and watch their programs to anyone who drops them a letter requesting it. The sports channels and many of the other commercially sponsored networks will do the same, as will a number of the public service channels such as C-SPAN. For the home viewer of satellite TV, it is probably a good idea to send a short note to the major programmers who provide these types of shows to get these permission letters sent to you—if only for your personal files.

A third regulatory issue presented to the home satellite viewer concerns the "unauthorized publication of communications" covered in

Section 605 of the Communications Act of 1934. Basically, this section prohibits any person who receives a satellite television, radio, or data signal from "divulging or publishing the existence, contents, substance, purport, effect or meaning thereof." This means that while it may be all right to watch, for example, the Cable News Network from the privacy of one's home, it might not be all right to invite the neighbors in, or to carry this channel on a projection television system in one's restaurant without first obtaining permission from the Turner Broadcasting System. Good common sense, of course, suggests that this would be true. Ted Turner is in the business of selling his CNN service to his advertisers. His contractual agreement is necessary so that he can determine how many people are actually watching at any given time. Likewise, the organization that delivers the signal to the CATV companies nationwide has to be paid for the expense of renting the CNN satellite transponder from RCA.

One possible way of becoming an instant mini-CATV system, and thus meeting all the tests for eligibility in picking up and paying wholesale rates for the pay TV channels, is actually to get together with a few neighbors and agree to share the satellite services on a paying basis.

By filing a simple form with your local city or county government, you can legally establish a DBA (or "doing business as") company name for your instant CATV company, perhaps calling it the CATV of XXX Street Company—which is simply you and your friends.

Armed with the name of your newly formed company, a checking account in the local bank under the company name, and some letterhead printed by a nearby printer, you should be able to write a formal letter to the pay-TV programmers requesting they provide your CATV company with service. For all practical purposes, you really are in the CATV business.

The use of video-cassette recorders to tape movies and other programming received directly from the satellites is still in question. A major lawsuit brought against the Sony Corporation by Walt Disney several years ago attempted to outlaw the use of video-cassette recorders for making home tapes of television shows. While the court ruled in Sony's favor, allowing home viewers who own video-cassette recorders legally to tape-record their own television programs from off-the-air television broadcasting stations, this case did not apply to the reception of nonbroadcasting signals, including the tape-recording of movie channels provided by the local CATV company. However, it appears that the common-sense grass-roots feeling in the United States will prevail, and the two million U.S. owners of video-cassette recorders will continue to tape-record any television programming that is technically possible.

Listening in on the tens of thousands of hidden audio and teletypwriter channels on the satellites clearly is covered by Section 605 of the Communications Act. The process of receiving these communications is somewhat similar to listening to mobile telephone and other radio signals using the popular scanning receivers. Millions of these receivers are now in use in the United States, and dozens of manufacturers sell them in hi-fi and stereo stores, electronics outlets, and even department stores.

Private radio channels that are picked up by these scanning monitors include ship-to-shore and marine communications, police, fire, coast guard, and FBI channels, mobile telephone and private business channels, and temporary communications links established for special events. The same restrictions that apply to the tactful use of these scanning receivers should apply also to monitoring the hidden channels on a communications satellite transponder. The individual who uses a scanning receiver has an obligation not to divulge any of the information or messages heard to any other person. That's why this regulation is called the Secrecy Act!

It is interesting to note that Section 605 of the Communications Act, as well as the entire Act itself, was written before television was invented, and well before the creation of the communications satellites. The section was originally designed to prevent any telephone company employee from listening in on private conversations, as was often possible in the days when telephone calls had to be placed through manual switchboards from city to city and to overseas points. The FCC of 1934 would be dazzled by the veritable paradise of telecommunications options available to the North American consumer today! Motion picture signals directly from space? Unheard of! Arthur C. Clarke was not to postulate his famous Clarke orbit theory in *Wireless World* until 1945, 11 years later. The Communications Act of 1934, while clearly an archaic document, has met the test of time by asserting that the American's right to own and operate radio reception equipment for private use cannot be arbitrarily abrogated by the federal or state governments. Thus, except in very isolated cases typically involving national security, the Secrecy Act has never been invoked in the almost 50 years of its existence. The FCC itself is moving more rapidly toward total deregulation of the U.S. airwaves, and given the predilection of the present individual-oriented Congress, it is clear that the rapid growth of home satellite television terminals will be nurtured rather than discouraged by the federal government.

Appendix 5 presents a more thorough analysis of the regulatory issues governing the home TV reception of satellite signals. It was prepared by Richard Brown, general counsel for SPACE.

Chapter 13
The New Satellites Are Coming

The Canadian Northern Territories Service

Broadcasting directly from space! For months the headlines have been announcing COMSAT General's new subsidiary, the Satellite Television Corporation. By the mid-1980s STC hopes to provide Americans with three nationwide television channels directly from space—picked up via 2-foot parabolic antennas. While the present TV satellite systems are used primarily to deliver television programs to CATV companies and local broadcast stations, the new STC satellite system will provide direct broadcast service (or DBS) to a tiny dish on one's rooftop.

Actually, the future is already here. Since the late 1970s the TeleSat Canada organization has been experimenting with its high-power 12-GHz ANIK satellite system. Recently the Canadian Broadcasting Corporation (CBC) began to use the satellites to deliver low-cost network television programming directly to isolated homes scattered throughout the northern provinces. Delivering both English and French programming, the Canadian ANIK DBS service, known as CBC North, uses small parabolic antennas roughly 3 feet each in diameter.

The CBC North project has been a great success, and thousands of Canadian viewers have been able to receive television for the first time, viewing perfect pictures transmitted from Montreal, Toronto, and Vancouver studios.

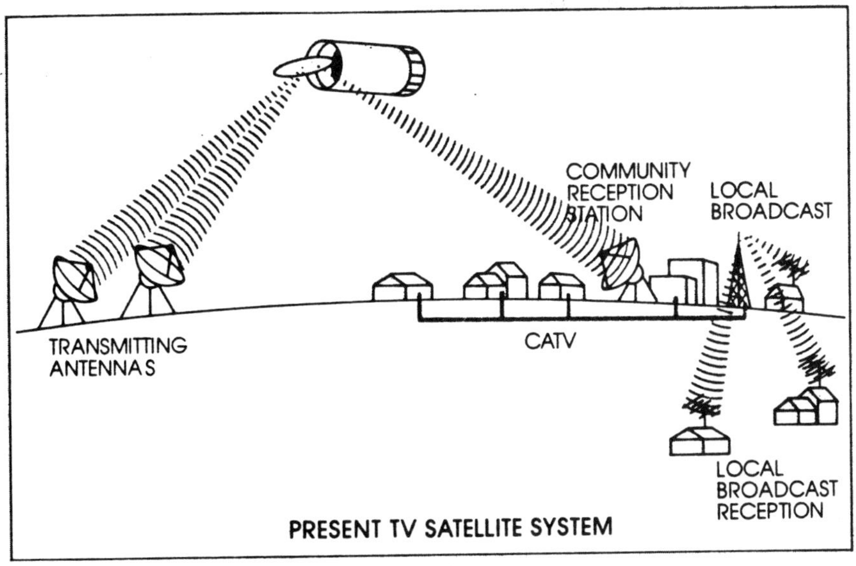

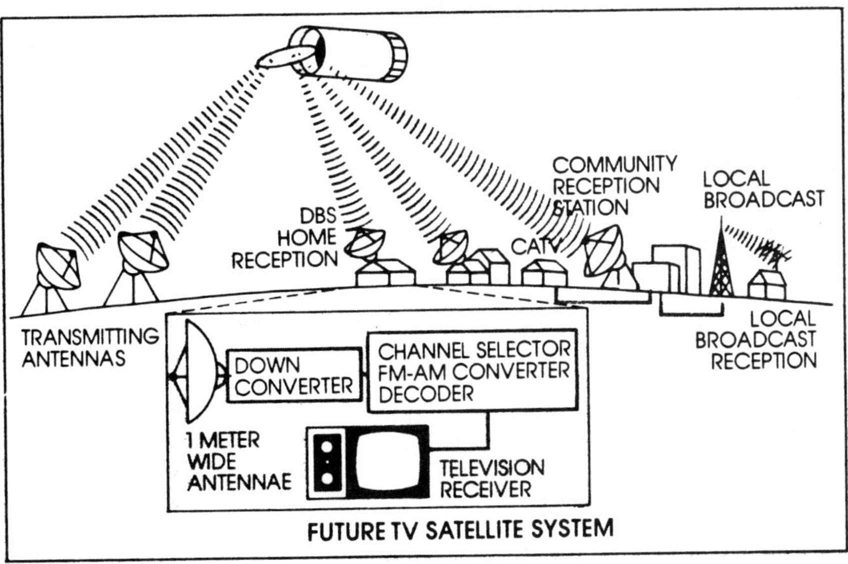

Television transmission via satellite. (Courtesy of Dept. of Communications, Ottawa, Canada)

The costs to deliver such a multichannel programming service are well within the range of an ordinary terrestial-based television network. The Canadian experience has shown that low-cost terminals can be easily installed in large quantities, and reliable nationwide television broadcasting can be fed from a single transmitter 22,300 miles in space,

Experimental 14-/12-GHz direct-broadcast, TVRO earth station. (Courtesy of TeleSat Canada)

thereby eliminating the hundreds of local TV broadcasting transmitters that would otherwise be required. Of course, the trade-offs are of some concern. The elimination of local transmission facilities potentially curtails the delivery of locally oriented news. However, it appears that for certain types of national programming, the creation of a high-power direct broadcasting satellite system ultimately will take place in every major country in the world, and the DBS projects are both technically and economically feasible.

Many other countries have begun to experiment with similar concepts, including the Japanese, whose national DBS television project was begun in the mid-1970s. Western Europeans are considering the establishment of a multinational intra-European television service operating in multiple languages across many borders.

The Canadian ANIK DBS service can be picked up throughout North America using a conventional home TVRO earth station, provided that a different receiver and feedhorn arrangement, tuned to the higher microwave frequencies used in the direct broadcasting TV system, replace the conventional GHz C-band satellite TV receiver.

(Courtesy of Paraframe Inc.)

(Courtesy of Paraframe Inc.)

The Canadian experiment has been a tremendous success in many ways unenvisioned by TeleSat Canada. A number of viewers from Central American and Caribbean nations, including Haiti, a French-speaking country, report excellent reception as far south as the equator for the CBC French programming!

DBS is alive and well—in Canada today!

The Satellite Television Corporation

On December 17, 1980, COMSAT, through its subsidiary the Satellite Television Corporation, applied on FCC Form 301 to construct and operate a new television broadcasting station. What made this application unique, however, was that the broadcasting transmitter would be located not a few miles down the street from the home TV receiver, but in the Clarke geosynchronous orbit. In addition, the television service would not use the ordinary UHF or VHF channels, but would broadcast over three new microwave frequencies in the 12-GHz band. These TV signals would be picked up by special microwave receivers and miniature microwave parabolic dish antennas to be installed by the Satellite Television Corporation on the viewer's premises.

The Satellite Television Corporation also made a working capital of $625 million available to fund the new project!

In operation, STC plans to launch four satellites to feed three channels of programming to the Eastern, Central, Mountain, and Pacific time zones with additional service to Alaska and Hawaii.

To determine what new national programs Americans would be interested in receiving, STC conducted an in-depth market survey using a number of nationally recognized consulting firms to interview the public. STC concluded that the public interest could best be served by providing a balanced combination of distinctive entertainment, informational, and educational programming on a diversified, multichannel schedule. As currently conceived, STC's program schedule will offer general and family entertainment, including movies, plays, concerts, night club acts, opera, and dance. In addition, STC will reserve broadcast time for the presentation of public affairs and special-interest educational programs. The company proposes to provide minority-oriented programming, children's programming for both preschool and school-age children, and quality educational and instructional programming designed for adults as well as children.

Additionally, some programs will be sold on a per-program or per-series basis. All of the programming will be supported directly by the viewers, who will pay a monthly subscription fee to receive the 24-hour-per-day programming transmitted by STC on a scrambled basis. In creating an instantaneous national narrowcast market, STC believes

System configuration. (Courtesy of Satellite Television Corporation)

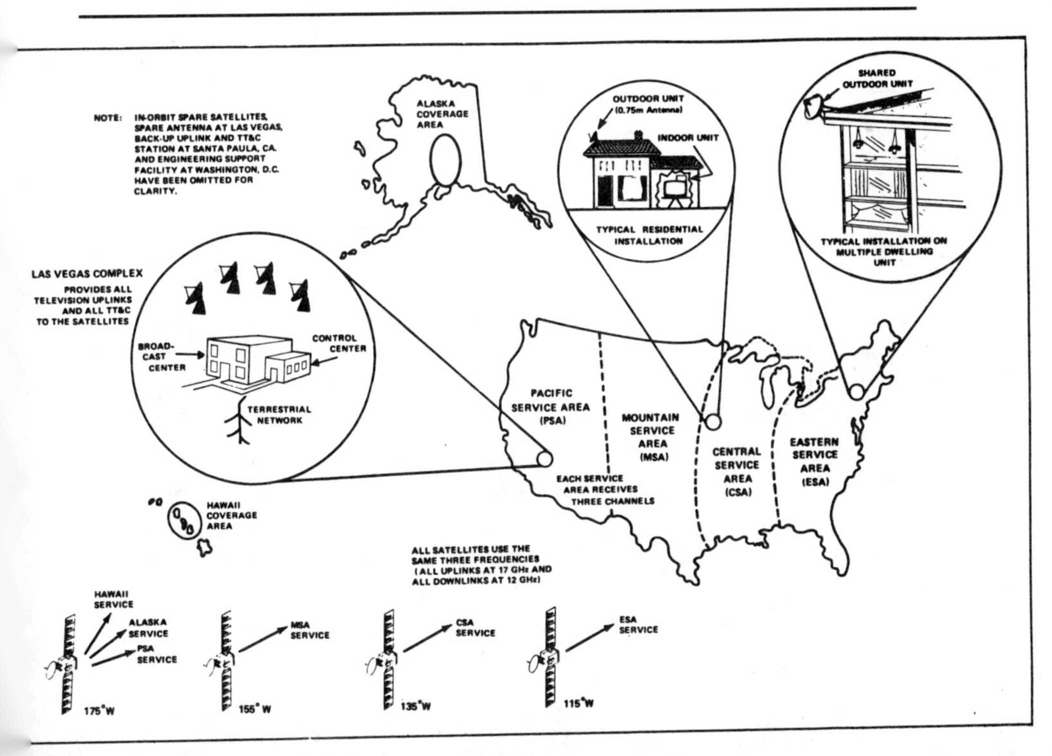

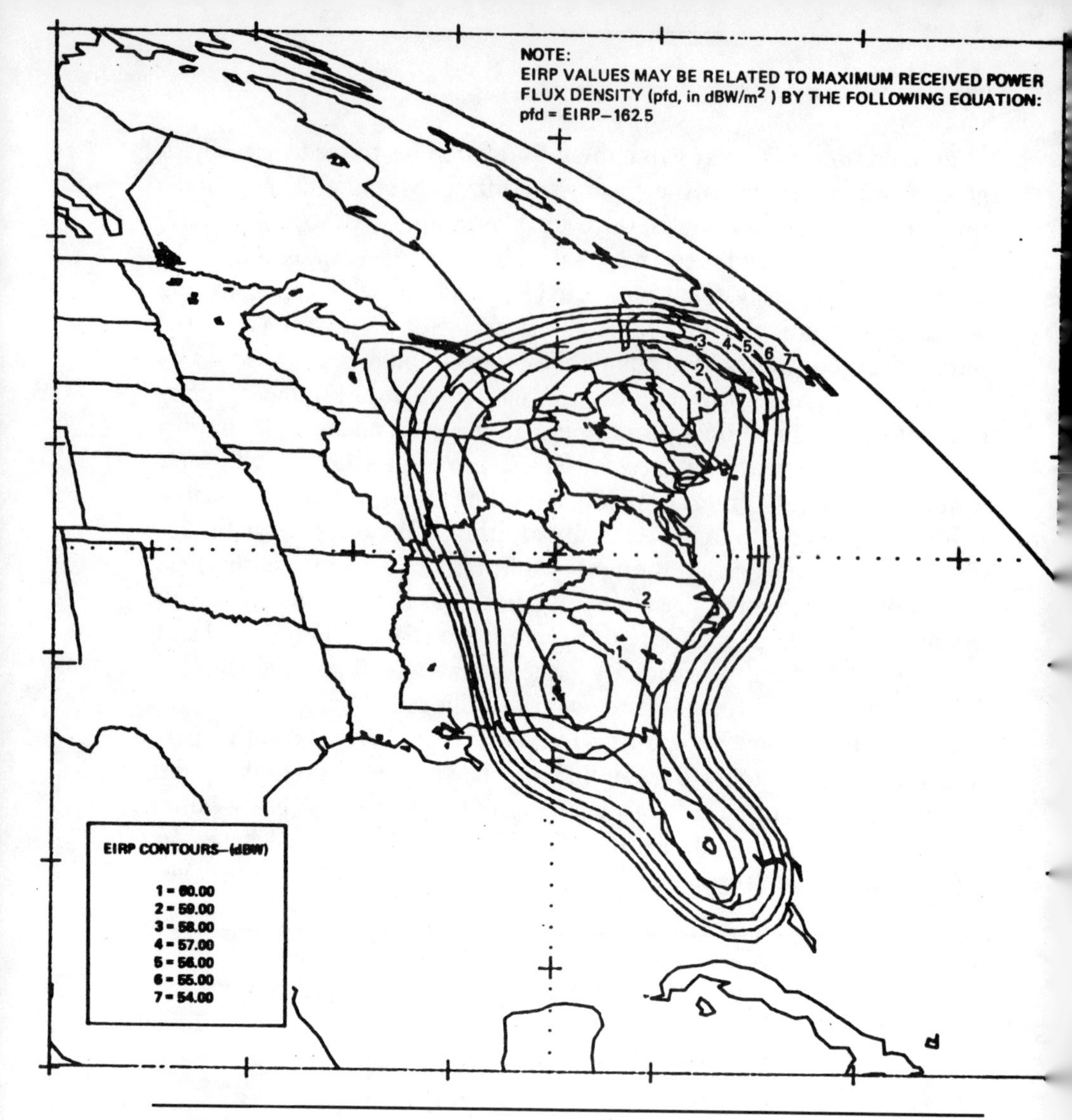

Computer predicted EIRP contours for ESA beam. (Courtesy of Satellite Television Corporation)

that small numbers of viewers will be willing to pay to see a particular series, thereby justifying the continual production of those shows.

The initial DBS system is divided into three channels. Channel A, called Superstar, will provide major motion pictures, popular concerts, theater specials, and family entertainment. Channel B, dubbed Spectrum, will offer film classics, children's programs, variety shows, performing arts and cultural attractions, and public affairs. Channel C, entitled Viewer's Choice, will feature adult education courses, regularly scheduled sports events, instructional sports activities, periodic major sporting events, theatrical productions, and special interest programming.

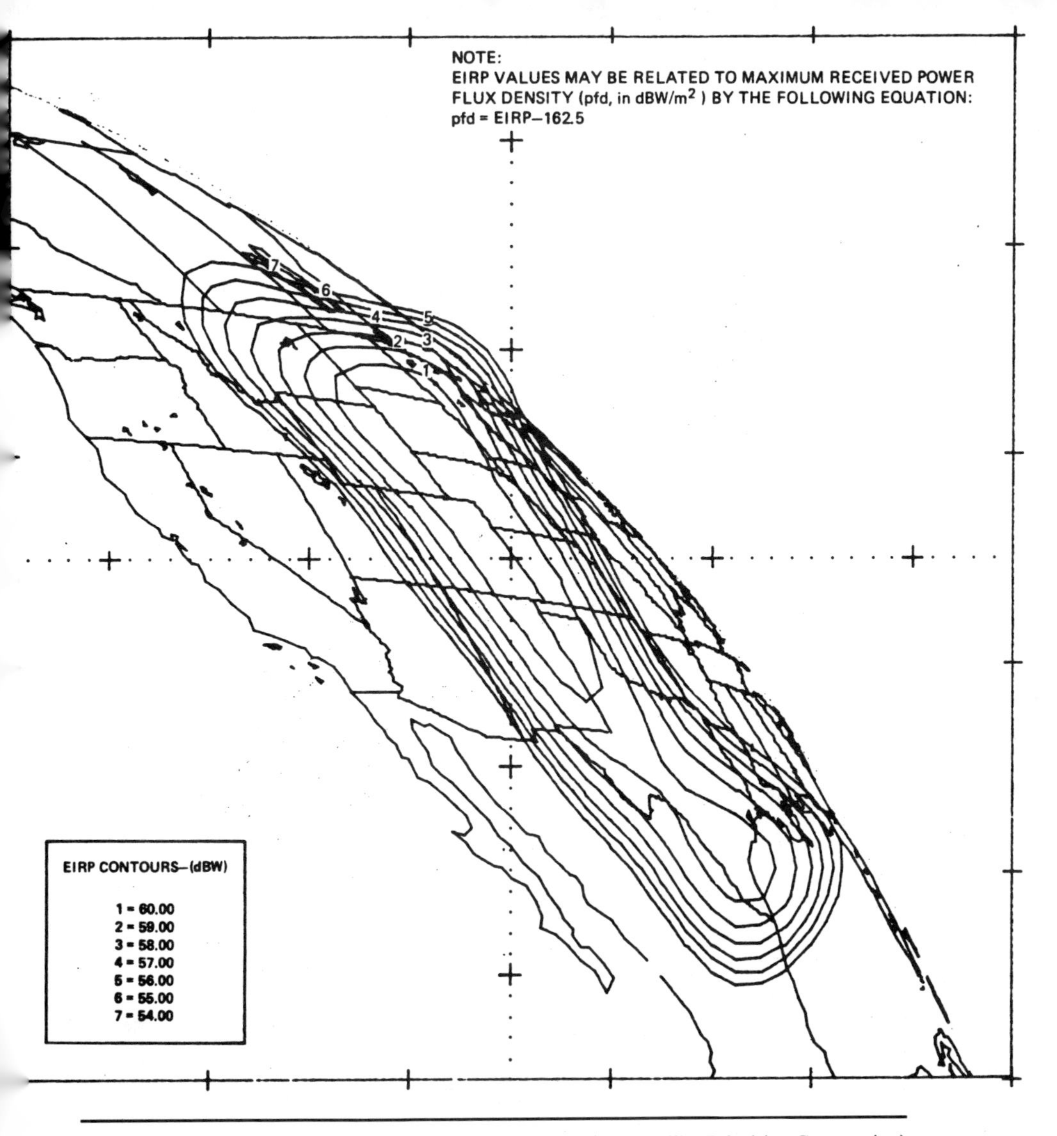

Computer predicted EIRP contours for MSA beam. (Courtesy of Satellite Television Corporation)

In analyzing the marketplace, STC has concluded that first-run pay TV movies are of greatest interest nationwide, and approximately 82 percent of the Channel A program material, or 138 hours per week, will be devoted to feeding a pay movie service. Eight percent of the Superstar channel will be reserved for sport specials, and 4 percent for "star" specials featuring a variety of performers. Theater specials, involving filmed or taped versions of major Broadway and off-Broadway productions, will account for a projected 2 percent of the total Channel A content.

Channel B, the Spectrum service, will cater to a variety of viewer interests ranging from children's programming and film classics to public affairs and the performing arts. STC believes that the broad spectrum of material on Channel B will widen the customer base and substantially enhance subscription revenues, thus providing new funds for additional special interest programming. Forty-five percent of the channel will be devoted to classic award-winning pictures; 19 percent will go to children's programming to both "entertain and instruct." STC plans to broadcast the children's series devoid of gratuitous violence and uninterrupted by advertising. Performing arts and cultural events will fill 17 percent of the channel's time, scheduling classical music, dance, drama, and similar shows. Public affairs programming—including the STC *Journal,* catering to the needs of minority, women, and other special interest groups; *Singled Out,* another program exploring a wide range of social problems; and *Transition,* a semiweekly series dealing with changes in life styles—will be the fourth most frequently broadcast programming service, accounting for 14 percent of the total Channel B programming week.

Channel C, Viewer's Choice, will include regularly scheduled lectures and discussions on major social, economic, political, and scientific issues of the day, and a series featuring the best in regional theatrical production. Channel C will also carry sports programs, including two regularly scheduled weekly series as well as periodic broadcasts of special events that will be sold on a per-program basis. As surveys have indicated, movies, sports, and cultural programming represent the most popular pay TV offerings. Thus, by counterprogramming these three genres on Channels A, B, and C, STC is expected to provide the greatest number of viewers with their first choice in programming. Channel C will resemble a "magazine stand of the air," affording a potpourri of interests to the public, including sports, photography, cooking, and other favorite pastime activities, with a series of "how to" tips on general subject areas such as car repair, home repair, home decorating, furniture building, and refurbishing.

Adult high school and continuing education classes will account for one-third of all hours available on Channel C during the week, and special interest programming will furnish over 20 percent of the network feeds. Professional classes suited to meeting continuing educational requirements of several professions, such as medicine, law, engineering, and the physical and social sciences, will fill 14 percent of the total air time on Channel C. Experimental theater series featuring the works of lesser known playwrights and local theater productions from throughout the United States and Canada will be aired on *Curtain Call,* which will present filmed or taped versions of live stage productions.

Taken together, the three channels represent an ambitious adventure on the part of COMSAT in its bid to broaden its corporate activity from that of a common carrier organization that simply rents its facilities for others to use, to that of a full-service television network providing innovative and original programming in direct competition with the four major television networks. All told, the Satellite Television Corporation proposes to broadcast about 387 hours of programming each week, of which over two-thirds will consist of general entertainment programs. Educational and instructional programs will account for an additional 20 percent of total air time on all three channels, whereas 7 percent of the broadcast week will be devoted to children. Public affairs will make up 6 percent of total air time during an average week.

The reaction to STC's proposal and FCC application was immediate and intense. The existing major TV networks cried foul play. So did a number of the common carriers. In its business plan STC proposes to bridge the gap between the common carrier and the broadcasting entity, thus further blurring an already confusing distinction. Many of the "old-line" corporations objected to the new and different competition, and STC's proposed DBS system will have an uphill battle in winning full approval as originally conceived.

Yet it appears clear that some form of DBS service ultimately will make its debut in the United States. Its feasibility will not be tested in court, however, but in the marketplace.

Direct broadcasting satellites, cable systems, subscription television stations, MDS, and video disks and tapes are close substitutes, and thus effective competitors. Therefore, commercial success of this new DBS technology will depend in large measure on the Satellite Television Corporation's ability to compete effectively not only in terms of the price charged for the service, but also in terms of its general attractiveness, relative to the programming distributed by other pay video sources. STC believes, of course, that its proposed program line-up will be entertaining, informative, imaginative, balanced, and highly diverse. On the other hand, the new cable television, MDS, and low-power TV networks also believe the same thing! This means that the consumer will ultimately benefit from the fallout as these new megabuck organizations come onstream with their heavy-hitting satellite-based television programming in the mid-1980s.

The Satellite Television Corporation plans to install tiny roof-mounted parabolic antennas at the customer's residence as part of a turnkey television broadcasting service. The typical equipment configuration consists of an outdoor unit (the ODU), which includes a 2- to 2½-foot-diameter receiving antenna and associated microwave elec-

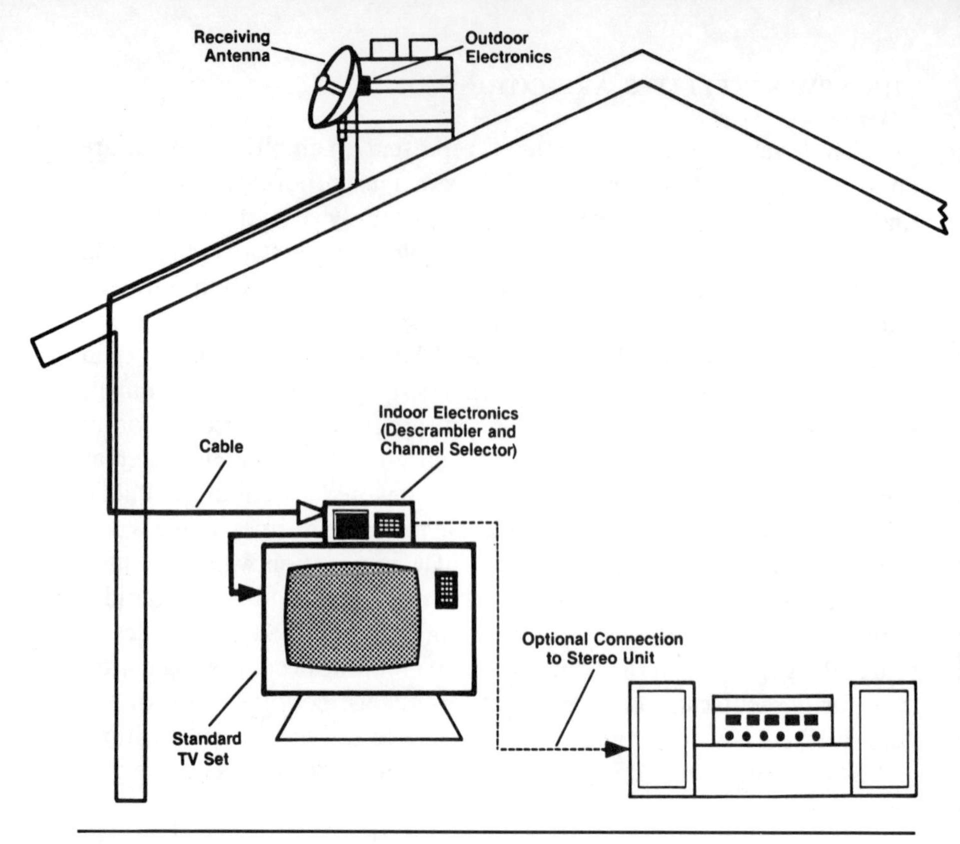

Typical residential installation. (Courtesy of Satellite Television Corporation)

tronics, and an indoor unit (IDU) connected to the outdoor unit with up to 100 feet of miniature low-loss coaxial cable. The indoor unit consists of a receiver and decoding electronics package with cables connecting the unit to both the subscriber's television set and, optionally, to an associated stereo system for reception of stereo television audio planned for the system.

Multiple-dwelling units such as apartment buildings and condo developments will have a single outdoor unit and a special intermediate-frequency distribution amplifier that will distribute the three separate channels to multiple indoor units located in the subscribers' homes.

One of the basic concepts of STC is that the customer will obtain programs on a subscription television basis, paying approximately $14 to $18 a month for a total subscription fee that includes the three basic channels of programming. Additional tiers or levels of television programming will be offered periodically in lieu of the regular programming ordinarily available on one of the channels. These special programs will be priced on a "pay-per-program" or "pay-per-series" basis, and in addition special service features, including stereo sound, second-language track, closed captioning for the deaf, and teletext service, will also be offered for somewhat higher fees.

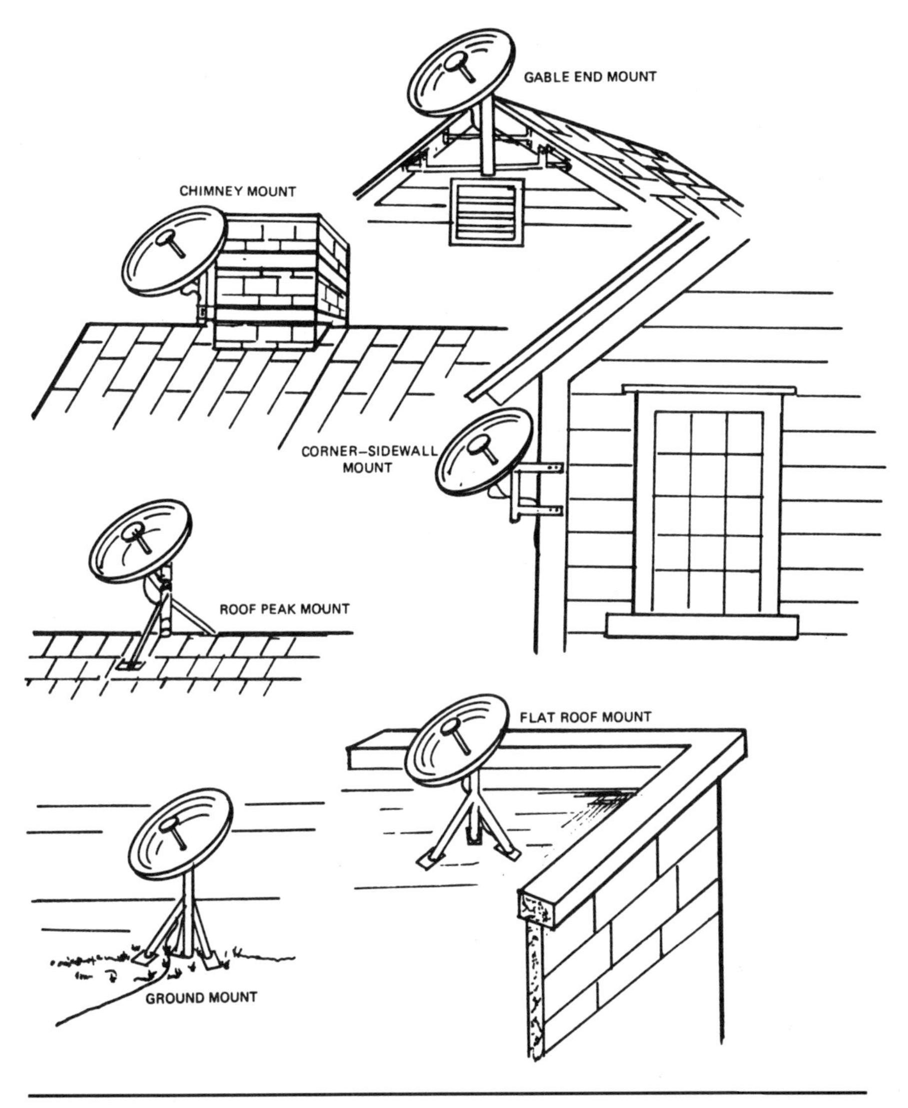

Home equipment outdoor unit—typical mounting configurations. (Courtesy of Satellite Television Corporation)

The home equipment costs, for the satellite antenna and electronics, are projected for rental at between $6 to $10 per month with a one-time charge of about $100 covering the installation of the antenna, mount, and cable, and the transfer of the title of this equipment to the subscriber. Optionally, the subscriber could elect to purchase the home equipment outright for perhaps $500 to $1000.

In either case, if a customer failed to continue the monthly subscription service, the Satellite Television Corporation central computer would send a signal to the subscriber's terminal to deactivate it, effectively removing it from receiving the three pay television channels. To have the equipment installed and service turned on, the customer need

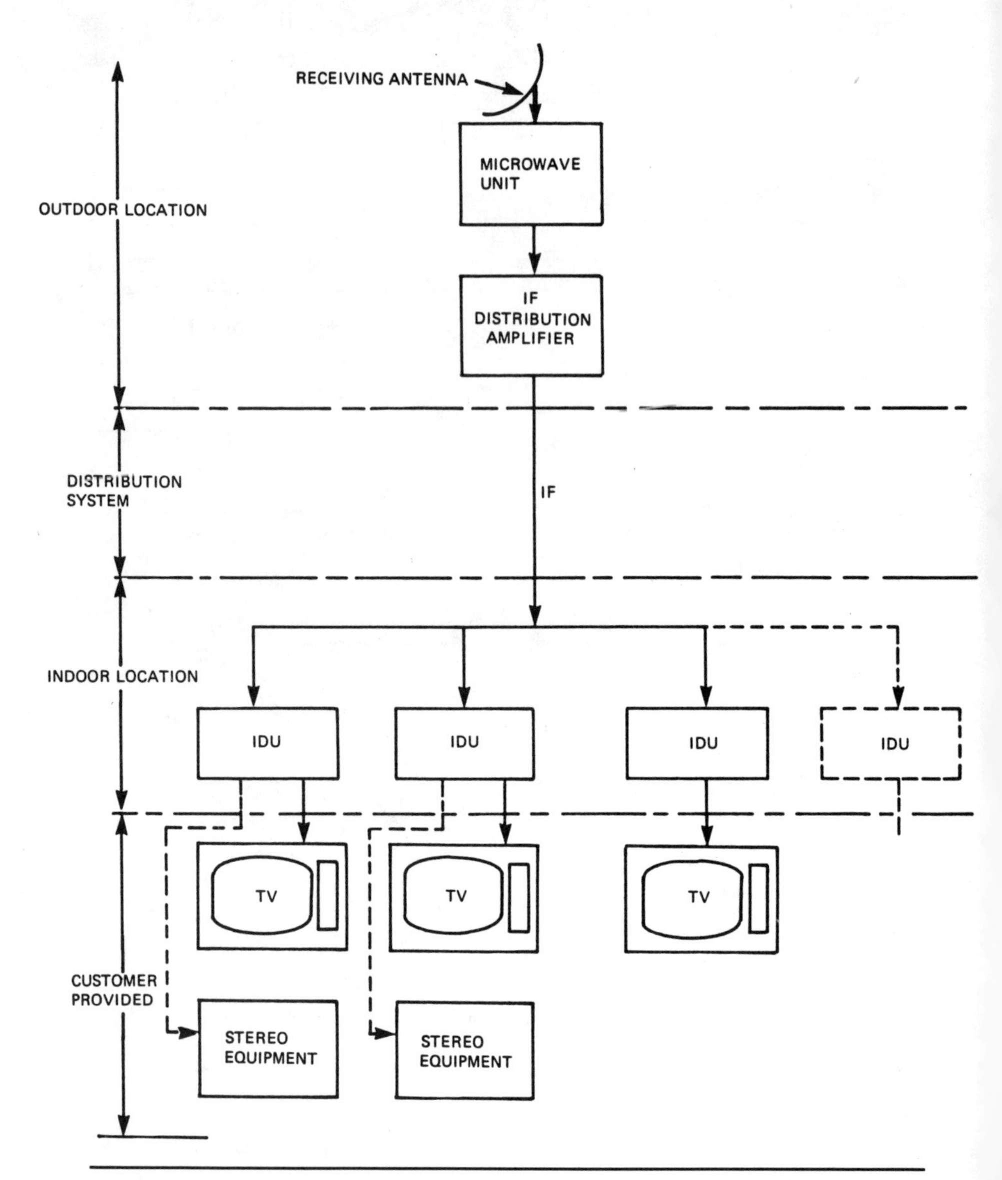

Multidwelling unit receiving equipment—basic elements. (Courtesy of Satellite Television Corporation)

only call a national toll-free telephone number for an STC representative or agent to appear the next day. Ultimately, the Satellite Television Corporation believes, every single-dwelling unit in the United States will be wired with its own roof-top 12-GHz satellite antenna, and will have a television outlet in each room similar to the modularized connectors installed by the telephone companies. Upon moving to a new location, a customer will need simply to unplug the indoor electronics unit and pack it away with the television set. Upon unpacking the equipment at a new residence across town, or across the country, the user can reconnect the home electronics unit simply by plugging it into the waiting home satellite outlet.

Although this whole concept may sound like a page from a science fiction novel, COMSAT is betting almost a billion dollars of its money that its dream of direct broadcast satellite transmission will come true. COMSAT was created by President John F. Kennedy and the U.S. Congress in 1962 as a private corporation to bring satellite communications to the world, and the company has an enviable track record of achieving the impossible!

In the meanwhile, the existing 4-GHz C-band satellite television services are growing more numerous every month, and the friendly skies of the Satellite Television Corporation's three pay TV channels may well be threatened by lower cost competitors, who will allow users to tune in, using "off-the-shelf" home TVRO earth terminals sold in the hundreds of thousands. By the time that the Satellite Television Corporation has its birds in the air, the cost of a typical C-band TVRO backyard satellite antenna may well be under a thousand dollars, and such antennas may be available in dozens of brands at the local department store!

Chapter 14
The World in Your Backyard

The Sky's the Limit

By the mid-1980s it is estimated that over 50 communications satellites will be circling the globe, providing 600 or more channels of news, sports, entertainment, religious, educational, and informational programs in English, French, Spanish, German, Russian, Portuguese, and Dutch! The RCA Corporation projects that, in the United States alone, by the year 1990 more than 41 communications satellites operated by half a dozen U.S. companies will be circling the earth. By the middle of the 1980s, in addition to Western Union, RCA, and COMSAT, new players such as AT&T and GTE will have joined the fray, along with Hughes Corp. and the Southern Pacific Communications Corporation.

As satellites get bigger and bigger, larger and larger launching vehicles are needed to put them into orbit. Luckily, the Space Shuttle has arrived on the scene in the nick of time, allowing gigantic supersatellites to be launched into geosyschronous orbits. During the mid-eighties the Space Transportation System (as the Space Shuttle is officially known) will be operating weekly flights carrying tens of thousands of pounds of satellites into orbit. The shuttle's hold can handle a package of equipment 60 feet long by 15 feet in diameter, capable of literally putting five Volkswagens into orbit, bumper to bumper!

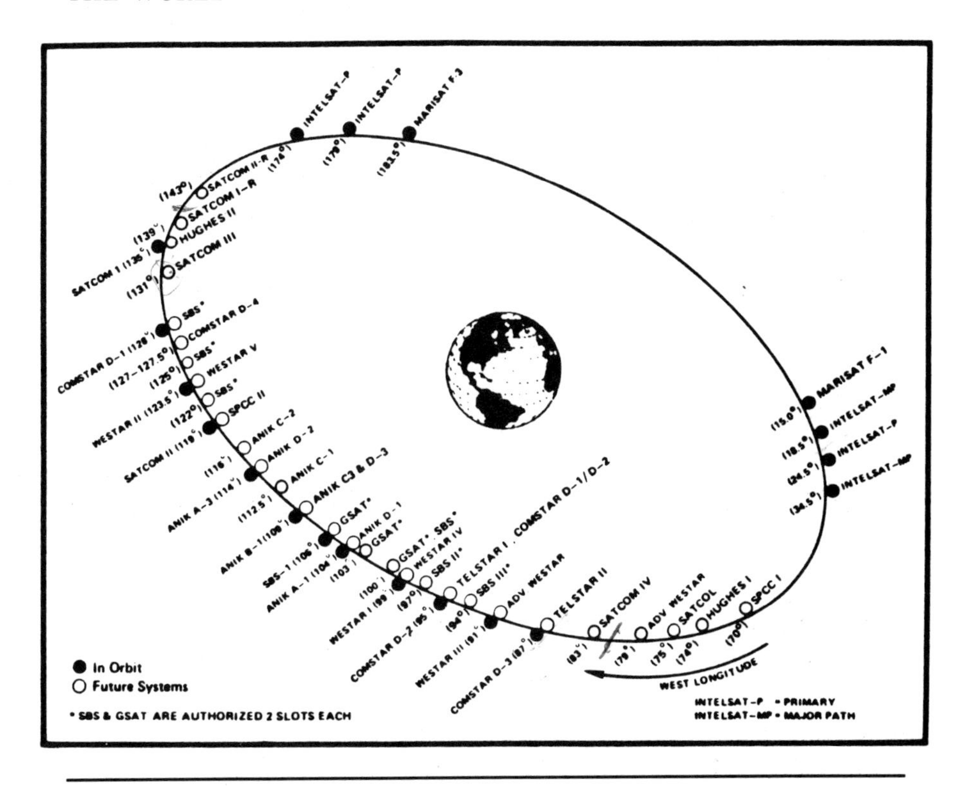

Where will the satellites be located? (Courtesy of COMSAT)

(Courtesy of RCA American Communications, Inc.)

PROJECTED SATELLITES IN ORBIT 1981-1990

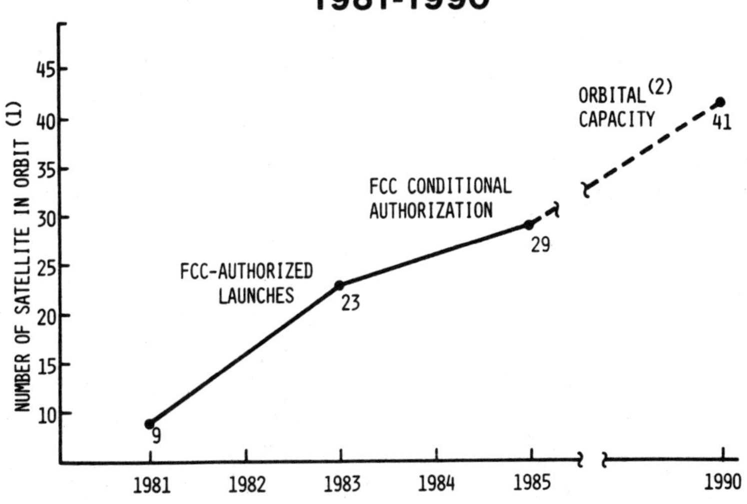

(1) HYBRID C/K-BAND SATELLITES COUNTED AS TWO

(2) ASSUME 3° SPACING FOR BOTH C-BAND AND K-BAND FOR ORBITAL CAPACITY

SATELLITE DEPLOYMENT ON THE ORBITAL ARC AUTHORIZED LAUNCHES

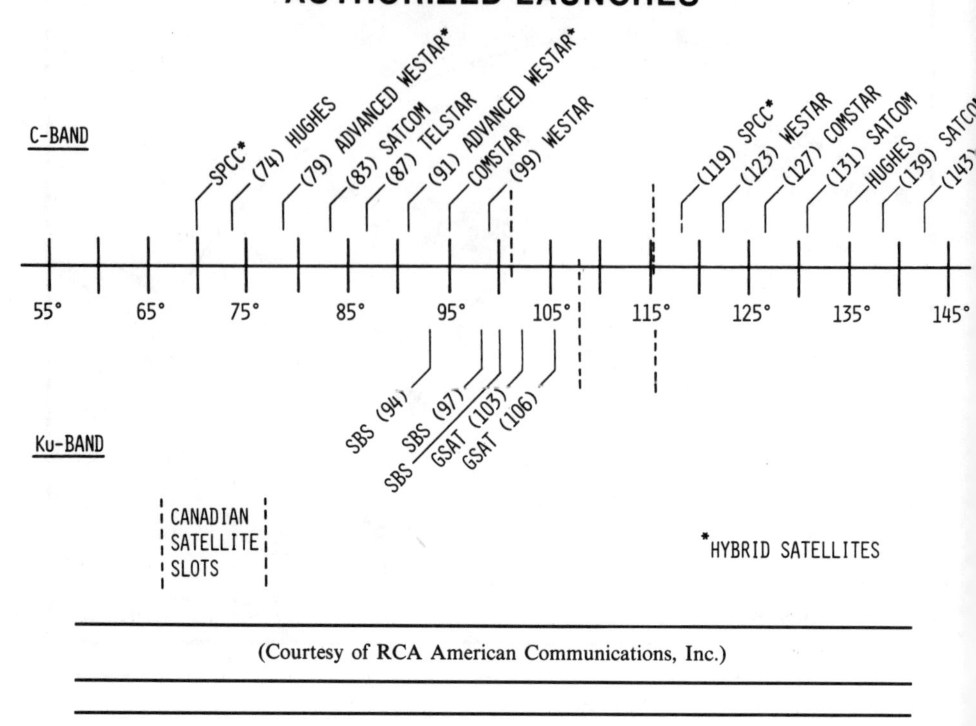

(Courtesy of RCA American Communications, Inc.)

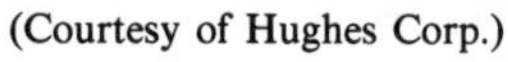
(Courtesy of Hughes Corp.)

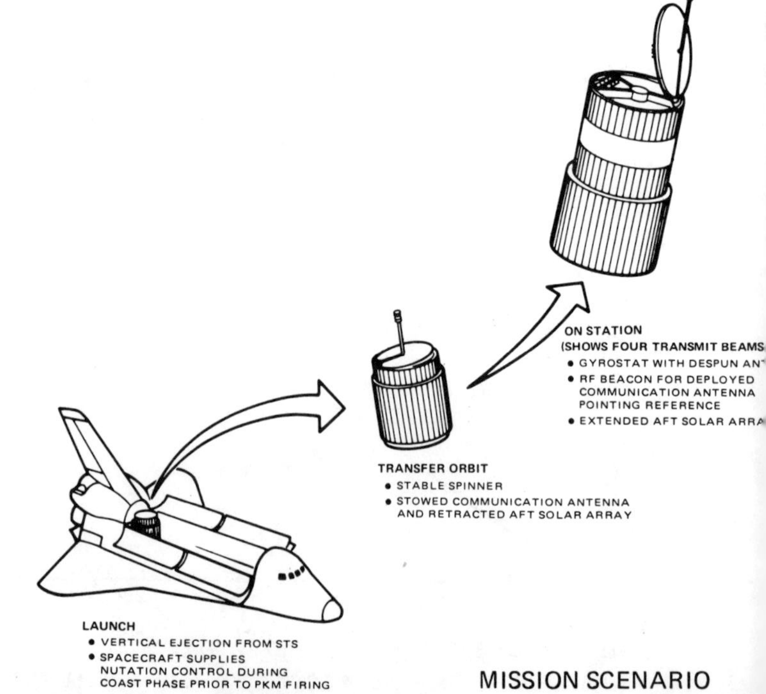

MISSION SCENARIO

In addition, newer and more powerful rockets are being developed both by the National Aeronautics and Space Administration (NASA) and the European Space Organization (ESO), using the French equatorial launching facilities located on Devil's Island in the Pacific.

As the conventional television satellites become bigger and bigger, their ability to radiate far more powerful signals will increase dramatically, allowing smaller and smaller TVRO home antennas to be used in picking up their television signals. Proposals on the drawing board for the late eighties call for satellites to carry hundreds of transponders. And new television bandwidth reduction techniques eventually will allow multiple TV signals to be squeezed onto a single transponder, permitting hundreds, or perhaps thousands, of television pictures to be carried by a single satellite. Given a reduction of orbital spacing in the Clarke belt from 4 degrees to 2 degrees between satellites, this recent FCC directive ultimately will allow upward of 30 or more satellites (both C-band and the higher frequency K-band) to be receivable from any point in North America.

Launch sequences. (Courtesy of Hughes Corp.)

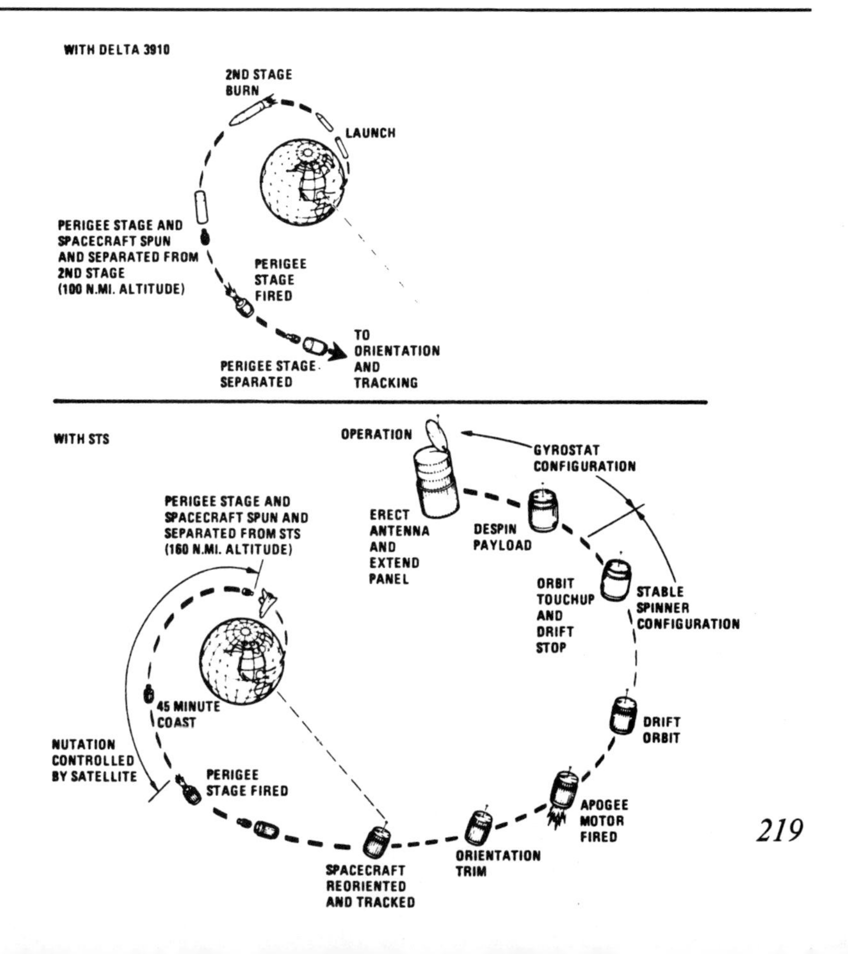

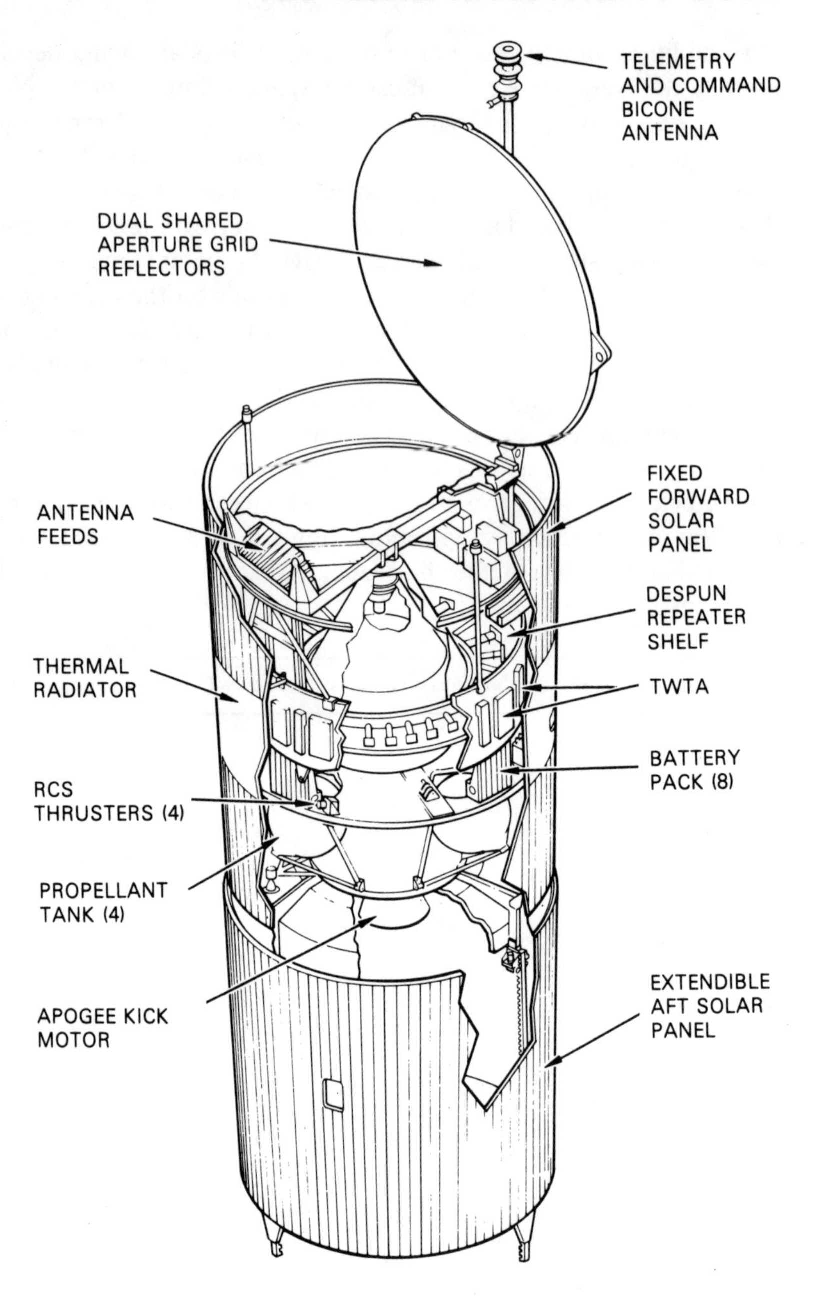

Location of WESTAR IV units. (Courtesy of Hughes Corp.)

Bigger satellites with giant antenna arrays also mean that smaller ground station uplink antennas can be used to transmit TV pictures from any point within the United States or Canada on an instantaneous basis for relay by television satellite nationwide. The 6- to 10-foot transmit antennas beaming up regional sports activities and special events to various independent and geographically regionalized television networks will become as popular in the late 1980s as the use of the local electronic news-gathering (ENG) eyewitness news TV feeds are today. Just about every television station in the United States and Canada owns at least one or more minicams, or portable television cameras, and relay equipment, often mounted on the roof of stationwagons and small panel vans. Through the use of these modern ENG video links, the local TV station can travel immediately to fast-breaking news events in the area. Several hundred television stations now operate their own commercial TVRO earth terminals for reception of network programming via satellite. Many public broadcasting stations own both transmit and receive earth stations for feeding noncommercial programming from region to region. By the end of the decade, thousands

Spacecraft—exploded view. (Courtesy of Hughes Corp.)

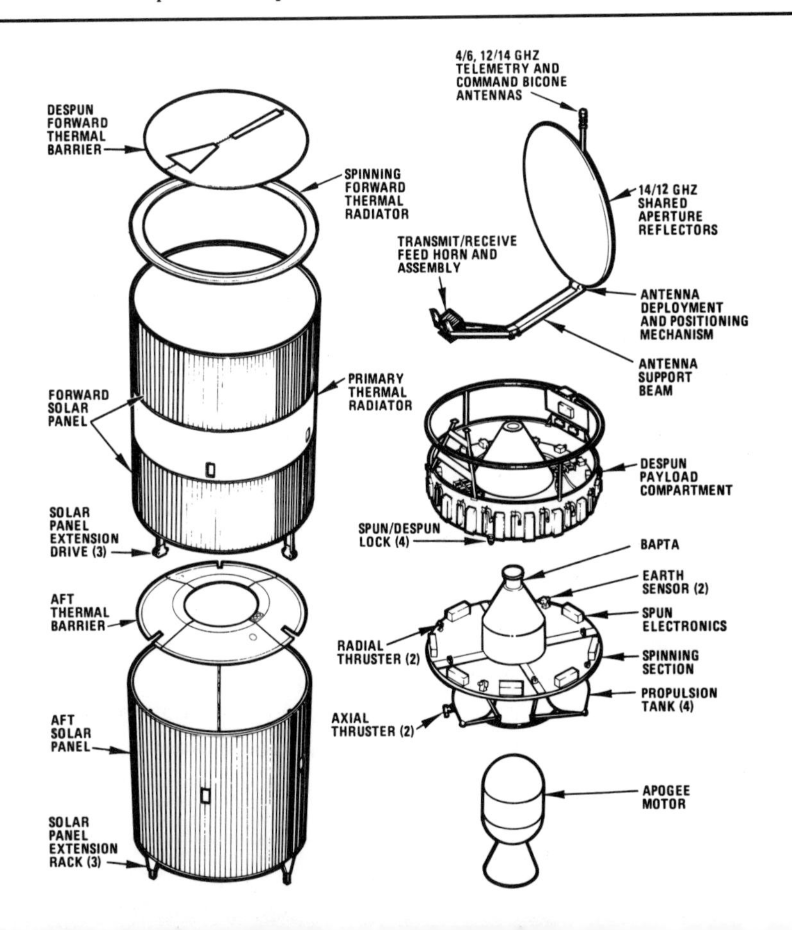

of television stations and other organizations will operate uplink earth terminals, thus creating an enormous diversity of narrowcast programming to audiences of as few as several thousand people scattered nationwide, who will receive the programs directly on their backyard home TVRO earth stations. By then, the ultimate promise of satellite technology will have been fulfilled. We will have created the multiprogram "magazine stand of the air," replacing the antiquated printing press technology with the instantaneous medium of the electronic image.

Where Do We Go from Here?

So, you've read this book and are dazzled by the prospect of tuning into the world through your own backyard satellite dish. Or perhaps you have known all along about this exciting new field, and are now ready to go out and buy one of these electronic genies yourself. Maybe you would like to meet other people who share the same dream and the same enchantment with this electronic wizardry of satellite television. Where do you go from here?

One thought is to join SPACE, the Society of Private and Commercial Earth Station Users. Membership in this nonprofit organization entitles you to receive the monthly newsletter, to participate in SPACE conferences, workshops, and trade shows, and to be involved on an intimate basis with the pioneers, movers, and shakers of the industry. SPACE members are mostly individual owners of satellite TV earth stations, not giant corporations. As a consumer organization, this is one of the best around.

The monthly *Coop's Satellite Digest* is a folksy, down-to-earth magazine written by Bob Cooper, Jr., one of the founders of the home satellite TV revolution. The magazine reviews new products and equipment of interest to the home satellite TV user, as well as the latest information on new programming and television feeds on the various birds. Often what you read in *Coop's* will appear months later in the commercial television publications.

Cooper also sponsors the Satellite Television Technology conferences and shows held three times a year in various parts of the country. At these home satellite TVRO happenings, all the leading manufacturers get together to show off their latest equipment, and informative seminars are run on every conceivable aspect of the exciting world of home satellite television reception. It is always a tremendous experience to attend one of these shows, and upon arriving at the hotel to be confronted with a vista of dozens and dozens of satellite antennas of all shapes and sizes pointed up to the various satellites in space. Considering that Bob Cooper himself put together one of the first backyard satellite television earth terminals as late as 1976, the skyrocket growth of the home satellite TV field has been phenomenal.

Another excellent source of information is the *SAT Guide,* sort of a giant monthly *TV Guide* of satellite television program listings. This magazine is also chock full of interesting articles and information on new programs, networks, and video feeds, and its orientation to the mini-CATV operator does not get in the way of its popularity with the home satellite TV subscriber.

Finally, the people at The Satellite Center in San Francisco publish a new home satellite TV newsletter. In addition they can give detailed information on setting up a mini-CATV system and LPTV station, as well as advice on what equipment to buy.

For more information on SPACE, and the conferences, magazines, and organizations mentioned, tune in to the appendices following this chapter. And don't hesitate to drop a note to Richard Brown or Taylor Howard at SPACE, or Robert Cooper, Jr., directly. You might also contact any of the other pioneers of the industry, including Portus Barlow, president of Downlink Inc., in Connecticut, or other "space pioneers," those home satellite TV manufacturers who helped kick off the revolution in the beginning. SPACE can put you in touch with all of them. Good luck and happy viewing of the world from your backyard!

Appendices

Introduction

This book was written for the interested but not overly technical satellite TV hobbyist and entrepreneur. One does not need to have a degree in electrical engineering to put together a top-notch satellite TV earth station—either for a home backyard installation or a commercial mini-CATV system.

As one digs into this exciting and explosive field, however, a number of somewhat technical questions are bound to arise. How, for example, do I find SATCOM F-3 in space—and point my antenna at it? How big an antenna do I really need? What is the meaning of those more obscure technical words? (My copy of *Webster's Dictionary* doesn't list "threshold extension" just yet. . . .) Where do I turn for further information? Who makes and sells satellite TV earth station equipment? How can I find out what programs are on the birds? What about government rules covering my mini-CATV system?

The purpose of this section of the book is to answer these and other questions that the reader will probably ask eventually. The author has attempted to present this information as accurately and completely as possible, by assembling these more detailed matters in a convenient reference format.

Appendix 1 is a glossary of terms mentioned in the book.

Appendix 2 presents a list of manufacturers, dealers, distributors, and suppliers of satellite television earth station equipment. Each entry in the list is keyed according to type of product and service provided. As is always the case in developing such a list, a few firms inadvertently may have been left out. Of course, inclusion on the list does not necessarily mean that the author endorses a given product.

Appendix 3 lists the present satellite TV programmers who are transmitting their signals via satellite. The appendix is divided into both a satellite-by-satellite listing and an alphabetical listing of programmers, with types of programs offered, addresses, and telephone numbers.

Appendix 4 is a bibliography of contacts, associations, books, and periodicals on the subject of satellite television. The appendix is further divided into sections covering the various reader interests: owning a home satellite terminal, starting a dealership, operating a mini-CATV system, etc.

Appendix 5 presents a summary of the Federal Communications Commission's rules and regulations governing the establishment and operation of a cable TV system or MATV apartment television system in the United States. It is must reading for anyone who is contemplating going into the business of operating a mini-CATV system. Appendix 5 also reviews the present copyright laws governing the retransmission of a television stations's programming by a CATV system. This section is also important to any reader considering starting a mini-CATV system. Appendix 5 was provided by Richard Brown, general counsel of SPACE, the Society of Private and Commercial Earth Station Users.

Appendix 6 discusses just how to find the birds in the heavens. The problem is twofold: accurately positioning the dish in relation to true north, and accurately aiming it at the correct point in space where the desired satellite is located. A complete Apple II satellite tracking program is included, along with a thorough analysis of the satellite tracking geometry.

Appendix 7 consists of the set of satellite footprint maps of effective signal coverage for various locations on the earth's surface. A discussion of antenna size versus location parameters is also included, along with a quick-guide rule-of-thumb table of minimum antenna sizes required for satellite TV reception from various places.

Appendix 1
Glossary

AC Voltage (Alternating-Current Voltage) The voltage found in one's house. Produced by a generator or oscillator, AC voltage oscillates periodically between some maximum positive and negative value as determined by its frequency, typically 60 cycles per second (or 60 Hz). AC voltages can be used as signals that can be sent along wires or transmitted via radio over great distances, when the frequency of oscillation is very large.

AFC (Automatic Frequency Control) A circuit used in a satellite television receiver that automatically locks onto the varying frequency (FM) signal, preventing the receiver from going out of tune due to temperature changes or component variations.

AGC (Automatic Gain Control) A circuit used in satellite television receivers to adjust the amplification of the receiver automatically so that the signal level will remain constant in the receiver, despite atmospheric variations or signal levels that may be different from transponder to transponder.

Alignment The process of tweaking an electronic circuit through the use of screwdriver-adjustable components to maximize the sensitivity and signal-receiving capability of the circuit.

AM (Amplitude Modulation) The process of varying a transmitter's power output in direct relationship to incoming signal amplitude, either audio or video. Amplitude-modulation techniques are more susceptible to impulse and natural noise (see **FM**). In the North American television system, the sound channel is transmitted using FM, while the picture is transmitted using AM. This is why during an electric storm the sound will remain static free while the visual pictures will be broken up and distorted with each flash of lightning. Unlike terrestrial television, satellite TV systems use the frequency-modulation scheme for both the sound and picture signals. The home satellite TV receiver must then convert these signals to the standard format required by the television set.

Analog Continuously variable, as opposed to variable in discrete steps (digital). Examples of signals that are inherently analog in nature include human speech, music, and television pictures. Satellite systems in use today primarily relay analog signals back to earth by FM techniques.

ANIK The Canadian domestic satellite system that transmits Canadian Broadcasting Corporation's (CBC) network feeds throughout the country. ANIK is also used heavily by the Canadian telephone companies to bring voice communications to major metropolitan centers and small communities throughout the country.

ANIK A-1 is located at 104 degrees west longitude. ANIK B-1 and B-2 use 12 transponders at C-band and four Ku-band transponders especially developed for direct broadcast to homes in northern Canada; co-located at 109 degrees west longitude. ANIK A-3 with 12 transponders is located at 114 degrees west longitude. All ANIK satellites are operated by TeleSat Canada, Ottawa. Future satellites in the ANIK series will include ANIK C (Ku-band) and ANIK D (C-band).

Antenna A metallic or metallic-coated device connected to a transmitter or receiver that amplifies the high-frequency AC signals that are propagated to or received from the atmosphere. Depending on their use and operating frequency, antennas can take the form of a single piece of wire or a sophisticated parabolic-shaped dish.

AP (Associated Press) A national news service that transmits both hourly audio news-wire feeds and photographs to newspapers and broadcast stations via satellite.

Apogee The point in an elliptical satellite orbit that is farthest from the surface of the earth. Geosynchronous satellites that maintain circular orbits around the earth are first launched into highly elliptical orbits with apogees of 22,300 miles. When the communications satellite reaches its initial apogee, small on-board thruster rockets are fired to nudge the satellite into its permanent circular orbit of 22,300 miles. (See **Geosynchronous.**)

Attenuation The process of reducing the level of a signal either intentionally by means of an attenuator or by a natural process. Attenuators are passive devices used in the satellite receiver. They are also heavily used in the CATV business to compensate for excessively high level signals for those homes nearest the CATV distribution amplifier. (See **Distribution Amplifier.**)

Azimuth Angle The angle of rotation (horizontal) that a ground-based parabolic antenna must be rotated through to point to a specific satellite in a geosynchronous orbit. The azimuth angle for any particular satellite can be determined for any point on the surface of the earth given the latitude and longitude of that point. It is defined with respect to due north for convenience.

Baseband The basic direct 6-MHz output signal from a television camera, satellite television receiver, or videotape recorder. Baseband signals can be viewed only on studio monitors. To display the baseband signal on a conventional television set, a remodulator is required to convert the baseband signal to one of the VHF or UHF television channels that the TV set can be tuned to receive. (See **NTSC.**)

Beta Format One of the two popular home consumer videotape recorder formats. The Beta system was developed by the Sony Corporation. It is incompatible with the more popular VHS system. (See **VHS Format.**)

Bird The actual communications satellite. A typical bird weighs several thousand pounds, has an average lifetime of seven years, and is parked into a circular orbit at an altitude of 22,300 miles above the earth. The satellite acts as an electronic mirror, retransmitting telephone, video, and data signals fed up to it by network control centers, and relaying these signals to the earth via geographically scattered receive-only satellite terminals. (See **Geosynchronous, TVRO.**)

Blanking (Retrace) Interval between picture frames. An ordinary television signal consists of 30 separate complete pictures, or frames, sent every second. They occur so rapidly that the human eye blurs them together to form an illusion of moving pictures. This is the basis for television and motion picture systems. The blanking interval is that portion of the television signal that occurs after one picture frame is sent and before the next one is transmitted. During this period of time special data signals can be sent that will not be picked up on an ordinary television receiver. (See **Teletext.**)

BNC Connector A standard medium-sized twist-lock coaxial cable connector used at higher frequencies throughout the television industry and by many satellite TV receivers. Other popular connectors include the larger screw-in Type N and the smaller miniature-type connectors.

Broadcast Bureau The department of the Federal Communications Commission that issues television and radio station licenses. (See **Federal Communications Commission.**)

Broadcasting The process of transmitting a radio or television signal via an antenna to multiple receivers that can simultaneously pick up the signal. It is different from cable television. (See **Narrowcasting, CATV.**)

Carrier The basic radio or television transmitter's signal (information carrier). The carrier is modulated by manipulating its amplitude or its frequency (shifting it higher or lower) in direct relation to the video signal coming to the television or audio signal originating from a microphone. Satellite carriers are frequency modulated. (See **FM.**) The term is also used in a nontechnical sense to describe operating agencies (satellite carriers).

Carrier Frequency The main frequency on which a radio station, television station, or microwave transmitter operates. AM radio stations operate in the frequency band from 535,000 cycles per second (kilohertz) to 1600 kHz. FM radio stations operate in the frequency band from 88 to 108 MHz. Terrestrial television station transmitters operate in the frequency band whose channels range from 54 MHz (Channel 2) to several hundred megahertz. Microwave and satellite communications transmitters operate in the band from 1 to 6 gigahertz (a gigahertz is one billion cycles per second). (See **Modulation.**)

Cassegrain Antenna The antenna principle that utilizes a subreflector at the focal point, which reflects energy to or from a feed located at the apex of the main reflector.

CATA The Community Antenna Television Association, a nonprofit organization composed of several thousand small cable television systems, some of which have as few as a dozen or so subscribers. (See Appendix 4.)

CATV Originally meant Community Antenna Television. Independent "mom and pop" companies in rural communities would build a large television receiving antenna on a nearby mountain to pick up the weak TV signals from a distant metropolis. These signals were amplified, modulated onto television channels, and sent along a coaxial cable strung from house to house. Now standing for Cable Television; most independent CATV companies have long since been purchased by national organizations that own multiple cable systems in rural and urban areas. (See **Multiple Service Operator.**)

CATV Convertor A small box that sits atop the television set and is connected to the incoming CATV cable. The convertor replaces the function of the television set tuner, allowing the subscriber to select the various television channels sent over the cable.

CATV Decoder Similar to a CATV convertor, the TV set decoder includes an additional module that descrambles pay television premium channels on the cable.

C-Band The 4–6-GHz frequency band used by the domestic and international communications satellites. To transmit signals to the earth, the 4-GHz (actually 3.7- to 4.2-GHz) band is used. (See **Microwave.**)

CCIR (Consultative Committee for International Radio) One of the two major arms of the International Telecommunications Union, a United Nations agency located in Geneva. The CCIR developed international standards for radio, television, and satellite television transmissions.

CCITT (Consultative Committee for International Telephone and Telegraph) Second major arm of the International Telecommunications Union. The CCITT developed international standards for telephone and telegraph communications, including certain engineering specifications for satellite transmissions. (See **CCIR.**)

Channel A frequency band in which a specific broadcast signal is transmitted. Channel frequencies are specified in the United States by the FCC. Television signals require a 6-MHz-wide frequency band to carry all the necessary picture detail. (Compare this with a typical high-fidelity audio channel of 15,000 Hz!)

Christian Broadcasting Network (CBN) The most highly watched satellite program network. CBN is carried on the RCA F-1 satellite. Plans are under way to change its name to Continental Broadcasting Network, Inc., and to expand its entertainment programming to reach more households. CBN is received by 10 million households on over 2000 cable systems.

Circular Polarization Utilized by INTELSAT satellites. Unlike domestic North American satellites that utilize vertical or horizontal polarization, the international INTELSAT satellites transmit their signals in a rotating corkscrew-like pattern as they are downlinked to earth. In some satellites, both right-hand rotating and left-hand rotating corkscrew-like signals can be transmitted simultaneously and on the same frequency, thereby doubling the capacity of the satellites to carry communications channels.

Clarke Orbit That circular orbit in space 22,300 miles from the surface of the earth at which geosynchronous satellites are placed. First postulated by the science-fiction writer Arthur C. Clarke in *Wireless World* in 1945. Satellites placed in these orbits, although traveling around the earth at thousands of miles an hour, appear to be stationary when viewed from a point on the earth, as the earth is rotating upon its axis at the same angular rate that the satellite is traveling around the earth. Modern techniques allow very tight control over satellite position in orbit, so that earth terminal antennas can be fixed (nontracking). (See **Station-keeping.**)

C/N (Carrier-to-Noise Ratio) The ratio of the power in a satellite signal carrier to the received noise as measured in decibels. The larger the ratio, the better the television picture looks. If C/N is below 7 dB, the picture quality is poor; above 11 dB, the picture quality is excellent. (See **Threshold Extension, dB.**)

Coaxial Cable A transmission line in which an inner conductor is surrounded by an outer conductor, or shield, and separated by a nonconductive dielectric, typically foam. Coax cables have the capacity to carry enormously high-frequency signals and thus are used by CATV companies for their multisignal distribution. Coax cables also connect the backyard satellite TV antenna to the TV-set-top satellite television receiver.

Common Carrier Any organization that operates communications circuits used by other people. Common carriers include the telephone companies as well as the owners of the communications satellites, RCA Americom Inc., Western Union Telegraph Corporation, and COMSAT General Corporation (which leases its COMSTAR satellites to AT&T).

Common Carrier Bureau The FCC division that regulates all activities of common carriers in the United States, including the service price tariffs issued by the common carriers themselves.

Communications Act of 1934 The federal law that established the Federal Communications Commission and that regulates all interstate wire and broadcast communications in the United States. Since satellite communication was not envisaged at that time, there is confusion today over whether the Act should apply equally to satellite communications also.

COMSAT The Communications Satellite Corporation, a New York Stock Exchange company established by an act of Congress in 1962. COMSAT launches and operates the international satellites for the INTELSAT consortium of countries, and through its COMSAT General subsidiary owns the MARISAT series of marine communications satellites and the three 24-transponder COMSTAR domestic satellites leased to AT&T for domestic telephone, data, and television use. COMSAT has also proposed the first Direct Broadcast Satellite (DBS) system for use in the United States. (See **DBS.**)

COMSTAR Three 24-transponder satellites designated COMSTAR 1 (D-1), COMSTAR 2 (D-2), and COMSTAR 3 (D-3). D-1 and D-2 are now co-located at 95.0 degrees west, and D-3 is at 86.9 degrees west.

Copyright Tribunal Established by the Copyright Act of 1976, the Copyright Tribunal collects fees from the cable television operators for their use of off-the-air broadcasting station signals. The Copyright Office collects these fees through a mandatory copyright license, and distributes the money to the motion picture and television producers and copyright owners. The use of satellite signals by private earth terminal owners has created problems in interpreting the Copyright Act.

dB (Decibel) A mathematical ratio developed originally by Alexander Graham Bell to express logarithmically the relationship between two different levels or signals. Typically used in the measurement of the power of a signal in a channel compared with the power of the noise in that channel; a 3-dB increase in signal is equivalent to a signal that is twice as strong. A hi-fi stereo set requires a signal-to-noise ratio of at least 60 dB to be considered acceptable by the average listener. A telephone channel can have a signal-to-noise ratio of only 30 dB to be considered acceptable. The signal-to-noise ratio of a good-quality television picture should be at least 45 dB.

dBi Decibels of gain, relative to isotropic. The term dBi is used properly to express the gain of an earth terminal antenna, although often a simpler form, dB, is used. The gain of the antenna, *G,* is used to compute the G/T of the earth terminal, which determines the ultimate signal quality of the received signal.

dBm The ratio in dB of the power of a signal as compared with a 1-milliwatt reference power.

DBS (Direct Broadcast Satellite) A new generation of satellites designed to operate in the Ku-band (12–14 GHz) The Canadian ANIK B satellite has four such transponders, which are used for direct reception by 3- to 5-foot antennas in an experimental program.

Higher frequency DBS satellites have the advantage of allowing smaller receive antennas to be used, but numerous difficulties need be overcome before the service can be introduced in any significant fashion. In the United States COMSAT has proposed the creation of a new DBS satellite system to be constructed and operated by its Satellite Television Corporation subsidiary in the mid- to late 1980s. Although not originally intended for this use, the existing C-band communications satellites essentially function as direct broadcast satellites when their signals are picked up by an 8-foot or larger backyard home satellite TV antenna.

DC Voltage (Direct-Current Voltage) Like the voltage produced by a battery, a signal that flows continuously in one direction on a wire. Because DC voltages do not produce alternating magnetic fields, they cannot be transmitted through space via an antenna.

Decibel (See **dB**)

Demodulator A satellite receiver circuit that extracts or demodulates the video and audio signals from the received microwave carrier.

Detector A demodulation circuit used in a satellite television receiver to recover the audio and video signals from the carrier. (See **Demodulator.**)

Digital Discretely variable. A digital signal has, typically, two states corresponding to two different signal conditions. For example, slow-speed computer terminals that operate over ordinary telephone lines use a modem (modulator and demodulator unit) to shift an audio tone between two frequencies to correspond to the "one" and "zero" computer logic states. Some satellite systems carry special digital signals for news services, etc., as well as the analog television signals. To decode these secret digital signals, a special digital demodulator must be used.

Dipole A type of antenna. The typical "rabbit ears" television antenna is a simple dipole. Other dipoles are sometimes used in the feedhorn assembly of a satellite television antenna.

Distribution Amplifier Wideband amplifier operating at the VHF television channel frequencies used by the CATV companies periodically to strengthen the weakened signals as they are transmitted down the

CATV company's cable network. Distribution amplifiers are also used in apartment and other MATV, or master antenna television, installations, and some models operate on baseband signals as well.

Distribution Center The central point from which the television signals are distributed from a television programming organization (such as HBO) to its network stations or receivers. The distribution center for a satellite television programmer will usually consist of a bank of videotape machines, studio facilities, and a 10-meter uplink (transmit) antenna for transmission of the signals directly to the desired satellite.

DOMSAT A domestic satellite. In the United States there are three major domestic satellite systems: the AT&T COMSTAR series, the RCA SATCOM series, and the Western Union WESTAR series. Among the eight satellites there is a total of 156 potential satellite transponders, each one of which can carry a television signal. There are over 60 transponders presently carrying video signals of one kind or other, including Picturephone Meeting Service teleconferences, network prefeeds, distribution feeds, and pay television programming. In addition, a new business communications satellite series, Satellite Business Systems (SBS), has been launched, which carries video as well as business data in the K-band.

Down-Convertor That portion of the satellite television receiver that converts the signals from the 4-GHz microwave range to (typically) the more readily used 70-MHz range. Down-convertors typically were located physically at the front end of the receiver, requiring bulky and expensive coaxial cable feeds from the antenna to the receiver. Newer designs have seen the down-convertor placed at the antenna itself, often in combination with the LNA, allowing miniature CATV-like coaxial cable to bring the satellite TV signal into the house. This type of design allows more flexibility in the location of the antenna with respect to the house.

Downlink The 4-GHz frequency range utilized by the satellite to retransmit the television signal down to earth for reception by the TVRO ground stations. (See **TVRO.**)

Earth Station The term used to describe the combination of antenna, low-noise amplifier (LNA), down-convertor, and receiver electronics used to receive a signal transmitted by a satellite. Earth station antennas vary in size from the 10-foot- to 20-foot-diameter size used for TV reception, to as large as 100 feet in diameter for international communications.

EIRP (Effective Isotropic Radiated Power) A measure of the power of a satellite television signal received on earth. EIRP is expressed in dBw or the ratio of the power of the signal as compared with a 1-watt reference signal. Satellite common carriers calculate EIRP signal strengths when the satellites are tested before launch, and these studies are used to produce footprint maps showing the varying satellite signal strength as expected to be received throughout the United States. At the boresight of a typical DOMSAT, usually aimed at the continental United States (or CONUS) center-country point, the satellite EIRP may be as high as 37 dBw or higher. In areas such as Southern California or Florida, the EIRP can drop to 30 dBw. There is a direct relationship between EIRP as received on the earth and the required antenna size. In areas whose signal strength is 34 dBw or above, good-quality satellite television reception can be accomplished with a 10-foot parabolic antenna and a 120-degree LNA, a rather inexpensive combination. The antenna size can be reduced even further, depending on the threshold characteristic of the receiver used and the performance margin desired.

Elevation Angle The upward tilt to a satellite antenna measured in degrees required to aim the antenna at the communications satellite. When aimed at the horizon, the elevation angle is zero. If it were tilted to a point directly overhead, the satellite antenna would have an elevation of 90 degrees.

FCC (Federal Communications Commission) The federal agency established by the Communications Act of 1934 to regulate all interstate telecommunications services in the United States. It is located at 1919 M Street, N.W., Washington, D.C. (See **Broadcast Bureau, Common Carrier Bureau.**)

Feed A term used to describe the transmission of video programming from a distribution center. Also used to describe the feed system of an antenna. The feed system may consist of a subreflector plus a feedhorn or a feedhorn only.

Feedhorn A satellite TV receive antenna component that collects the signal reflected from the main surface reflector and channels this signal into the low-noise amplifier (LNA).

Feedline The transmission line, typically coaxial cable, from the satellite antenna to the receiver.

Field Strength Meter A test instrument used in the cable television and broadcast industries to measure the power of a signal in a transmission line or from an antenna.

FM (Frequency Modulation) A process of varying the frequency of a sinusoidal "carrier" wave in relation to an incoming signal, thereby modulating the carrier with the signal. When the modulated carrier is transmitted by an antenna, an FM radio or television signal results. (See **AM.**)

FM Threshold That point at which the input signal power is just strong enough to enable the receiver demodulator circuitry successfully to detect and recover a good-quality television picture from the incoming video carrier. Using threshold extension techniques, a typical satellite TV receiver will successfully provide good pictures with an incoming carrier noise ratio of 7 dB. Below the threshold a type of random noise, called "sparklies," begins to appear in the video picture.

Footprint A map of the signal strength showing the EIRP contours of equal signal strengths as they cover the earth's surface. Different satellite transponders on the same satellite will often have different footprints of signal strength. Footprint maps are available from the common carriers, and from both SST and CATA. (See Appendix 4.) The accuracy of EIRP footprint or contour data improves with the operational age of the satellite, although the actual EIRP levels decrease slowly as the satellite ages.

Frequency The property of an alternating-current signal measured in cycles per second or hertz. In general, the higher the frequency of a signal, the more capacity it has to carry information, the smaller an antenna is required, and the more susceptible the signal is to absorption by the atmosphere and physical structures. At microwave frequencies, radio signals take on a line-of-sight characteristic, and require highly directional and focused antennas to be seen successfully.

Frequency-Agile The ability of a satellite TV receiver to select or tune all 12 or 24 channels (transponders) from a satellite. Receivers not frequency-agile are dedicated to a single channel, and are most often used in the CATV industry. Frequency agility can be via continuously variable tuning or discrete-step (channel selection) tuning.

Frequency Coordination A computerized service utilizing an extensive database to analyze potential microwave interference problems that arise between organizations using the same microwave band. As the same C-band frequency spectrum is used both by the communications satellites and the terrestrial toll telephone network, a CATV company contemplating the installation of a TVRO earth station will often obtain a frequency coordination study to determine if any potential interference might be received at the desired earth station location.

Frequency Modulation (See **FM**)

Frequency Reuse A technique to expand the capacity of a given set of frequencies or channels by separating the signals either geographically or through the use of polarization techniques. Satellites are spaced approximately 4 degrees apart in their orbital positions. The beam width of a typical antenna operating in the 4-GHz range is only about 2 degrees; a TVRO earth station pointed at one satellite will not detect any signal from another satellite 4 degrees away, even though the two satellites are operating in the same frequency band. The newer 24-transponder satellites use vertical and horizontal polarization techniques to reuse the same frequencies on the same satellite, allowing 24 television signals to be squeezed into the space normally capable of carrying only 12. (See **Polarization.**)

Gain A measure of the increase in level or strength of a signal, usually measured in decibels. Gain is realized in an antenna by virtue of its signal concentration properties and in an amplifier because of its elevation of a signal to a higher level.

Geostationary (See **Geosynchronous**)

Geosynchronous The Clarke circular orbit above the equator, which for a planet the size of the earth is 22,300 miles. (See **Clarke Orbit.**)

Ghost An unwanted, weaker reflected copy of the picture that interferes with the main picture itself. In VHF and UHF receptions, ghosts are caused by the receiver antenna detecting both the direct line-of-sight signal from the transmitter antenna and one or more weaker signals bounced off nearby buildings, mountains, etc. The main technical advantage of cable TV and satellite TV is the elimination of such ghosts.

Gigahertz (GHz) One billion cycles per second. Signals operating above 1 gigahertz are known as microwaves, and begin to take on the characteristics of visible light.

Global Beam An antenna downlink pattern used by the INTELSAT satellites, which effectively covers one-third of the globe. Global beams are aimed at the center of the Atlantic, Pacific, and Indian oceans by the respective INTELSAT satellites, enabling all nations on each side of the ocean to communicate with the satellites. Because they transmit to such a wide area, global beam transponders have significantly lower EIRP outputs at the earth's surface compared with a U.S. domestic satellite system, which covers just the continental United States. Therefore, earth stations receiving global beam signals need to be much larger in size (typically 30 feet to 100 feet in diameter).

G/T (Gain-over-Noise Temperature) A relationship, in decibels, between the gain of a TVRO antenna system and the surrounding ambient noise. As this number increases, the picture quality increases. G/T can be raised by using an LNA with a lower noise temperature or increasing the size of the receive antenna.

Guard Channel Television channels are separated in the frequency spectrum by being spaced several megahertz apart. This unused space serves to prevent the adjacent television channels from intefering with each other.

Headend The master distribution center of a CATV system in which the incoming television signals from space and distant broadcast stations are received, amplified, and remodulated onto television channels for transmission down the CATV coaxial cable.

Heliax A special type of thick coaxial cable often used to connect TVRO antennas to satellite receivers because of the cable's low-loss transmission characteristics at microwave frequencies.

Hertz Cycles per second. An AC signal or carrier completes a sinusoidal geometric pattern periodically in time. This phenomenon is known as its frequency, and its repetition rate is measured in hertz.

Home Box Office The original premium pay TV cable television programmer, owned by Time-Life, Inc. HBO transmits first-run movies and special shows over several domestic satellites daily.

Integrated Circuit (IC) A miniaturized solid-state device that replaces thousands of discrete-component circuit elements. ICs can be fabricated to function as microcomputer circuits, stereophonic amplifiers, and television receivers.

INTELSAT The international satellite consortium that owns and operates the global satellite communications system. INTELSAT is headquartered in Washington, D.C., and its U.S. member, COMSAT, operates the international satellite system under management contract. INTELSAT's satellites carry telephone and television transmission to all parts of the world.

International Telecommunications Union (ITU) The United Nations organization, headquartered in Geneva, responsible for all international telecommunications coordination. The ITU predates the UN by almost a century, having been founded by Napoleon in the mid-19th century. (See **CCITT, CCIR.**)

INTERSPUTNIK The Soviet Union's rapidly expanding international satellite network of stationary satellites used to carry television and telephone communications between communist countries.

Kelvin (K°) The temperature measurement scale used in the scientific community. Zero degrees Kelvin represents absolute zero, and corresponds to minus 459 degrees Farenheit. Thermal noise characteristics of LNAs are measured in degrees Kelvin, with respect to room temperature. (See **Noise Temperature.**)

Latitude An angular measurement of a point on the earth above or below the equator. The equator represents zero degrees, the north pole plus 90 degrees, and the south pole minus 90 degrees.

License An operating permit issued by the FCC to run a radio, television, or satellite transmitter. Voluntary licenses can also be obtained from the FCC by mini-CATV companies for their 15-foot or larger TVRO earth stations. While not legally required, the issuance of a

TVRO license by the FCC guarantees that no future microwave transmitter can be constructed in the path of the TVRO antenna, thereby minimizing the possibility that future terrestrial interference will occur.

LNA (Low-Noise Amplifier) A very sensitive preamplifier used at the feedhorn of the TVRO satellite antenna to strengthen the very weak satellite signal. The most important parameter of the LNA is its noise temperature as described in degrees Kelvin. In general, the lower the noise temperature, the better the signal quality will be. Typical LNAs have noise temperatures of 120°K. There is a limited but useful trade-off between the noise temperature of the LNA used and the size of the satellite-receive antenna. The higher the noise temperature, the bigger an antenna required. Conversely, the G/T of an earth terminal can be improved by lowering the LNA noise temperature.

Longitude An angular measurement of a point on the surface of the earth in relation to the meridian of Greenwich (London). The earth is divided into 360 degrees of longitude, beginning at the Greenwich mean. As one travels west around the globe, the longitude increases.

Low-Power TV (LPTV) A new television service proposed by the FCC in October 1980. LPTV broadcasting stations typically radiate between 100 and 1000 watts of power, covering a geographic radius of 10 to 15 miles. Upward of 5000 LPTV stations ultimately will be licensed for operation, many of which will obtain their programming from new satellite television networks now being formed. (See **National Entertainment Television.**)

Master Antenna Television (MATV) The master distribution coaxial cable television antenna system found in a modern apartment building, condominium complex, or hotel. Really a mini-CATV system in itself, an MATV system can easily be modified, with the addition of a TVRO earth station, to pick up and distribute first-run movies and other satellite programming directly to the MATV users.

MDS (Multipoint Distribution System) A common carrier licensed by the FCC to operate a broadcast-like omnidirectional microwave transmission facility within a given city. MDS carriers often pick up satellite pay TV programming and distribute it via their local MDS transmitter to specially installed antennas and receivers in hotels, apartment buildings, and individual dwellings throughout the area.

MDS Pirate Owner of an unauthorized receiver. Although the MDS system is similar to a broadcasting station, it is actually licensed by the FCC Common Carrier Bureau. Therefore, the signal cannot legally be received and divulged without the permission of the MDS operator. A number of enterprising electronics firms have sprung up nationwide to manufacture unauthorized MDS receivers, prompting the use of the term "pirate" for the owner of such a receiver. MDS should not be confused with satellite television, which is a totally different service, falling under different regulations and using different frequencies and technologies.

Microwave A radio or television signal whose carrier is oscillating at a frequency of 1 GHz or higher. (See **Gigahertz.**)

Microwave Interference Occurs when a TVRO earth station aimed at a distant satellite picks up a second, often stronger, signal from a local telephone terrestrial microwave relay transmitter. Microwave interference can also be produced by nearby radar transmitters as well as the sun itself. Relocating the TVRO antenna by only several feet may completely eliminate the microwave interference.

Modulation The process of manipulating the frequency or amplitude of a radio or television carrier in relation to an incoming video or audio signal. (See **AM, Carrier, FM.**)

Modulator A device that modulates a carrier. Modulators are found as components in broadcasting transmitters and in satellite transponders. Modulators are also used by CATV companies to place a baseband video television signal onto a desired VHF or UHF channel. Home videotape recorders also have built-in modulators that enable the recorded video information to be played back using a television receiver tuned to VHF Channel 3 or 4.

Multiple Service Operator (MSO) A major cable television owner. The top 50 MSOs own cable television franchises throughout the United States. Among the big ten are Teleprompter Corporation, Time-Life/ATC, Warner-Amex Cable Communications, Times–Mirror Cable Television (*Los Angeles Times*), and Viacom International.

Narrowcasting The concept of delivering a television program to a very small audience market share. By carrying 20 or more channels of television, CATV companies have begun to threaten the established television networks with their narrowcasting capabilities. LPTV stations threaten to erode the market share of the major television networks as dozens of new and alternative satellite-based television services turn on in the eighties. (See **National Entertainment Television.**)

National Association of Broadcasters (NAB) A trade association headquartered in Washington, D.C., whose members are the major radio and television stations in the United States. The NAB holds an annual convention and trade show that attracts visitors from around the world to see the latest technology of the broadcasting industry.

National Cable Television Association (NCTA) A trade association whose members are most of the cable television companies in the United States. It is headquartered in Washington, D.C. Like the NAB, the NCTA also sponsors its own annual trade show and convention.

National Entertainment Television (NET) The first satellite-based U.S. television network to orient its programming to the new low-power television broadcast stations. The San Francisco organization will provide its first-run movie programming to any TVRO satellite owner or organization.

Network Television programming and transmission arrangement consisting of a program creation and distribution center, telecommunications channels (either terrestrial, microwave, or satellite video circuits), and viewer delivery mechanisms (typically broadcasting station affiliates or CATV companies). There are over 30 television networks in operation today whose programming can be picked up by satellite. (See Appendix 3.)

Noise Temperature The noise temperature of a low-noise amplifier (LNA) is expressed in degrees Kelvin with respect to room temperature.

NTSC (National Television Standards Committee) The television system in use in the United States today, in which 30 separate still pictures are transmitted per second using an electron beam to scan 525 horizontal lines per picture on the face of the picture tube. (See **PAL, SECAM.**)

Occasional-Use Transponder A satellite transponder used for television transmission on a periodic basis, typically rented by the satellite common carrier to the programmer for an hourly fee.

PAL (Phase Alternation System) A European color television system incompatible with the U.S. television system. PAL television picture formats have 625 lines per frame, providing a higher resolution television picture. U.S. and Canadian satellites utilize the NTSC color television transmission system. INTELSAT satellites often use the PAL system. Although a U.S. black-and-white television monitor can view PAL TV pictures, a U.S. color television monitor cannot.

Parabolic Antenna The most frequently found satellite TV antenna. It takes its name from the shape of the dish described mathematically as a parabola. The function of the parabolic shape is to focus the weak microwave signal hitting the surface of the dish into a single focal point in front of the dish. It is at this point that the feedhorn is usually located, or a subreflector that reflects the energy back into a feedhorn.

Pay TV The concept made popular by Home Box Office and the CATV companies in which a special decoder box is installed adjacent to the viewer's television set to enable the scrambled pay TV channel to be watched. The viewer pays a per-month or per-movie charge for the rental of the box. Pay TV services are also provided by MDS operators and STV television stations that transmit their pay TV signals over the air in a scrambled mode.

Perigee The point in an elliptical satellite orbit that is closest to the surface of the earth. (See **Apogee.**)

Polarization A technique used by the satellite designer to increase the capacity of the satellite transmission channels by cleverly reusing the satellite transponder frequencies. In linear polarization schemes half of the transponders beam their signals to earth in a vertically polarized mode; the other half horizontally polarize their downlinks. Although the two sets of frequencies overlap, they will not interfere with each other. To receive these signals on earth successfully, the TVRO earth station must be outfitted with a properly polarized feedhorn to select the vertically or horizontally polarized signals as desired. In most home TVROs the feedhorn is simply rotated 90° to change from one to another. In deluxe TVRO installations the feedhorn will have the capability of receiving the vertical and horizontal transponder signals simultaneously, and routing them into separate LNAs for delivery to two or more satellite television receivers. Unlike the U.S. and Canadian domestic satellites, the INTELSAT series use a technique known as left-hand and right-hand circular polarization. (See **Frequency Reuse.**)

Power The strength of a signal as measured in watts. The output power of a typical satellite transponder is only 5 watts, equivalent to shining a small flashlight at the surface of the earth from a point 22,300 miles in space. Combined with a satellite antenna gain of 30 dBi, this would result in an EIRP of approximately 36 dBw toward the earth.

Private Terminal A television receive-only earth terminal (or earth station) used by anyone other than a CATV system.

Program Control Tones Special musical tones transmitted by a programmer's network control center before and after each program. Program control tones are used by the CATV companies automatically to switch earth station receivers and associated equipment at their feed ends.

Programmer A supplier of television programs transmitted via satellite. Programmers include Home Box Office, Cable News Network, Christian Broadcasting Network, and the dozens of other users of satellite transponders who provide daily video feeds to CATV companies.

Protected-Use Transponder A satellite transponder provided by the common carrier to a programmer with a built-in insurance policy. If the protected-use transponder fails, the common carrier guarantees that the programmer will switch over to another transponder, often preempting some other nonprotected programmer from the other transponder. Protected-use transponders cost more to lease than nonprotected-use transponders.

Radio-Frequency Spectrum Those frequency bands in the electromagnetic spectrum that range from several hundred thousand cycles per second (very low frequency) to several billion cycles per second (microwave frequencies).

Receiver An electronic device that detects and decodes a radio or television signal. The typical home television set is a receiver designed to detect UHF and VHF television signals and display them on a built-in television screen. The typical satellite television receiver detects signals in the microwave range that are picked up by a satellite antenna and amplified by an LNA. The satellite television receiver provides its output as a baseband video signal to a TV monitor or as a VHF television channel that can be fed into an ordinary television set.

Registered TVRO A satellite television receive-only earth station for which a license has been issued by the FCC. (See **License.**)

Reuters The British-based news service similar to the Associated Press and United Press International. Reuters operates a high-speed global stock exchange and electronic news wire system called Monitor, and transmits this signal to thousands of users throughout the United States using an entire satellite transponder on the SATCOM F-3 satellite.

RF Adaptor An add-on modulator that interconnects the output of the satellite television receiver to the input (antenna terminals) of the user's television set. The RF adaptor converts the baseband video signal coming from the satellite receiver to a radio-frequency (RF) signal that can be tuned in by the television set on VHF Channel 3 or 4.

SATCOM The series of 24-transponder domestic satellites owned and operated by RCA American Communications Inc. SATCOM F-3 (known as the cable bird) is located at 131 degrees west longitude. SATCOM 2 is positioned at 119 degrees west longitude. The original cable bird, SATCOM F-1, is sitting at 135 degrees west longitude, and is now being used for occasional video feeds. SATCOM birds operate using 12 vertically and 12 horizontally polarized transponders. (Note: Changes in satellite positions will occur as more satellites are launched in the next few years.)

Satellite A smaller body revolving in orbit around a larger body in space. The moon is a natural satellite of the earth. Man-made satellites travel in a variety of orbits around the earth, as the earth rotates on its axis. All major communications satellites are placed into geosynchronous orbits around the earth. (See **Geosynchronous.**)

Satellite Receiver A wideband FM receiver operating in the microwave range, converting the incoming C-band RF signal to a standard baseband video signal.

Satellite Television Corporation (STC) A new organization formed by COMSAT to provide three nationwide pay TV network channels directly to each subscriber's home via 2½-ft.-in-diameter rooftop-mounted TVRO satellite dishes. STC has proposed to begin operation around 1986, using the 12–14-GHz frequency spectrum.

Satellite Television Technology (STT) The publishing and seminar organization founded by Bob Cooper, Jr., the father of the home satellite TV business. STT specializes in home satellite television information, is located in Oklahoma, and holds home satellite TV conferences and trade shows three times a year in various parts of the United States.

Satellite Terminal A receive-only satellite earth station consisting of an antenna (typically parabolic in shape), a feedhorn, a low-noise amplifier, a down-convertor, and a satellite receiver.

SECAM A color television system developed by the French and used in the Francophile countries and the U.S.S.R. SECAM operates with 625 lines per picture frame, but is incompatible in operation with the European PAL system.

Secrecy Act That section of the Communications Act of 1934 that protects the privacy of communications transmitted over private common carrier communications circuits, such as microwave and satellite systems. The Secrecy Act prevents an unauthorized person from receiving and divulging communications not intended for that individual. However, simple reception of the signal does not appear to violate the Secrecy Act, nor does the Secrecy Act clause apply to broadcast signals.

Signal Information sent through electronic means from a transmitter to a receiver, by either an electric circuit or radio transmission technique.

Signal-to-Noise Ratio (S/N) The measure in decibels of the power or strength of a signal as measured in watts compared with the power of the noise surrounding the signal. Just as signal-to-noise ratios are important in high-fidelity stereophonic systems, so a high signal-to-noise ratio means a better, clearer, snow-free television picture. The S/N of the television signal output from a satellite earth terminal receiver is determined directly by the C/N ratio at the input and the C/N threshold of the receiver.

Single-Channel-per-Carrier (SCPC) A special voice or audio service offered by the satellite common carriers, used to transmit several dozen audio signals over a single satellite transponder. Although many audio channels could be handled (a transponder has the capacity to carry 2000 voice conversations normally), each audio signal is transmitted with more power, thereby allowing much smaller receive-only antennas to be used at the earth stations. Using SCPC techniques, the American news wire services are transmitting audio news feeds to radio stations nationwide equipped with antennas only 3 to 4 feet in diameter.

Slot Term used for the longitudinal angular position in the geosynchronous orbit into which a communications satellite is parked. Above the United States, communications satellites are typically positioned in slots that are spaced 3 to 5 degrees apart.

Snow A form of noise picked up by a television receiver, caused by a weak signal. Snow is characterized by alternate dark and white dots randomly appearing on the picture tube. To eliminate snow, a more sensitive receive antenna must be used, or a lower noise temperature LNA must be used.

Society of Private and Commercial Earth Station Users (SPACE) The Washington, D.C., nonprofit association of individuals and mini-CATV organizations that own satellite terminals. SPACE has been active in encouraging the development of home satellite television reception, and is lobbying Congress to have this category of satellite service recognized as a legitimate segment of the industry.

Space Shuttle The U.S. space transportation system that will carry most major satellite packages into near-earth orbit over the next 20 years. As the space shuttle has only enough thrust to reach an orbit of several hundred miles, an additional booster rocket system will be carried in the space shuttle hold along with the communications satellite. Upon its release from the space shuttle hold, the booster rocket will "kick" the communications satellite into its 22,300-mile geosynchronous orbit.

Sparklies A form of satellite television "snow" caused by a weak signal. Unlike terrestrial VHF and UHF television snow that appears to have a softer texture, sparklies are sharper and more angular noise "blips." As with terrestrial reception, to eliminate sparklies either the satellite antenna must be increased in size, or the low-noise amplifier must be replaced with one that has a lower noise temperature.

Spherical Antenna The second major form of satellite television earth station antenna. The spherical antenna (unlike the parabolic antenna) has the ability to simultaneously to see, and thus receive, several satellites in orbit. Because of this feature, spherical antennas are becoming increasingly popular for the home satellite TV user who may wish to switch from satellite to satellite easily and conveniently. (See **Parabolic Antenna.**)

Spot Beam A focused transponder pattern used by the INTELSAT satellites in a limited geographical area. Spot beams are also used by the U.S. domestic satellites to deliver certain transponder signals to Hawaii, Alaska, and Puerto Rico.

Subcarrier A second signal piggybacked into a main signal to carry additional information. In satellite television transmission the video picture is transmitted over the transponder's main carrier. The corresponding audio is sent via an FM subcarrier, which is located at a frequency slot above the video baseband. Some satellite transponders carry as many as four special audio or data subcarriers, whose signals may or may not be related to the main transponder's programming.

Subreflector A small reflector used at the focal point of a parabolic dish antenna to reflect signal energy into the feed beam.

Subscription Television (STV) A broadcasting television station transmitting pay television (usually first-run movies) in a scrambled mode. Over 50 STV stations are presently in operation or planned for cities throughout the United States.

Superstation A term originally used to describer Ted Turner's WTBS Channel 17 UHF station in Atlanta, Georgia. With the addition of several other major independent television stations whose signals are also carried by satellite, the term superstation has come to mean any regional television station whose signal is picked up and retransmitted by satellite to cable companies nationwide.

Synchronization (Sync) The process of orienting the transmitter and receiver circuits so that information sent in relation to a precise instant of time by the transmitter will be perfectly related to that same instant by the receiver. Home television sets are "synchronized" by an incoming sync signal with the television cameras in the studios 60 times per second. Horizontal and vertical hold controls on the television set are used to "fine tune" synchronize the receiver to the incoming television picture.

Teletext An "electronic newspaper of the air" consisting of several hundred "pages" of 24-character by 20-line English text displays. Teletext signals can be transmitted simultaneously over a satellite transponder using a data subcarrier. Several satellite programmers are now transmitting teletext signals, and the national television networks have begun to implement their own teletext systems. A special teletext decoder is required to capture and display the teletext pages as they are transmitted.

Terrestrial TV Ordinary VHF (very high–frequency) and UHF (ultrahigh-frequency) television transmission limited to an effective range of 100 miles or less. Terrestrial TV transmitters operate at frequencies between 54 MHz and several hundred megahertz, far lower than the 4 billion hertz microwave frequencies used by satellite transponders.

Threshold Extension A technique used by satellite television receivers to improve the carrier-to-noise ratio threshold of the receiver by approximately 3 dB. When using small receive-only antennas (10 to 12 feet), a receiver equipped with the threshold extension feature can make the difference between obtaining a good picture or a poor picture because the C/N ratio is above the threshold point.

Translator An unattended television repeater usually operating on a UHF channel. A translator picks up a distant city's broadcast TV station and retransmits the picture locally on another channel. (See **Low-Power TV.**)

Transmitter An electronic device consisting of oscillator, modulator, amplifier, and other circuits that produce a radio or television electromagnetic wave signal for radiation into the atmosphere by an antenna.

Transponder A combination receiver, transmitter, and antenna package physically part of a communications satellite. Transponders have a typical output of 5 watts, operate over a frequency band with a 36-MHz bandwidth in the 4- to 6-GHz microwave spectrum. Communications satellites typically have between 12 and 24 on-board transponders.

TT&C Stations (Telemetry Tracking and Command) The master earth station operated by the common carrier that monitors the on-board operation of the satellite and directs the satellite electronics and positioning equipment. Each satellite is controlled by its own TT&C station using a separate set of telemetry and command frequencies and transponders, unrelated to the communications satellite function.

Tuner That portion of a receiver that can variably select under user control a desired signal from a group of signals in a frequency band. By twisting the tuner knob on a conventional VHF television set, the viewer is selecting one of 12 different incoming channel frequencies that range from 54 MHz (Channel 2) to several hundred megahertz (Channel 13).

Turnkey An installation of a satellite earth station in which all steps are performed by the manufacturer or dealer.

TV Receiver A sophisticated electronic device consisting of a VHF and UHF tuner, audio and video amplifiers, and television display monitor. The "everything in one" TV receiver will probably go the way of the hi-fi console of the 1950s, as more and more modular video equipment such as videotape recorders, home computers, projector television sets, and satellite receivers become popular.

TVRO (Television Receive-Only Terminal) (See **Satellite Terminal**)

Tweaking The process of adjusting an electronic receiver circuit to optimize its performance.

Twin Lead The old-fashioned flat antenna wire used in CATV systems in the 1950s. Twin lead has since been replaced with coaxial cable, capable of carrying far more television channels with less distortion over longer distances.

Ultrahigh Frequency (UHF) The set of television frequencies starting at 470 MHz designated as Channels 14 through 70.

United Press International (UPI) A U.S. news agency similar to the Associated Press, which operates both audio and data news-wire feeds via satellite transponder subcarriers.

Unprotected-Use Transponder A transponder rented to a programmer on an as-is basis by the common carrier. If the transponder fails, the programmer has no recourse to demand replacement or restoral of service by the common carrier. (See **Protected-Use Transponder.**)

Uplink That set of frequencies used by the satellite distribution center's earth station to transmit the video signal up to the satellite.

Vertical Blanking (See **Blanking**)

Very High Frequency (VHF) That frequency band reserved for television channels 2 through 13, beginning at 54 MHz and extending up to 216 MHz in frequency.

VHS Format One of the two most popular home videotape recorder systems. VHS is supported by over a dozen manufacturers, and has obtained approximately 75 percent of the market share in North America. (See **Beta Format.**)

Video-Cassette Recorder (VCR) Originally developed for commercial use, and since purchased by almost two million people in the United States. There are two mutually incompatible schemes: the Beta (Sony) system and the VHS (Japan Victor Corp.) system.

Video Monitor A television set without the tuner. A video monitor only accepts a baseband video signal, not a VHF or UHF television broadcast signal. Video monitors usually have higher resolution and display quality than conventional television sets.

Waveguide A metallic hollow pipe microwave conductor, typically rectangular or oval in shape, used to carry microwave signals into and out of microwave antennas.

WESTAR The U.S. domestic satellite system owned by Western Union. There are three WESTAR birds, currently located at 99 degrees west, 123.5 degrees west, and 91 degrees west longitude, respectively. Each WESTAR has the capacity to carry 12 television channels using its 12 vertically polarized transponders.

Wind Loading The pressure placed upon a satellite earth station antenna caused by the wind. Well-designed parabolic antennas should be able to operate in 40-mph winds without noticeable picture degradation, and to be able withstand winds of up to 125 mph.

World Administrative Radio Conference (WARC) An international meeting coordinated by the ITU at which the world's countries determine which frequencies will be allocated for what services and to whom. The use of domestic communications satellites located in orbits above the United States is determined by the WARC. A key meeting of WARC is scheduled for 1983, at which time assignment of frequencies and orbital arc usage will be determined for direct broadcast satellite (DBS) systems.

Appendix 2

Manufacturers and Suppliers of Satellite Television and CATV Equipment

A list of manufacturers and suppliers of equipment for both the commercial CATV and the home satellite TVRO markets is presented in the following. Not all manufacturers could be included, of course, as the industry is constantly changing. The information is divided into six major categories, depending on the type of equipment supplied. These categories are as follows:

Categories

1. Cable television and commercial equipment suppliers (for mini-CATV systems)
2. Complete satellite home terminal systems (antenna, LNA, receiver, modulator, installation)
3. Satellite antennas (both parabolic and spherical)
4. Satellite receivers (both home and mini-CATV use)
5. Low-noise amplifiers for satellite antennas
6. Miscellaneous equipment (including feedhorns, antenna mounts, CATV modulators)

Note: Unless otherwise indicated, firms listed are manufacturers. CATV manufacturers and other commercial manufacturers tend to charge more for their products than home satellite TV manufacturers, although quality is often (but not always) higher. Some dealers also manufacture one or more of their own components, such as antennas and antenna mounts. Installing dealers are noted whenever possible, although most dealer/distributor organizations have some ability to install equipment in their area. The major installing dealers will travel nationwide, and are so noted.

Following the breakdown of principal suppliers into categories, an alphabetical listing of manufacturers and suppliers includes brief descriptions of their products, and their addresses and telephone numbers.

Category 1. CATV and Commercial Equipment Suppliers

Andrew Corporation
Arvin
Avcom
Blonder-Tongue
Bogner Broadcast Equipment
California Microwave
Delta-Benco-Cascade
Emcee
Forth Worth Tower
Gardiner Communications
Harris Corporation
Hubbard-Payne
Hughes Corporation
Jerrold Systems
Magnavox CATV Systems
Microdyne
Microwave Associates
North Supply
Oak Communications
RCA CATV Systems
Rigel Systems
Rockwell International (Collins Transmission Systems)
Scientific-Atlanta
Scientific Communications
Sylvania CATV Transmission Systems
Tocom
Toner Cable
United States Tower Corporation
Westinghouse Electric Corporation

Category 2. Complete Satellite Home Terminal Systems Dealers, Distributors, and Installing Dealers

A-B Electronics
Advanced Electronics
Anixter-Mark
Antenna Development and Manufacturing
ATV Research
Avcom

Birdview Satellite Communications, Inc.
Black and White Enterprises (Downlink)*
Channel Master
Channel One*
Comtech Telecommunications
Earthstar Corporation
Echosphere Corporation
Gardiner Communications Corporation
Hamilton Systems
Hero Communications
Helfer's Antenna Service
Home Satellite Television Systems
H & R Communications (Starview Systems)*
Interstar Signals*
KLM Electronics
Lindsay Specialty Products
Robert A. Luly
MacLine
Microdyne Corporation
MicroLink Systems
Microwave Associates
Microwave General
Mid-America Video
Mini-Casat
National Microtech*
Norsat Systems
Orlando Antenna
Ramsey Electronics (Sat-Tec Systems)
SATELCO (Thunder Compute)
Satelinc
Satellite Supplier
Satellite Systems Unlimited
Satellite Technology Services Corporation
Satellite Television Systems
Satellite Video Systems
SAT Finder Systems (Rielco Electronics)
SAT Share, Inc.
SAT-TEC
SAT Vision
SED Systems
Spacecom Industries
Starpath
Starvision Systems
Alan Swan
Telemetry Communications and Instrumentation
Third Wave Communications*
Tri-Star General
Via Cable
Wilson Microwave Systems, Inc.*

*Major supplier of home TVRO satellite terminal systems.

Category 3. Satellite TVRO Antennas

Andrew Corporation
Anixter-Mark
Antenna Development and Manufacturing Company
Antenna Technology Corporation
Arvin (Applied Technology Group)
Black and White Enterprises (Downlink)*
Cayson Electronics
Channel One
Comm/Plus
Comtech Antenna Corporation
Discom Satellite Systems, Inc.
Forth Worth Tower
Gardiner Communications
H & R Communications (Starview Systems)
Hamilton Systems
Harrell's Southside Welding
Hastings Antenna Co., Inc.
Hero Communications
Home Satellite Television Systems
KLM Electronics
Lindsay Specialty Products
Robert A. Luly and Associates
MacLine
McCullough Satellite Systems†
Microwave Associates
Microwave General
Mid-America Video
National Microtech
Orlando Antenna
Paraframe
Prodelin
Ramsey Electronics (Sat-Tec Systems)
SATELCO (Thunder Compute)†
Satellite Supplies†
Satellite Technology Services Corporation
Satellite Television Systems
Satellite Video Systems*
SAT Vision
SED Systems
Alan Swan†
Tri-Star General*
Via Cable
Wagner Industries†

Category 4. Satellite Receivers

Arunta Engineering Co.
Automation Techniques
Avcom

*Spherical and parabolic antennas.
†Spherical antennas; all others are parabolic.

Robert M. Coleman
Comm/Plus
Comtech Data
R.L. Drake Company
Earth Terminals (Washburn)
Gardiner Communications
Gillaspie & Associates, Inc.
H & G Systems
H & R Communications (Starview Systems)
Hi-Tek Satellite Systems
Howard Engineering
International Crystal Manufacturing Company (ICM)
Interstat Corp.
KLM Electronics
Merrimac Industries
Microdyne Corporation
Microwave Associates
Mid-America Video
Newton Electronics
Norsat Video
Ramsey Electronics (Sat-Tec Systems)
Rohner and Associates
SATELCO (Thunder Compute)
Satellite Supplies
Satellite Technology Services Corporation
Satellite TV Tech. Int.
SAT-TEC
Scientific Communications
SED Systems
Sigma International
Standard Communications
Telcom Industries
Toner Cable

Category 5. Low-Noise Amplifiers (LNAs)

Amplica
Avantek
Dexcel
Gardiner
Microwave Associates
Real World System
Simcomm Labs

Category 6. Miscellaneous Equipment

Chaparral (feedhorns)
H & G (mounts)
ICM (signal purifiers)
KLM Electonics (AZ/EL satellite-locator tool)
Merrimac Industries (down-convertors)
Microwave General (motorized antenna mounts)

Quantum Associates (motorized antenna mounts)
Satellite Communications (stereo demodulators)

Other manufacturers also supply specialty components. Check with the manufacturers and suppliers listed under Category 3.

Alphabetical Listing of Manufacturers and Suppliers

Supplier: A-B Electronics, 1782 West 32nd Place, Hialeah, FL 33012; (305) 887-3203
Product: Parabolic antenna systems.
Category: 2

Supplier: ADM (Antenna Development & Mfg. Inc.), 2745 Bedall Ave., Poplar Bluff, MO 63901; (314) 785-5988
Product: Parabolic 11- and 13-foot antennas.
Category: 3

Supplier: Advanced Electronics, 5085 Arville St., Las Vegas, NV 89118; (800) 634-6047
Product: Complete systems. Dealer.
Category: 2

Supplier: Amplica Inc., 950 Lawrence Dr., Newbury Park, CA 91320
Product: Low-noise amplifiers. Principal manufacturer.
Category: 5

Supplier: Andrew Corp., 10500 West 153rd St., Orland Park, IL 60462; (312) 349-3300
Product: 3-meter, 4.5-meter, 8-meter, and 10-meter antennas.
Category: 1, 3

Supplier: Anixter-Mark, Inc., 2180 S. Wolf Rd., Des Plaines, IL 60018; (312) 298-9420
Product: Complete 10–16-foot parabolic antennas. CATV-oriented manufacturer.
Category: 3

Supplier: Antenna Technology Corp., 895 Central Florida Parkway, Orlando, FL 32809; (305) 851-1112

Product: Antennas—SIMULSAT 5, commercial stretched parabolic multisatellite antenna.
Category: 3

Supplier: Arunta Engineering Co., Department CSP, P.O. Box 15082, Phoenix, AZ 85060; (602) 956-7042
Product: Satellite receivers.
Category: 4

Supplier: Arvin (Applied Technology Group), 4490 Old Columbus Rd. NW, Carroll, OH 43112; (614) 756-9211
Product: SAT-weather Receiver, which converts National Oceanic and Atmospheric Administration (of the Department of Commerce) weather pictures of the earth's hemisphere into color video pictures. The company's weather satellite receiver provides current weather pictures, continuous video output, and realistic color.
Category: 1, 4

Supplier: ATV Research, 13th and Broadway, Dakota City, NE 68731; (402) 987-3771
Product: Parabolic antenna systems.
Category: 2

Supplier: Automation Techniques, P.O. Box 52481, Tulsa, OK 74152; (918) 582-7545
Product: Frequency-agile TVRO satellite receiver.
Category: 4

Supplier: Avantek, 3175 Bowers Ave., Santa Clara, CA 95051; (408) 727-0700
Product: Microwave semiconductors; amplifiers, wideband and narrowband; scan and lock oscillators; microwave modules; digital microwave radio relay systems; earth station assemblies and subsystems; CATV test equipment; transmission network engineering, low-noise and power amplifications, CATV receiver technology, telecommunications systems analysis; high-performance receive-only CATV antenna dish; low-noise amplifiers. Principal manufacturer.
Category: 1, 4, 5, 6

Supplier: AVCOM of Virginia Inc., Satellite Systems Div., 500 Research Rd., Richmond, VA 23235; (804) 794-2500

Product: Receivers—COM3, COM3R, COM3R remote-control unit; satellite video receiver—PSR3; antennas; low-noise amplifiers.
Category: 2, 3, 4, 5, 6

Supplier: Birdview Satellite Comm., Inc., P.O. Box 963, Chanute, KS 66720; (316) 431-0400
Product: Complete home terminal systems.
Category: 2

Supplier: Blonder–Tongue Laboratories Inc., 1 Jake Brown Rd., Old Bridge, NJ 08857; (201) 679-4010
Product: CATV products—single-channel antennas, head-in processor components, low-noise channel preamplifiers, custom-built channel converters; equipment housings and accessories; CATV indoor distribution amplifiers; CATV line powered amplifiers and audio and video modulator products; products for smaller CATV systems. Major CATV manufacturer.
Category: 1, 6

Supplier: Bogner Broadcast Equipment Corp., 401 Railroad Ave., Westbury, NY 11590; (516) 997-7800
Product: Antennas—low- and medium-power TV broadcast antennas, UHF TV low-power broadcast antennas, MDS receiving antennas. LPTV specialists.
Category: 6

Supplier: California Microwave Inc., 455 W. Maude Ave., Sunnyvale, CA 94086; (408) 732-4000
Product: Telecommunications and satellite communications products; small satellite terminals, low-density applications. Major commercial manufacturer.
Category: 1, 3, 4

Supplier: Cayson Electronics, Rte. 3, Box 160, Fulton, MI 38843; (601) 862-2132
Product: 10-foot parabolic fiberglass antenna, main dish mount, feed mount with rotor motor, feedhorn.
Category: 3

Supplier: Channel Master (Div. of Avnet, Inc.), Ellenville, NY 12428; (914) 647-5000

Product: Complete home terminal systems.
Category: 2

Supplier: Channel One Inc., Willarch Rd., Lincoln, MA 01773; (617) 259-0333
Product: Earth link system. Popular home TVRO systems. Installing national dealer.
Category: 2

Supplier: Chaparral Communications Co., P.O. 832, Los Altos, CA 94022; (415) 941-1555
Product: Superfeed—designed for use with parabolic antennas. Most popular feedhorn for home TVROs.
Category: 3, 6

Supplier: Robert M. Coleman; Rte. 3, Box 58-A, Travelers Rest, SC 29690
Product: Satellite television receiver—Starview. Coleman and Taylor Howard have reengineered and simplified all the circuits involved in a TVRO receiver. Produces circuit boards that make TVRO construction easier. LNA units.
Category: 4, 5

Supplier: Comm/Plus (Communications Plus), 3680 Cote Vertu, The Bazaar Center, St. Lawrence, Que., Canada HR41P8 (represents Technimat Canada, Inc.); (514) 337-7255
Product: Technimat 692A model earth satellite receiver; antennas.
Category: 3, 4

Supplier: Comtech Antenna Corp., 3100 Communications Rd., P.O. 428, St. Cloud, FL 32769; (305) 892-6111
Product: Satellite earth terminals, 3-meter and 5-meter parabolic antennas, tropospheric scatter antennas, reflectors, antenna feeds, special application antennas, general-purpose antennas, feed systems, and feedhorns.
Category: 3, 6

Supplier: Comtech Data Corp., 613 S. Rockford Dr., Tempe, AZ 85281; (602) 968-2433
Product: TVRO satellite receivers
Category: 4

Supplier: Comtech Telecommunications Corp., 135 Engineers Rd., Smithtown, NY 11787; (516) 231-5454
Product: Complete TVRO home satellite terminals. Dealer.
Category: 2

Supplier: Delta-Benco-Cascade Ltd., 124 Belfield Rd., Rexdale, Ont., Canada M9W1G1; (416) 241-2651
Product: Cable television and miscellaneous equipment.
Category: 1, 6

Supplier: Dexcel Inc., 2285C Martin Ave., Santa Clara, CA 95050; (408) 727-9833
Product: Low-noise amplifiers. Principal manufacturer. New miniaturized satellite receivers.
Category: 4, 5

Supplier: Discom Satellite Systems, Inc., 4201 Courtney, P.O. Box 8699, Independence, MO 64054; (816) 836-2828
Product: Satellite TVRO antennas.
Category: 3

Supplier: Downlink Inc., P.O. Box 33, Putnam, CT 06260; (203) 928-7955
Product: Parabolic and spherical antennas, low-noise amplifiers, receivers, complete satellite home terminals. Leading national installing dealer and distributor.
Category: 2, 3, 4, 5

Supplier: R.L. Drake Co., 540 Richard St., Miamisburg, OH 45342; (513) 866-2421
Product: TVRO receivers.
Category: 4

Supplier: Earthstar Corp., 16012 Cottage Grove, South Holland, IL 60473; (312) 755-5400
Product: Complete TVRO home satellite terminals. Dealer.
Category: 2

Supplier: Earth Terminals, Inc., P.O. Box 636, Fairport, NY 14450; (716) 223-7457
Product: Receivers. Washburn TVRO home satellite systems.
Category: 4

Supplier: Echosphere Corp., 5315 S. Broadway, Littleton, CO 80120; (303) 797-3231
Product: Complete satellite systems and components. Dealer.
Category: 2

Supplier: EMCEE Broadcast Products, Div. of Electronics, Missiles and Communications Inc., P.O. Box 68, White Haven, PA 18661; (800) 233-6193
Product: LPTV transmitters, MDS transmitters.
Category: 1

Supplier: Fort Worth Tower Co., Inc., 1901 East Loop 820 South, P.O. Box 8597, Ft. Worth, TX 76112; (817) 457-3060
Product: Towers for cable television systems, 12-foot, 16-foot, 7-meter parabolic antennas. Major CATV manufacturer.
Category: 3

Supplier: Gardiner Communications Corp., 1980 South Post Oak Rd., Suite 2040, Houston, TX 77056; (713) 961-7348
Product: 3-meter, 5.6-meter parabolic antennas, low-noise amplifiers, power supplies, receivers, CATV modulators, power dividers. Major CATV manufacturer.
Category: 1, 2, 3, 4, 5, 6

Supplier: Gillaspie & Associates, Inc., 950 Benicia Ave., Sunnyvale, CA 94086; (408) 730-2500
Product: Satellite receivers, down-convertors.
Category: 4

Supplier: H & G Systems, 15109 Chicago Rd., Dolton, IL 60419
Product: TVRO home satellite receivers.
Category: 4

Supplier: H&R Communications, Inc., Rte. 3, Box 103G, Pocahontas, AK 72455; (501) 647-2001
Product: 10-, 12-, 13-, 16-, 20-foot parabolic antennas (manufactured by Starview Systems subsidiary), TVRO systems.
Category: 2, 3, 5

Supplier: Hamilton Systems, 1101 E. Chestnut, Suite A, Santa Ana, CA 92701; (714) 543-5217

Product: Complete TVRO home terminals. Dealer.
Category: 2

Supplier: Harrell's Southside Welding, Old Highway 7 North, Route 2, P.O. Box 46, Grenada, MS 38901; (601) 226-4081
Product: Satellite TVRO antennas.
Category: 3

Supplier: Harris Corp., Satellite Communications Div., P.O. Box 1700, Melbourne, FL 32901; (305) 725-2070
Product: 20-foot, 11-meter TVRO and uplink parabolic antennas. Satellite receivers. CATV equipment.
Category: 1, 3, 4, 5

Supplier: Hastings Antenna Company, Inc., 847 West First, Hastings, NE 68901; (402) 463-3598
Product: Satellite TVRO antennas.
Category: 3

Supplier: Helfer's Antenna Service, 23 Brookside Pl., Pleasantville, NY 10570; (914) 769-2588
Product: TVRO complete home terminal systems, antennas, LNAs, receivers. Nationwide installing dealer.
Category: 2, 3, 4, 5

Supplier: Hero Communications (Div. of A-B Electronics), 1782 W. 32nd Pl., Hialeah, FL 33012; (305) 887-3203
Product: TVRO systems. Dealer.
Category: 2, 3, 4, 5, 6

Supplier: Hi-Tek Satellite Systems, Division of Gillaspie & Assoc., Inc., 177 Webster St., Suite A455, Monterey, CA 93940; (408) 372-4771
Product: Satellite receivers, LNAs, down-convertors
Category: 4, 5, 6

Supplier: Home Satellite Television Systems, Div. of Rieco TV Service Inc., 6541 E. 40 St., Tulsa, OK 74145; (918) 664-4466
Product: Complete satellite home terminal systems, remote-control antenna package. Dealer.
Category: 2, 3

Supplier: Hoosier Electronics, Inc., P.O. Box 3300, Terre Haute, IN 47803; (812) 238-1456
Product: Complete systems. Dealer.
Category: 2

Supplier: Howard Engineering, P.O. 48, San Andreas, CA 95249
Product: *Howard Terminal Manual* distributed by Satellite Television Technology, circuit boards for the Howard receiver. For the advanced electronics experimenter.
Category: 4

Supplier: Hubbard-Payne, 4811 Clinton Highway, Suite 113, Knoxville, TN 37912; (615) 688-6261
Product: Professional-quality, small parabolic antennas for broadcast and CATV use.
Category: 1

Supplier: Hughes Corp. Microwave Communications Products, P.O. Box 2999, Torrance, CA 90509; (213) 534-2146
Product: 4.5-, 5-, 6-meter antennas with various feed options; satellite video receivers. Complete satellite home terminal systems. Major CATV manufacturer.
Category: 1, 2, 3, 4, 6

Supplier: International Crystal Manufacturing Co. (ICM), 10 N. Lee, Oklahoma City, OK 73102; (405) 236-3741
Product: Satellite receiver ICMTV 4300, satellite receiver signal purifier. Leading home satellite receiver manufacturer.
Category: 4, 6

Supplier: Intersat Corporation, 2 Hood Drive, St. Peters, MO 63376; (314) 278-2178
Product: Satellite receivers.
Category: 4

Supplier: Interstar Satellite Systems, 21708 Marilla St., Chatsworth, CA 91311; (213) 882-6770
Product: Complete distributor of TVRO home satellite terminals and equipment.
Category: 2, 3, 4, 5, 6

Supplier: Jerrold Electronics, Inc., Hatboro, PA 19040; (215) 674-4800
Product: CATV-type modulators, convertors, amplifiers. Major CATV manufacturer.
Category: 6

Supplier: KLM Electronics Inc., P.O. Box 816, Morgan Hill, CA 95037; (408) 779-7363
Product: TVRO home satellite systems, TVRO receivers. Major home satellite receiver manufacturer.
Category: 2, 4, 6

Supplier: Lindsay Specialty Products Ltd., 50 Mary St. W., Lindsay, Ont., Canada K9V4S7; (705) 324-2196
Product: 8-, 10-, 12-, 15-foot, 7-, 11-meter parabolic antennas, circular feedhorns.
Category: 3, 6

Supplier: Robert Luly Assoc., P.O. Box 2311, San Bernardino, CA 92405; (714) 888-7525
Product: Patented unique, fold-up ultralightweight parabolic antennas; wind load kit and polar mounts for Luly antennas. Geodesic domes for antennas. Complete TVRO systems using Luly antennas.
Category: 2, 3

Supplier: Mac Line Inc., W. 3125 Seltice Blvd., Coeur d'Alene, ID 83814; (208) 765-0909
Product: 12-foot sectionalized parabolic antenna, complete package systems
Category: 2, 3

Supplier: Magnavox CATV Systems, Inc., 133 W. Seneca St., Monduis, NY 13104; (315) 682-9105
Product: CATV systems. All equipment required from convertors/decoders to line amplifiers.
Category: 1

Supplier: McCullough Satellite Systems, Inc., P.O. Box 57, Salem, AR 72576; (501) 895-3167
Product: 8-, 10-, 12-foot spherical multisatellite antennas. Major home and mini-CATV antenna manufacturer.
Category: 3

Supplier: Merrimac Industries, Inc., 41 Fairfield Pl., West Caldwell, NJ 07006; (201) 575-1300
Product: Receivers, down-convertors.
Category: 4, 6

Supplier: Micro-Link Systems, Inc., 6012 Dixie Dr., Dayton, OH 45414; (513) 898-9964
Product: Complete TVRO home satellite systems. Dealer.
Category: 2

Supplier: Microdyne Corp., P.O. Box 7213, Ocala, FL 32672; (904) 687-4633
Product: Series of 12-, 16-, 20-foot parabolic antennas, satellite receivers, complete system packages, modulators, LNAs. Major CATV manufacturer.
Category: 2, 3, 4, 5, 6

Supplier: Microwave Associates Communications (MA/Com), 63 Third Ave., Burlington, MA 01803; (617) 272-3100
Product: 4.6-meter parabolic antenna, 80-degree, 100-degree, 110-degree, 120-degree LNAs. TVRO receivers: VR-3X, VR-4X, VR-3XT. Major commercial manufacturer.
Category: 1, 2, 3, 4, 5, 6

Supplier: Microwave General, 2680 Bayshore Frontage Rd., Mountain View, CA 90019
Product: Complete TVRO home terminals. 10-, 13-, 16-, 20-foot parabolic antennas. TVRO receivers. Motorized antenna mounts.
Category: 2, 3, 4, 6

Supplier: Mid-America Video Inc., 324 Pershing Blvd., P.O. Box 511, N. Little Rock, AR 72115; (501) 753-3555
Product: 3-, 4-, 5-, 6-meter parabolic antennas, tunable home TVRO receivers. Major home receiver manufacturer. Dealer.
Category: 2, 3, 4, 5

Supplier: Mini-Casat, Route 3, Box 160, Fulton, MS 38843; (601) 862-2132
Product: Consumer-oriented antenna and receiver systems.
Category: 2

Supplier: MUNTZ Electronics, Inc., 7700 Densmore, Van Nuys, CA; (213) 782-7511
Product: Consumer-oriented TVRO home systems.
Category: 2

Supplier: National Microtech Inc., P.O. Box 417, Grenada, MS 38901; (800) 647-6144
Product: Complete home terminal systems, 3-, 4-, 4.5-, 5-, 6-meter parabolic antennas, TVRO receivers, 80-degree, 90-degree, 100-degree, 110-degree, 120-degree LNAs. Major national distributor.
Category: 1, 2, 3, 4, 5, 6

Supplier: Newton Electronics, Inc., 2218 Old Middlefield Way, Mountainview, CA 94043; (415) 967-1473
Product: Satellite receivers.
Category: 4

Supplier: Norsat Systems, Box 232, Surrey, B.C., Canada V3T4W8; (604) 585-9428
Product: Video receivers, home model Norsat 2001-B, industrial model Norsat 2001-C, options include two stereo-audio demodulators; 3.7-meter antenna kit, 3-meter, 4.5-meter parabolic antennas; 100-degree, 120-degree LNAs.
Category: 2, 3, 4, 5, 6

Supplier: North Supply Co., 10951 Lakeview Ave., Lenexa, KS 66219; (913) 888-9800
Product: 3.66-meter, 5-meter parabolic antennas; satellite earth station receivers, TV modulators, CATV distribution equipment, cables. Major CATV/telephone company distributor.
Category: 1, 3, 4, 6

Supplier: Oak Communications Inc., CATV Div., Crystal Lake, IL 60014; (815) 459-5000
Product: Satellite signal incription system ORION (provides signal security to satellite users). CATV decoders, STV decoders. Major CATV/STV manufacturer.
Category: 1, 6

Supplier: Orlando Antenna Co., 2356 W. Oakridge Rd., Orlando, FL 32809; (305) 851-8332

Product: 3-, 4-, 5-meter fiberglass parabolic antennas
Category: 3

Supplier: Paraframe Inc., Box 423, Monee, IL 60449
Product: 12-foot and larger stressed parabolic antenna kits, plus antenna plan design manuals. Major home satellite antenna manufacturer.
Category: 2, 3

Supplier: Prodelin Inc. (Div. MA/Com), Box 131, Hightstown, NJ 08520; (609) 448-2800
Product: Parabolic earth station antennas—6-, 8-, 10-, 12-, 15-foot diameters. 10-foot (3-meter)-diameter segmented antenna. Major home satellite antenna manufacturer.
Category: 3

Supplier: Quantum Associates, Inc., Box 18, Alpine, WY 83128; (307) 654-2000
Product: Satellite-selector, remote-control motorized antenna mounts.
Category: 6

Supplier: Ramsey Electronics Inc. (Sat-Tec Systems), 2575 Baird Rd., Penfield, NY 14526; (716) 586-3950
Product: 12-foot spherical antennas; 85-degree, 120-degree LNAs; R2B, R3, and R4 tunable receivers. Major home satellite receiver manufacturer.
Category: 2, 3, 4, 5

Supplier: RCA Community Television Systems, 7355 Fulton Ave., No. Hollywood, CA 91605; (213) 764-2411
Product: CATV modulators, equipment.
Category: 1

Supplier: Real-World Systems, 128 Cross House Rd., Greenside, Sheffield, England S303RX
Product: Kits and circuit boards for LNA units for advanced experimenters.
Category: 5

Supplier: Rigel Systems, 2974 Scott Blvd., Santa Clara, CA 95050; (408) 727-3628

Product: Parabolic antennas for small CATV systems.
Category: 1

Supplier: Rockwell International, Collins Transmission Systems Div., P.O. Box 10462, Dallas, TX 75207; (214) 996-5340
Product: Satellite video receivers, Models SVR4A1, SVR4T1, SVR4F. Major commercial manufacturer.
Category: 4

Supplier: Rohner & Associates, 501 N. Elm St., West Liberty, IA 52776; (319) 627-4819
Product: Dual FET LNA boards and LNA kits; TVRO receivers.
Category: 4, 5, 6

Supplier: SATELCO, Div. of Thunder Compute, 5540 W. Pico Blvd., Los Angeles, CA 90019; (213) 931-6274
Product: Complete satellite home systems with dish sizes at 10, 16, and 20 feet. Dealer.
Category: 2

Supplier: Satelinc, Riverside Drive, Chestertown, NY 12817; (518) 494-4151
Product: Complete home terminal systems.
Category: 2

Supplier: Satellite Communications, Inc., 8101 Gallia St., Wheelersburg, OH 45694; (614) 574-8121
Product: Complete satellite TVRO home terminals. Dealer. Manufacturer of stereo-receiver demodulators.
Category: 2, 6

Supplier: Satellite Supplies, Box 278, Aldergrove, B.C., Canada VOX1A0
Product: TVRO satellite receivers with antenna-mounted downconvertors.
Category: 4

Supplier: Satellite Systems Unlimited, P.O. Box 43, Conway, AR 72032; (501) 327-6501
Product: Complete home terminal systems.
Category: 2

Supplier: Satellite Technology Services, Inc., 11684 Lilburn Park Rd., St. Louis, MO 63141; (800) 325-4058
Product: Complete satellite TVRO home terminals. Dealer.
Category: 2

Supplier: Satellite Television Systems/Star, P.O. Box 51837, Lafayette, LA 70505
Product: 10- and 12-foot sectionalized fiberglass parabolic antennas. Complete systems.
Category: 2, 3

Supplier: Satellite Television Technology International, Inc., P.O. Box G, Arcadia, OK 73007; (405) 396-2574
Product: Satellite receivers.
Category: 4

Supplier: Satellite Video Systems, P.O. Box 673, Cabot, AR 72023; (501) 843-7358
Product: Satellite TVRO home terminals. Dealer.
Category: 2

Supplier: SAT Finder Systems (Div. of Rieco Electronics), 6541 E. 40th St., Tulsa, OK 74145; (718) 664-4466
Product: Complete TVRO home terminal systems. Dealer.
Category: 2

Supplier: SAT Share, Inc., 5301 Hollister, Suite 100, Houston, TX 77040; (713) 460-9900
Product: Complete home terminal systems.
Category: 2

Supplier: SAT-TEC, P.O. Box 10101, Rochester, NY 14610
Product: Low-cost TVRO receiver kits for the advanced electronics experimenter.
Category: 2, 4

Supplier: SAT Vision, P.O. Box 1490, Miami, OK 74354; (918) 542-1616
Product: 3.3-meter parabolic antenna.
Category: 3

Supplier: Scientific-Atlanta Inc., 3845 Pleasantdale Rd., Atlanta, GA 30340; (404) 449-2000

Product: 3-, 4-, 5-, 6-, 7-, 10-meter parabolic antennas. 80-, 100-, 120-degree LNAs, tunable receivers, fixed crystal receivers, CATV decoders, complete receiving terminals, CATV-type modulators. Major commercial manufacturer.
Category: 1, 2, 3, 4, 5, 6

Supplier: Scientific Communications Inc., 3425 Kingsley Rd., Garland, TX 75041; (214) 271-3685
Product: TVRO satellite receivers, LNAs, CATV supplier.
Category: 4, 5

Supplier: SED Systems, Ltd., P.O. Box 1464, 2414 Coyle Ave., Saskatoon, Sask., Canada 57K 3P7; (306) 664-1825
Product: 10-, 12-, 15-foot parabolic antennas. Complete systems. Dealers.
Category: 2

Supplier: Sigma International, Inc., 617 N. Scottsdale Rd., Scottsdale, AZ 85257; (602) 994-3435
Product: Basic satellite receiver modules.
Category: 4

Supplier: Simcomm Labs, Box 60, Kersey, CO 80644; (303) 352-1020
Product: LNA for small earth stations Model LNA-60.
Category: 5

Supplier: Space Com Industries, P.O. Box 66, Cape Girardeau, MO 63701; (314) 334-2011
Product: Complete TVRO home terminals. Dealer.
Category: 2

Supplier: Standard Communications, P.O. Box 92151, Los Angeles, CA 90009; (213) 532-5300
Product: Satellite TVRO receivers. Major commercial manufacturer.
Category: 4

Supplier: Starpath Inc., Rte. 1, Box 188, Fisk, MO 63940; (314) 967-3201
Product: Complete TVRO home terminals. Dealer.
Category: 2

Supplier: Starvision Systems, 611 N. Wymore Road, Winterpark, FL 32789; (305) 628-5458
Product: Complete home terminal systems.
Category: 2

Supplier: Swan Antenna Co., 614 Cimarron, Stockton, CA 95210; (209) 948-5254
Product: 12-, 14-, 16-foot multisatellite spherical antennas.
Category: 3

Supplier: Sylvania CATV Transmission Systems, 10841 Pellicano Dr., El Paso, TX 79935; (915) 591-3555
Product: CATV equipment. Major CATV manufacturer.
Category: 1

Supplier: TCI Telemetry Communications and Instrumentation Corp., 411 N. Buchanan Circle #3, Pacheco, CA 94553; (415) 676-6102
Product: Complete TVRO home terminals. Dealer.
Category: 2

Supplier: Telecom Industries Corp., 27 Bonaventura Dr., San Jose, CA 95134; (408) 262-3100
Product: Satellite TVRO receivers.
Category: 4

Supplier: Third Wave Communications Corp., 2600 Gladstone, Ann Arbor, MI 48104; (313) 996-1483
Product: Completely satellite home terminal systems. Major nationwide installing dealer.
Category: 2

Supplier: TL Systems, 3001 Redhill Ave., Esplanade V, Suite 207, Costa Mesa, CA 92626; (714) 556-9830
Product: TVRO home satellite receivers.
Category: 4, 6

Supplier: TOCOM, 3301 Royalty Row, Box 47066, Dallas, TX 75247; (214) 438-7691
Product: CATV equipment, convertors. Major CATV manufacturer.
Category: 1

Supplier: Toner Cable Equipment Inc., 969 Horsham Rd., Horsham, PA 19044; (800) 523-5947
Product: Satellite TVRO receivers, LNAs, modulators. Major CATV manufacturer.
Category: 1, 4, 5, 6

Supplier: Tri-Star General, 4810 Van Epps Rd., Brooklyn Heights, OH 44131; (216) 459-8535
Product: Complete satellite home terminals. Dealer.
Category: 2

Supplier: United States Tower Co. (USTC), P.O. Drawer S, Afton, OK 74331; (918) 257-4257
Product: Unique multisatellite spherical TVRO antennas. Major CATV manufacturer.
Category: 3

Supplier: VIA Cable Inc., Box 552, Ingram, TX 78025; (512) 367-5714
Product: 3-, 4-, 5-meter parabolic antennas, complete systems.
Category: 2, 3

Supplier: Wagner Industries, P.O. Box 559, Alva, OK 73717; (405) 327-1877
Product: Eagle spherical section antenna, 3-, 3.6-, 4.8-meter sizes.
Category: 3

Supplier: Western Satellite, P.O. Box 22959, Wellshire Station, Denver, CO 80222
Product: Complete TVRO home satellite systems and for shared use.
Category: 2

Supplier: Westinghouse Electric Corp., Electronic Products and Overseas Div., 1111 Schilling Rd., Hunt Valley, MD 21030; (301) 667-1000
Product: Ground-mounted 3.1-meter antenna, portable 1.2-meter antenna, roof-mounted 4.6-meter antenna; receivers. Major commercial manufacturer.
Category: 1, 3, 4, 6

Supplier: Wilson Microwave Systems, Inc., 4286 S. Polario Ave., Las Vegas, NV 89103; (800) 634-6898
Product: Complete systems and components. National distributor.
Category: 2

Appendix 3
Satellite Video Programming Sources

This appendix is divided into two sections. First, each satellite is presented on a transponder-by-transponder basis, listing the programming carried and program supplier.

Next, the programmers are listed, in alphabetical order, along with a description of the programming carried, satellite transponder, address and telephone number, and programming cost for CATV and mini-CATV/MATV systems (when known).

Guide to Transponder Numbers versus Satellites

The RCA SATCOM and ATT/GTE COMSTAR satellites use 24 transponders, consisting of 12 vertically and 12 horizontally polarized transponders. RCA numbers its transponders from 1 to 24. ATT/GTE uses 1–12V and 1–12H.

Western Union's WESTAR and TeleSat Canada's ANIK satellites use 12 vertically polarized transponders, numbered 1–12. On a 24-channel tunable receiver, these transponders would actually correspond to Channels 1, 3, 5, 7, etc., as only the vertical channels are being transmitted. The following table should help to clarify this.

APPENDIX 3

TABLE A3–1.
Satellite Downlink Transponder Frequencies versus Satellite Transponder Designations

CATV Standard Transponder Designation	Downlink Frequency (MHz)	Satellite:			
		SATCOM	COMSTAR	WESTAR*	ANIK*
1	3720	1	1V	1	1
2	3740	2	1H		
3	3760	3	2V	2	2
4	3780	4	2H		
5	3800	5	3V	3	3
6	3820	6	3H		
7	3840	7	4V	4	4
8	3860	8	4H		
9	3880	9	5V	5	5
10	3900	10	5H		
11	3920	11	6V	6	6
12	3940	12	6H		
13	3960	13	7V	7	7
14	3980	14	7H		
15	4000	15	8V	8	8
16	4020	16	8H		
17	4040	17	9V	9	9
18	4060	18	9H		
19	4080	19	10V	10	10
20	4100	20	10H		
21	4120	21	11V	11	11
22	4140	22	11H		
23	4160	23	12V	12	12
24	4180	24	12H		

*All transponders horizontally polarized

Part I: Listing by Satellite

RCA SATCOM F-3
Satellite Transponder Assignments
"Cable Net I"
Location: 131 Degrees West Longitude

TRANSPONDER	PROGRAM SERVICE
1	Nickelodeon—Warner Amex's children's programming. Arts—ABC's cultural network from New York's Lincoln Center, other locations.

2	PTL Television Network—24 hours per day of religious programming from Charlotte, North Carolina. WAME-AM, a Charlotte radio station, is also carried on a subcarrier.
3	WGN—United Video Retransmission of WGN-TV in Chicago. WFMT-FM, Chicago's classical music station, and Seeburg's record company audio service are also carried as subcarriers on WGN's video signal, as well as Dow Jones Stock Market data.
4	Spotlight—24 hours per day of major movies, specials, and classic films.
5	The Movie Channel—Warner Amex's national pay television first-run movie programming, 24 hours per day. East Coast feed.
6	WTBS—Southern Satellite System retransmission of Ted Turner's WTBS Atlanta Channel 17. In addition to Channel 17 video and audio, SSS also carries on a subcarrier "UPI Newstime." This service provides news photos with voice-over news reports 24 hours daily. Three other special audio subcarriers also carry other feeds on WTBS' video signal.
7	Entertainment and Sports Programming Network (ESPN)—National CATV sports programming. An advertiser-sponsored network of 24-hour-per-day sports events.
8	Christian Broadcasting Network (CBN)—24 hours per day of religious programming from Virginia Beach, Virginia.
9	USA Network—Prime-time coverage of Madison Square Garden sports, Calliope children's programming, and The English Channel.
	BET—Black Entertainment Television featuring black performers in movies, specials.
10	Showtime—National premium entertainment. West Coast feed. Movies, specials, plays, nightclub acts.
11	The Music Channel (Modern Television Network—MTV)—Rock and popular music groups on stage.
12	Showtime—National premium entertainment. East Coast feed.
13	Home Box Office—National pay TV service. West Coast feed.

14	Cable News Network—24 hours a day of news, sports news, weather, futures from Atlanta and over a dozen international news bureaus.
15	CNN-2—Continuously updated 30-minute "news wheel" of headline news.
16	ACSN—Appalachian Community Services Network of daytime, educational, and public-service-oriented programming.
	AETN (American Educational Television Network) —Continuing education courses for professionals.
	Christian Media Network—Evening commercially sponsored films, gospel sings, interviews.
	National Jewish Television—Sunday program feeds.
	Window on Wall Street—Friday analysis show at 6 p.m. New York time.
17	WOR-TV—24-hour-per-day superstation Channel 9 from New York carried by Eastern Microwave.
18	Reuters—Alpha/Numeric Monitor Data Display System during the business day.
	Galavision—Spanish International Network's premium Spanish-speaking movie service with sports and specials from Latin America and Spain.
19	C-SPAN—Daytime gavel-to-gavel coverage from the House of Representatives.
	Times-Mirror Pay TV Channel—Remainder of day.
20	Cinemax (HBO)—24 hours a day national programming. East Coast feed. Special programming differs from HBO movie service.
21	Home Theater Network (HTN)—Evening G and GP pay TV first-run movie service.
	The Preview Channel—PR/Advertising distribution service for new TV pilots, film clips, etc.
22	Home Box Office—National pay TV–West Coast feed.
	Beta—Women's Network (ABC Video Enterprises).
	Modern Satellite Network (MSN)—Five hours of daytime commercially sponsored shows.
23	Cinemax (HBO)—24 hours a day national pay programming. West Coast feed. (See transponder 20.)
24	Home Box Office—National pay TV service. East Coast feed. (See transponder 13.)

Note: Multiple audio networks and radio station feeds are carried as special audio subcarriers on SATCOM F-3's transponders as follows:

TRANSPONDER	AUDIO/DATA SERVICE	FORMAT
2	Satellite Radio Network	6.2-MHz subcarrier
3	WFMT–Chicago	5.8-MHz subcarrier
	Seeburg Music	7.6-MHz subcarrier
	Dow Jones Cable News	Vertical blanking interval (text)
	Electronic Program Guide	Vertical blanking interval (text)
6	North American Newstime	6.2-MHz subcarrier
	The Women's Channel	7.4-MHz subcarrier
	Moody Bible Network	5.6-MHz subcarrier
	Reuters News View	Vertical blanking interval (text)
	UPI Cablenews Wire	Vertical blanking interval (text)
	Consumer News	Vertical blanking interval (text)
	View Weather	Vertical blanking interval (graphics)
	SSS Cable Text	Vertical blanking interval (color text)

ATT/GTE COMSTAR D-2
Satellite Transponder Assignments
"Cable Net II"
Location: 95 Degrees West Longitude

TRANSPONDER	PROGRAM SERVICE
1 (1V)	Message (telephone) traffic.
2 (1H)	Occasional transmissions: teleconferencing, sporting events, news, and network television feeds.
3 (2V)	Message (telephone) traffic.
4 (2H)	Occasional transmissions: teleconferencing, sportings events, news, and network television feeds.
5 (3V)	Message (telephone) traffic.
6 (3H)	Bravo (performing and cultural arts).
7 (4V)	National Christian Network (religious programming).

	Escapade (R-rated movies). Family Radio Network (transmitted as an audio subcarrier simultaneously with video programming).
8 (4H)	Occasional transmissions.
9 (5V)	Occasional transmissions.
10 (5H)	Occasional transmissions: teleconferencing, sporting events, news, and network television feeds.
11 (6V)	Occasional transmissions: teleconferencing, sporting events, news, and network television feeds.
12 (6H)	Message (telephone) traffic.
13 (7V)	Cinemax—HBO's complimentary entertainment service. East Coast feed. Same as SATCOM F-3, transponder 20.
14 (7H)	Message (telephone) traffic.
15 (8V)	Occasional transmissions: teleconferencing, sporting events, news, and network television feeds.
16 (8H)	Message (telephone) traffic.
17 (9V)	Trinity Broadcasting Network (TBN) (24 hour-per-day religious programming).
18 (9H)	Home Box Office (HBO)—First-run movie service. East Coast feed.
19 (10V)	Occasional video transmissions.
20 (10H)	Message (telephone) traffic.
21 (11V)	Occasional transmissions: teleconferencing, sporting events, news, and network television feeds.
22 (11H)	Warner Amex feeds.
23 (12V)	Occasional transmissions: teleconferencing, sporting events, news, and network television feeds. United Video feeds.
24 (12H)	Message (telephone) traffic.

Note: Multiple audio networks are carried as special audio subcarriers on COMSTAR D-2's transponders as follows:

TRANSPONDER	AUDIO SERVICE	FORMAT
6	Stereo TV (Bravo) audio	6.8-MHz subcarrier
7	Family radio—east	5.8-MHz subcarrier
	Family radio—west	6.0-MHz subcarrier

Western Union WESTAR 3
Satellite Transponder Assignments
Location: 91 Degrees West Longitude

TRANSPONDER	PROGRAM SERVICE
1	Audio networks.
2	Hughes Television Network—Major sporting events to broadcast TV stations. BLAIRSAT—Commercial "spots" delivery to TV stations via satellite.
3	XEW-TV, Mexico City. Special tariff. Feeds to U.S. TV stations.
4	Audio networks.
5	Wold Communications—Occasional transmissions: sporting events, news, and network feeds. Private Screenings—Hard-core, sexploitation R-rated movies.
6	CBS Cable—Original programming of the arts.
7	Financial News Network—Live business and financial news from New York (daytime). Select TV—Pay per view, first-run movie programming (evening).
8	SIN (Spanish International Network).
9	SPN (Satellite Program Network)—Classic movies channel; other general-interest programs.
10	ABC Network Contract Channel—Live network feeds from ABC-TV and private news "prefeeds."
11	CNN (Cable News Network) Contract Channel—News feeds, occasional transmissions, sporting events.
12	Eternal Word Television Network—Family entertainment (evening); occasional transmissions (daytime).

Note: WOLD Communication also utilizes WESTAR 3 for its entertainment network service including the Merv Griffin Show. Associated Press operates audio networks to over 700 affiliated radio stations on WESTAR 3, covering news, sports, and feature programming.

APPENDIX 3

Western Union WESTAR 1
Satellite Transponder Assignments
Location: 99 Degrees West Longitude

TRANSPONDER	PROGRAM SERVICE
1	Occasional transmissions: sporting events, news, and network feeds.
2	Occasional transmissions: sporting events, news, and network feeds.
3	Occasional transmissions: sporting events, news, and network feeds.
4	Message (telephone) traffic.
5	Occasional transmissions: sporting events, news, and network feeds.
6	Occasional transmissions: sporting events, news, and network feeds.
7	Message (telephone) traffic.
8	PBS (Public Broadcasting) Schedule A programming and "prefeeds."
9	PBS (Public Broadcasting) Schedule B programming and "prefeeds."
10	Message (telephone) traffic.
11	PBS (Public Broadcasting) Schedule C programming and "prefeeds."
12	PBS (Public Broadcasting) Schedule D programming and "prefeeds." Central Educational Network—Occasional transmissions: sporting events, news, and network feeds.

Note: Neighborhood TV Corporation, a project of Allstate Venture Capital Corporation (Sears), plans to rebroadcast KUSK-TV, country and western television service from Prescott, Arizona, to 130 LPTV stations throughout the United States on WESTAR 1 when low-power TV is approved by the FCC.

National Public Radio feeds 12 independent audio channels on transponder 6.

Mutual Broadcasting Systems and radio and other high-power audio channels also operate on WESTAR 1, allowing small antennas (3–5 feet) to be used, and Commodity News Service (CNS) delivers market data news service to subscribers via 2-foot dishes.

Western Union WESTAR 2
Satellite Transponder Assignments
Location: 123.5 Degrees West Longitude

TRANSPONDER	PROGRAM SERVICE
2	Wold Communications—Occasional networking. Independent News Network (INN) nightly news feeds to affiliates. Occasional transmissions: sporting events, news, and network feeds.
4	Occasional transmissions: sporting events, news, and network feeds.
9	Occasional transmissions: sporting events, news, and network feeds.
	Other transponders: nonvideo message (telephone) traffic and backup transponder capability.
12	Occasional transmissions: sporting events, news, and network feeds.

RCA SATCOM F-2
Satellite Transponder Assignments
Location: 119 Degrees West Longitude

TRANSPONDER	PROGRAM SERVICE
1	ABC, CBS, NBC, PBS feeds to Anchorage, Alaska.
2	Occasional transmissions: sporting events, news, and network feeds.
5	Occasional transmissions: sporting events, news, and network feeds.
8	NBC Network Contract Channel—Live and taped network feeds, news "prefeeds." New York City–Hollywood link. Now subleased to Wold Communications.
9	Armed Forces Satellite Network—Assorted independent and network programming feed to Armed Forces TV stations in Alaska and Cuba.
13	NASA Contract Channel—Video coordination between Houston, Cape Canaveral, Washington. Space Shuttle video.
16	Audio networks (see note).
23	Alaska Satellite Television Project—Assorted net-

	work and independent programming feed to TV stations in Alaska.*
	Other transponders: nonvideo message (telephone) traffic and backup transponder capability. Up to 14 transponders are used, on average, by RCA Alascom telephone and video feeds to Alaska from the Continental United States.

Note: NBC, CBS, ABC, National Black Radio, RKO Radio, AP radio and Teletype news, UPI radio and Teletype news, Voice of America's four radio networks, all are carried on audio subcarriers on transponder 16.

ATT/GTE COMSTAR D-3
Satellite Transponder Assignments
Location: 87 Degrees West Longitude

TRANSPONDER	PROGRAM SERVICE
20	Occasional transmissions: sporting events, news, and network feeds.
	Other transponders: NBC, ABC, CBS, NET, Wold Communications network video feeds. This satellite will eventually expand to become ATT's "Network Bird," carrying 24 video channels, as ATT swings over the network television feeds from terrestrial microwave circuits to satellite systems, beginning in early 1983.

TeleSat Canada ANIK B-1/B-2
Satellite Transponder Assignments
Location: 109 Degrees West Longitude

TRANSPONDER	PROGRAM SERVICE
4	Occasional transmissions: sporting events, news, and CBC/CTV network feeds.
6	CBC North—Assorted CBC network programming from Toronto and Montreal.
7	Occasional transmissions: sporting events, news, and CBC/CTV network feeds.

*Audio often found on transponder 3 subcarrier.

8	CBC (French Channel)—French-language CBC programming from Montreal.
9	CBC (English Channel 1)—English CBC programming from Toronto.
10	CBC (English Channel 2)—English CBC programming from Toronto.
	Other transponders: nonvideo message (telephone) traffic and backup protection for video feeds.
12	Same as transponder 10.

TeleSat Canada ANIK A-2/A-3
Satellite Transponder Assignments
Location: 114 Degrees West Longitude

TRANSPONDER	PROGRAM SERVICE
1	Occasional transmissions: sporting events, news, and CBC/CTV and BCTV (British Columbia Television) network feeds.
3	Daily live coverage of the Canadian House of Commons from Ottawa (with French translation).
4	CHCH-TV Hamilton, Ontario Television.
7	Daily live coverage of the Canadian House of Commons from Ottawa (standard English).
8	CHCT-TV Television.
10	CITV-TV Edmonton Television.
12	CTV North—Assorted CTV network programming from Toronto.
	Other transponders: nonvideo message (telephone) traffic and backup protection for video feeds.
	Four additional transponders provide regional TV superstation coverage to CATV systems nationwide from Toronto, Vancouver, Hamilton, Ontario, and Edmonton, Alberta. Service is jointly offered by Canadian Satellite Communications, Inc. (CANCOM).
	A consortium of 125 Canadian CATV operators, called Pay Television Network, Ltd. (PTNL), owns transponders on ANIK for development of a national first-run-movie pay TV service.

Part II: Listing of Major Programmers

Alpha Repertoire Television System (ARTS)
ABC Video Enterprises/Warner Amex Satellite Entertainment
1211 Avenue of the Americas
New York, NY 10036
(212) 944-4250

SATCOM III, transponder 1
Performing arts and visual arts programming commercially sponsored.

American Educational Television Network (AETN)
2172 DuPont Drive
Irvine, CA 92715
(714) 955-3800

SATCOM III, transponder 6
Educational programming including continuing education seminars sponsored by professional associations.

Appalachian Community Service Network (ACSN)
1200 New Hampshire Avenue NW, Suite 240
Washington, DC 20036
(202) 331-8100

SATCOM III, transponder 16
Educational programming for adult and post-secondary education, including college-level courses, for credit and continuing education seminars.

BETA
ABC Video Enterprises/Hearst Company
959 Eighth Avenue
New York, NY 10019
(212) 262-3315

SATCOM III, transponder 22
Commercially sponsored women's interest and service-oriented features on careers, fashion, and finance.

Black Entertainment Television (BET)
Suite 300, Prospect Place
3222 N Street, NW
Washington, DC 20007
(202) 337-5260

SATCOM III, transponder 9
Variety of programming featuring black performers in dominant or leading roles, including sports, music specials, films. Some commercials.

Blairsat
John Blair Company
717 Fifth Avenue
New York, NY 10022
(212) 752-0400

WESTAR III, transponder 2
Satellite distribution of TV commercials to major television stations.

Bonneville Satellite Corporation
130 Social Hall Avenue
Salt Lake City, UT 84111
(801) 237-2597

WESTAR I, transponder 5
Syndication and ad-hoc networking for independent TV stations and affiliates.

Cable News Network (CNN)
1050 Techwood Drive
Atlanta, GA 30318
(404) 898-8500

SATCOM III, transponder 14; CNN-2 SATCOM III, transponder 15
24-hour-per-day news network created by Ted Turner, with news bureaus in Washington, D.C., New York City, Los Angeles, and a dozen other cities worldwide.

Canadian Broadcasting Corporation (CBC)
354 Jarvis
Toronto, Ont., Canada
(416) 925-3311

ANIK B-2, ANIK A-3
Canadian commercial government-owned television network operates both English- and French-language services. Carries popular U.S. TV shows.

Canadian Television Corporation (CTV)
42 Charles East
Toronto, Ont., Canada
(416) 928-6000

ANIK B-2, ANIK A-3
Canadian commercial independent television network. Carries popular U.S. TV shows.

CBS Cable
1211 Avenue of the Americas
New York, NY 10036
(212) 975-3541

WESTAR III, transponder 6
Fine arts programming from New York. Original "up market" plays, cultural events, musical presentations. A subsidiary of CBS-TV. Commercially sponsored.

Christian Broadcasting Network (now known as CBN Satellite Network)
Continental Broadcasting Network, Inc.
Pembroke Four
Virginia Beach, VA 23463
(804) 424-7777

SATCOM III, transponder 8
Twenty-four hours per day of Christian programming including the *700 Club* and other enormously popular entertainment shows and musical events. CBN is the most-watched CATV satellite network.

Christian Media Network (CMN)
8120 Penn Avenue, South
Bloomington, MN 55431
(612) 884-4540

SATCOM III, transponder 16
Commercially sponsored Christian programming of films, gospel sings, interview shows.

Commodity News Service (CNS)
2100 West 89th Street
Kansas City, KS 66206
(913) 642-7373

WESTAR I
Multi-teletypewriter news network carried via audio subcarrier to newspapers, brokerage houses, and radio stations nationwide.

C-SPAN (Cable Satellite Public Affairs Network)
400 North Capital Street, Suite 155
Washington, DC 20001
(202) 737-3220

SATCOM III, transponder 19
Daily live coverage of the U.S. House of Representatives. Special events coverage includes National Press Club speeches.

Dow Jones Cable News
P.O. Box 300
Princeton, NJ 08540
(609) 452-2000

SATCOM III, transponder 3 (sent as a subcarrier of the WGN-TV signal, which is the main user of transponder 3 on SATCOM III)
News wire service. Consumer and investor audience programming emphasizing developments in energy, inflation, taxes, interest rates, housing, and securities markets.

ESPN, Entertainment and Sports Programming Network, Inc.
ESPN Plaza
Bristol, CN 06010
(203) 584-8477

SATCOM III, transponder 7, and COMSTAR D-2, transponder 3H (transponder 6)

Sports programming 24 hours per day including over 1200 annual hockey, basketball, NCAA collegiate games, etc.

Eternal Word Television Network
5817 Old Leeds Road
Birmingham, AL 35210
(205) 956-9537

WESTAR III, transponder 12 (transponder 23)
Family and religious-oriented entertainment, music, and comedy.

Financial News Network (FNN)
2525 Ocean Park
Santa Monica, CA 90405
(213) 450-2412

WESTAR III, transponder 7 (transponder 13)
Live, daily business news and financial/stock reports.

Galavision (Division of SIN National Spanish Network)
250 Park Avenue
New York, NY 10017
(212) 953-7550

SATCOM III, transponder 18 evenings only (transponder used by Reuters Monitor news service during the day)
Entertainment network with Spanish-language programming, with top movies, specials, and shows from Mexico, South America, and Spain.

Home Box Office (HBO)
Time and Life Building, Rockefeller Center
New York, NY 10020
(212) 484-1241

SATCOM III, transponders 13, 20, 23, and 24 and COMSTAR D-2, transponder 9H (transponder 18)
First-run movies, special entertainment events (Vegas, Hollywood musicals, shows).
HBO's Cinemax series on transponders 20 and 23 complement regular HBO movies on other transponders.

Home Theater Network (HTN)
465 Congress Street
Portland, ME 04101
(207) 774-0300

SATCOM III, transponder 21
First-run movies (G/PG only) entertainment programming for the family.

Hughes Television Network
4 Pennsylvania Plaza
New York, NY 10001
(212) 563-8900

WESTAR III, transponder 2 (transponder 3)
National specialty sports television network bringing major sporting events, such as golf tournaments, etc., to broadcast television stations.

Independent News Network (INN)
Telemine
888 Seventh Avenue
New York, NY 10106
(212) 489-7231

WESTAR II, transponder 2 (transponder 3)
Live news feeds to leading independent TV stations nationwide from independent New York City anchor station.

Independent TV News Association
220 East 42nd Street
New York, NY 10017
(212) 490-1212

WESTAR III, transponder 2 (transponder 3)
Live nightly news feeds to over 25 independent TV stations at 6:44 p.m. (EST) via Wold Communications transponder.

Modern Satellite Network (Division of Modern Talking Picture Service, Inc.) (MSN)
45 Rockefeller Plaza
New York, NY 10111
(212) 765-3100

SATCOM III, transponder 22
Daytime programming featuring entertainment and information for the general consumer. Sponsored by paid programs and commercials.

Modern Television (MTV), "The Music Channel"
Warner Amex Satellite Entertainment
1211 Avenue of the Americas
New York, NY 10036
(212) 944-4250

SATCOM III, transponder 11
24 hours per day of rock music, concerts, music-oriented films.

The Movie Channel
Warner Amex Corporation
75 Rockefeller Plaza
New York, NY 10019
(212) 484-6826

SATCOM III, transponder 5
An entertainment network featuring specials and G-, PG-, and R-rated films, 24 hours per day.

National Christian Network (NCN)
1150 West King Street
Cocoa, FL 32922
(305) 632-1000

COMSTAR D-2, transponder 4V (transponder 7)
Religious programming similar to CBN but greater nondenominational selections.

National Jewish Television (NJT)
Jason Films
2621 Palisades Avenue
Riverdale, NY 10463
(212) 549-4160

SATCOM F-3, transponder 16
Religious programming for the Jewish faith.

Nickelodeon
Warner Amex Satellite Entertainment Company
1211 Avenue of the Americas, 15th floor
New York, NY 10036
(212) 944-4255

SATCOM III, transponder 1
Programming for children has won several Emmys. Includes old-time

adventure serials, comic strips, new and original stories. In addition, a new service for adults and young people called ARTS (Alpha Repertory Television Service) includes performing visual and lively fine arts.

People That Love Television Network (PTL)
Charlotte, NC 28279
(704) 542-6000

SATCOM III, transponder 2
Twenty-four hours per day of Christian programming similar to CBN with talk, variety shows, and children's programming.

The Preview Channel
73 Market Street
Venice, CA 90291
(213) 392-9575

SATCOM III, transponder 21
Advertising/PR channel promotes TV pilots, previews, film clips, and movie trailers for a fee paid by producer.

Public Broadcasting Service (PBS)
475 L'Enfant Plaza
Washington, DC 20024
(202) 488-5000

WESTAR I, transponders 8, 9, 10 (occasionally 11)

Rainbow Programming Services
1355 South Colorado Boulevard, Suite 100
Denver, CO 80222
(303) 753-1822

COMSTAR D-2, transponder 4V (transponder 7)
Escapade service, an adult-oriented service with specials and R-rated films as late-night offering. Has joined forces with Playboy to provide the new Playboy channel.
COMSTAR D-2, transponder 3H (transponder 6)
Bravo service is a performing and cultural arts programming service for "up market" audience from Carnegie Hall and worldwide locations. Includes opera, jazz, modern dance, ballet.

RCTV (Rockefeller Center Television)
30 Rockefeller Plaza
New York, NY 10012
(212) 757-2007

Satellite pending, COMSTAR D-2
Mystery, adventure, drama, and children's and performing arts entertainment. Commercially sponsored. The Entertainment Channel features children's programs, sophisticated television series from action to comedy, Broadway shows, and international films.

Reuters, Ltd.
1700 Broadway
New York, NY 10036
(212) 730-2740

SATCOM III, transponder 18
News programming during the business day fed to special Reuters Monitor high-speed data terminals. Not a video television picture. Requires special decoder system.
SATCOM III, transponder 6
News-view, financial, sports, and general alphanumeric news wire service carried on subcarrier.

Satellite Programming Network (SPN)
P.O. Box 45684
Tulsa, OK 74145
(918) 481-0881

WESTAR III, transponder 9 (transponder 17)
Carries 24-hour-per-day talk shows, movies, corporate PR events, syndicated programs.

Satellite Syndicated Systems (SSS)
P.O. Box 45684
Tulsa, OK 74145
(918) 481-0881

SATCOM III, transponder 6 carries WTBS, transponders 14 and 15 carry CNN, and WESTAR III, transponder 9 carries SPN.
Satellite Syndicated Systems is a common carrier that retransmits several signals nationwide. In addition, SSS has a wholly owned subsidiary called Southern Satellite Systems, Inc., which originates its own programming. SSS carries entertainment, news, super TV station WTBS, entertainment on SPN, news on CNN, Cable Text, and The Women's Channel. (See each service for details.)

Satori Productions
250 West 57th Street
New York, NY 10019
(212) 581-8450

WESTAR III, transponder 5 (transponder 9)
Private Screenings service. Disco, late-night adult entertainment, R-rated (not X) movies, six hours per week.

Select-TV
4755 Alla Road
Marina Del Rey, CA 90291
(213) 827-4400

WESTAR III, transponder 7 (transponder 13)
STV pay TV first-run movie service (evenings only).

Showtime Entertainment, Inc.
1211 Avenue of the Americas
New York, NY 10036
(212) 575-5175

SATCOM III, transponders 10 and 12
Entertainment programming, including specials, Broadway and off-Broadway plays, original pay TV programming, and first-run films.

Spanish International Network (SIN)
250 Park Avenue
New York, NY 10017
(212) 953-7500

WESTAR III, transponder 8 (transponder 15)
Spanish-language entertainment programming including variety shows, drama, nightly news, sports. Live feeds from Puerto Rico, Spain, Mexico, South America.

TeleFrance USA
1960 Broadway
New York, NY 10023
(212) 877-8900

WESTAR III, transponder 9 (transponder 17)
French-oriented entertainment, news, and movies, broadcast three hours per day via SPN's transponder. Provided as part of Satellite Programming Network (SPN) service.

Times-Mirror
2951 28th Street, Suite 2000
Santa Monica, CA 90405
(213) 450-6488

SATCOM F-3, transponder 4
Spotlight pay TV movie and entertainment specials. Service primarily available to Times–Mirror CATV companies.

Trinity Broadcasting Network (TBN)
P.O. Box A
Santa Ana, CA 92711
(714) 832-2950

COMSTAR D-2, transponder 9V (transponder 17)
Religious programming 24 hours a day.

UPI Newstime
220 East 42nd Street
New York, NY 10017
(212) 682-0400

SATCOM III, transponder 6
24-hour all-news service that uses data subcarrier of WTBS-TV signal carried on transponder 6. Requires special modem decoding equipment to pick up the nonvideo digital signal.

USA Network
208 Harristown Road
Glenrock, NJ 07452
(201) 445-8550

SATCOM III, transponder 9
Programming features Thursday night baseball, Madison Square Gardens sports, NHL, NBA, and NASL games. Also, Calliope children's films and The English Channel cultural programming are carried.

UTV Cable Network
22-08 Route, P.O. Box 487
Fairlawn, NJ 07410
(201) 891-4242 or 794-3660

SATCOM IV, transponder 23
Programming for general audience, commercially sponsored.

Vidsat, Group W Productions
Westinghouse Broadcasting
310 Parkway View Drive
Pittsburgh, PA 15205
(412) 928-4700

WESTAR I
Satellite program distribution to Westinghouse affiliates for Group W Productions.

WGN-TV Superstation Channel 9
United Video
5200 South Harvard, Suite 215
Tulsa, OK 74135
(800) 331-4806

SATCOM III, transponder 3
United Video is the common carrier that retransmits superstation WGN-TV Channel 9 from Chicago. WGN provides 24-hour-per-day classic movies, top regional and national sports, and syndicated TV shows including Phil Donahue live. Also carries Chicago's WFMT classical music and fine arts programming on a subcarrier of WGN, as well as Seeburg Corporation's Lifestyle audio music programming on a second subcarrier.

Window on Wall Street (WWS)
6515 Sunset Boulevard
Hollywood, CA 90028
(213) 469-2790

SATCOM III, transponder 16
Friday one-hour analysis of week's activities on Wall Street.

Wold Communications
11661 San Vicente Boulevard
Los Angeles, CA 90049
(213) 820-2668

WESTAR III, transponder 5 (transponder 9) and others
Entertainment network featuring feeds to independent television stations, spotlighting the *Merv Griffin Show,* among others.

The Woman's Channel
Southern Satellite System
9252 South Harvard
Tulsa, OK 74136
(918) 481-0881

SATCOM III, transponder 6
Color slow-scan general-interest women's programming carried as subcarrier of WTBS.

WOR-TV Superstation Channel 9
Eastern Microwave, Inc.
P.O. Box 4872
Syracuse, NY 13221
(315) 455-5955

SATCOM III, transponder 17
New York's independent Channel 9 superstation featuring a variety of entertainment, news, sports, and movies, 24-hour programming, is carried by Eastern Microwave, Inc., another satellite "resale" common carrier similar to Satellite Syndicated Systems (which carries WTBS) and United Video (which carries WGN).

WTBS Superstation Channel 17
1050 Techwood Drive NW
Atlanta, GA 30318
(404) 898-8500

SATCOM III, transponder 6
A fabulous film library coupled with sports coverage of several Turner-owned teams including the Atlanta Braves; 24 hours per day. WTBS Channel 17 in Atlanta is retransmitted via satellite by Syndicated Satellite Systems, a "resale" common carrier formed for this purpose.

Note: WESTAR and COMSTAR satellite listings include corresponding CATV industry-standard transponder assignments as noted in parentheses after transponder number.

Appendix 4

Bibliography, Contacts, Further Information

Associations and Contacts

American Radio Relay League, Inc. (ARRL)
225 Main Street
Newington, CT 06111
(203) 666-1541

Cable Television Administration and Marketing Society (CTAM)
1725 K Street, NW, Suite 1103
Washington, DC 20026
(202) 296-4219

Cable Television Information Center
2100 M Street, NW
Washington, DC 20037
(202) 872-8888

California Community Television Association (CCTA)
3636 Castro Valley Boulevard, No. 10
Castro Valley, CA 94546
(415) 881-0211

Community Antenna Television Association (CATA)
4209 N.W. 23rd, Suite 106
Oklahoma City, OK 73107
(405) 947-7664

Copyright Royalty Tribunal
Washington, DC 20003
(202) 653-5775

Electronic Industries Association (EIA)
2001 I Street, NW
Washington, DC 20006
(202) 457-4900

Cable Television Bureau or
Public Information Office
Federal Communications Commission
1919 M Street, NW
Washington, DC 20554
(202) 632-9703

National Association of Broadcasters (NAB)
1771 N Street, NW
Washington, DC 20006
(202) 293-3500

National Association of Television Program Executives (NATPE)
Box 5272
Lancaster, PA 17601
(717) 626-4424

National Cable Television Association (NCTA)
918 16th Street, NW
Washington, DC 20006
(202) 457-6700

National Federation of Local Cable Programmers
c/o Miami Valley Cable Council
3200 Far Hills Avenue, Room 109
Kettering, OH 45429
(513) 298-7890

Satellite Television Technology
P.O. Box G
Arcadia, OK 73007
(405) 396-2574

Society of Private and Commercial Earth Station Users (SPACE)
1920 N Street, NW, Suite 510
Washington, DC 20036
(202) 887-0605

Society of Cable Television Engineers, Inc.
1900 L Street, NW
Washington, DC 20036
(202) 293-7841

The (Home) Satellite Center
680 Beach Street, Suite 428
San Francisco, CA 94109
(415) 673-7000

The Public Service Satellite Consortium
1660 L Street, NW
Washington, DC 20036
(202) 331-1154

APPENDIX 4

Magazines and Periodicals

Consumer and Electronics Hobbyist-Oriented Magazines

1. *Coop's Satellite Digest*
   Satellite Television Technology
   P.O. Box G
   Arcadia, OK 73007
   (405) 396-2574

   Published monthly, 1-year subscription = $50. A "must buy." Highly recommended. Specifically written for the home satellite TV enthusiast.

2. *Home Video*
   P.O. Box 2651
   Boulder, CO 80322

   Published monthly, 1-year subscription = $13.50.

3. *Video Magazine*
   Reese Publishing Company, Inc.
   235 Park Avenue South
   New York, NY 10003

   Published monthly, 1-year subscription = $18.

4. *Video Action*
   Video Action Incorporated
   21 West Elm Street
   Chicago, IL 60610

   Published monthly, 1-year subscription = $17.

5. *Video Buyers Guide*
   Hampton International Communications Incorporated
   P.O. Box 1592
   149 Hampton Road
   Southhampton, NY 11968
   (516) 283-3260

   Published annually, cost per issue = $2.95.

6. *Videography Magazine*
   United Business Publications Incorporated
   475 Park Avenue South
   New York, NY 10016

   Published monthly, 1-year subscription = $12.

7. *Videoplay Magazine*
   C.S. Tepfer Publishing Company, Inc.
   51 Sugarhollow Road
   Danbury, CT 06810
   (203) 743-2120

   Published bimonthly, 1-year subscription = $6.

8. *Video Review*
   CES Publishing
   325 East 75th Street
   New York, NY 10021
   (212) 794-0500

   Published monthly, 1-year subscription = $14.95.

9. *Popular Electronics*
   Ziff-Davis Publishing Company
   1 Park Avenue
   New York, NY 10016

   Published monthly, 1-year subscription = $14.

10. *QST*
    American Radio Relay League, Inc.
    225 Main Street
    Newington, CT 00611

    Published monthly, 1-year subscription (nonmembers) = $12.

11. *Radio-Electronics*
    Gernsback Publications, Inc.
    200 Park Avenue South
    New York, NY 10003
    (212) 777-6400

    Published monthly, 1-year subscription = $13.

12. *Popular Science*
    380 Madison Avenue
    New York, NY 10017
    (212) 687-3000

    Published monthly, 1-year subscription = $13.94.

13. *Popular Mechanics*
    224 West 57th Street
    New York, NY 10019
    (212) 262-5700

    Published monthly, 1-year subscription = $9.97.

Satellite Magazines and Periodicals

1. *Coop's Satellite Digest*
   Satellite Television Technology
   P.O. Box G
   Arcadia, OK 73007
   (405) 396-2574

2. *SAT Guide*
   Commtek Publishing Company
   P.O. Box 1700
   Haley, ID 83333
   (208) 788-4936

3. *Satellite News Newsletter*
   Phillips Publishing, Inc.
   7315 Wisconsin Avenue
   Washington, DC 20014

4. *Satellite Communications Magazine*
   3900 South Wadsworth
   Denver, CO 80235
   (303) 988-4670

5. *Satellite Technology Reports*
   The Satellite Center
   680 Beach Street, Suite 428
   San Francisco, CA 94109
   (415) 673-7000

Cable-TV-Oriented Magazines

1. *C-ED Magazine*
   Titsch Publishing, Inc.
   2500 Curtis Street
   P.O. Box 5400-TA
   Denver, CO 80217

   Published monthly, 1-year subscription = $20. The cable TV engineering management magazine.

2. *Cable Age*
   Television Editorial Corporation
   1270 Avenue of the Americas
   New York, NY 10020

   Published monthly, 1-year subscription = $25. Programming and financing CATV system coverage.

3. *CableVision*
   Titsch Publishing, Inc.
   2500 Curtis Street
   P.O. Box 5400-TA
   Denver, CO 80205
   (303) 573-1433

   Published weekly, except the week of December 28, 1981. 1-year subscription = $54. The industry "Bible."

4. *Community Antenna Television Journal*
   TPI Inc.
   4209 Northwest 23rd, Suite 106
   Oklahoma City, OK 73107

   Published monthly, 1-year subscription = $14. An applied "how-to" journal chock full of useful information for the mini-CATV operator.

5. *Home Video Report Newsletter*
   Knowledge Industry Publications, Inc.
   2 Corporate Park Drive
   White Plains, NY 10604
   (914) 694-8686

   Published weekly, 1-year subscription = $175.

6. *Multichannel News*
   Fairchild Publications
   1762 Emerson Street
   Denver, CO 80218
   (303) 832-4141

   Published weekly, newspaper format. 1-year subscription = $18.50.

7. *SAT Guide*
   Commtek Publishing Company
   P.O. Box 1700
   Haley, ID 83333
   (208) 788-4936

   Published monthly, 1-year subscription = $36. A "must-subscribe," lists all major programming on all satellites each month. Sort of a giant "TV Guide of the Sky."

8. *TVC*
   Cardiff Publishing Company
   (303) 988-4670

   Cardiff Publishing Circulation Service Center
   P.O. Box 1077
   Skokie, IL 60077

   Published monthly, 1-year subscription = $18.

9. *VIEW*
   Macro Communications Corporation
   150 East 58th Street
   New York, NY 10022
   (212) 826-4360

   Published monthly, 1-year subscription = $36. The magazine of cable TV programming.

Broadcasting-Oriented Magazines

1. *Broadcast Communications*
   Globecom Publishing Ltd.
   4121 West 83rd Street, Suite 132
   Prairie Village, KS 66208

   Published monthly, 1-year subscription = $36.

2. *Broadcast Engineering*
   Intertec Publishing Corporation
   9221 Quivira Road
   P.O. Box 12901
   Overland Park, KS 66212

   Published monthly, $3 per issue (free to industry subscribers).

3. *Broadcast Management/Engineering (BME)*
   Broadband Information Services Inc.
   295 Madison Avenue
   New York, NY 10017
   (212) 685-5320

   Published monthly, 2-year subscription = $24 (free to industry subscribers).

4. *Broadcasting*
   Broadcasting Publications, Inc.
   1735 DeSales Street, NW
   Washington, DC 20036

   Published 51 Mondays per year. Combined issue at year-end. 1-year subscription = $50.

5. *Broadcasting Systems and Operation*
   BSO Publications Ltd.
   P.O. Box 141 High Street Wivenhoe
   Colchester, CO79EA, U.K.

   Published monthly, 1-year subscription = 15 British pounds per year, by surface mail, 25 British pounds by airmail.

6. *Television/Radio Age*
   Television Editorial Corporation
   1270 Avenue of the Americas
   New York, NY 10020
   (212) 757-8400

   Published every other week, 1-year subscription = $65.

7. *Video News Newsletter*
   Phillips Publishing Inc.
   7315 Wisconsin Avenue
   Washington, DC 20014
   (301) 986-0666

   Published weekly, 1-year subscription = $167.

8. *Video Systems*
   Intertek Publishing Corporation
   9221 Quivira Road
   P.O. Box 12901
   Overland Park, KS 66212

   Published monthly, 1-year subscription = $24.

Video Programming and Production-Oriented Magazines

1. *Educational and Industrial Television*
   C.S. Tepfer Publishing Company, Inc.
   51 Sugarhollow Road
   Danbury, CT 06810
   (203) 743-2120

   Published monthly, 1-year subscription = $15.

2. *Television International*
   Television International Ltd.
   P.O. Box 2430
   Hollywood, CA 90028
   (213) 876-2219

   Published bimonthly, 1-year subscription = $20.

3. *Video Systems*
   Intertek Publishing Corporation
   9221 Quivira Road
   P.O. Box 12901
   Overland Park, KS 66212

   Published monthly, 1-year subscription = $24.

4. *View*
   The Magazine of Cable TV Programming
   Macro Communications Corporation
   150 East 58th Street
   New York, NY 10022
   (212) 826-4360

   Published monthly, 1-year subscription = $36.

Commercial and Dealer-Oriented Magazines

1. *Audio Video International*
   Dempa Publications Inc.
   380 Madison Avenue
   New York, NY 10017

   Published monthly, 1-year subscription = $20.

2. *Video Retailer*
   The National Video Clearing House, Inc.
   100 LaFayette Drive
   Syosset, NY 11791
   (516) 364-3686

   Published monthly, free.

3. *Video Retailing*
   Audio Mirror Publications, Inc.
   60 East 42nd Street
   New York, NY 10017
   (212) 682-7320

   Published monthly, 1-year subscription = $15.

CATV-Listings Magazines

1. *Cable Monthly*
   Cable Communications Media
   203 East Broad Street
   Bethlehem, PA 18018
   (215) 865-6600

   One-year subscription: $12.

2. *Cable Plus*
   Satellite Publications, Inc.
   4649 Westgrove
   Dallas, TX 75248
   (214) 931-2086

   One-year subscription: $18.

3. *On Cable*
   On Cable Publications
   P.O. Box 359
   Norwalk, CT 06856
   (203) 866-6256

   One-year subscription: $12.

4. *SAT Guide*
   Commtek Publishing Company
   P.O. Box 1700
   Hailey, ID 83333
   (208) 788-4936

   One-year subscription: $36.

   Complete listings for all services in RCA F1 satellite as well as several services on WESTAR and COMSTAR satellites. Recommended.

5. *TV Guide*
   Attn: Mr. J. Haynes
   Radnor, PA 19088
   (800) 523-7933

   One-year subscription: $20.
   Request local edition that carries satellite programming. If none is available, request special Texas satellite edition.

6. *VIP Cable Guide*
   Video International Productions, Inc.
   5757 Alpha Road, Suite 816
   Dallas, TX 75240
   (214) 934-2231

Books on Satellites and Home Satellite TV

AIAA Communications Satellite Systems Conference, Washington, D.C., 1966 (New York: Academic Press, 1966).

AIAA Communications Satellite Systems Conference, Los Angeles, Calif., 1970 (Cambridge, Mass.: MIT Press, 1971).

AIAA Communications Satellite Systems Conference (4th), Washington, D.C., 1972 (Cambridge, Mass.: MIT Press, 1974).

AIAA Communications Satellite Systems Conference (5th), Los Angeles, Calif., 1974 (New York: American Institute of Aeronautics and Astronautics, 1976).

AIAA Communications Satellite Systems Conference (6th), Los Angeles, Calif., 1970 (New York: American Institute of Aeronautics and Astronautics, 1977).

Bakan, Joseph D., and Chandler, David L. *The Independent Producer's Handbook of Satellite Communications* (New York: Shared Communications Systems, 1980).

Chander, R., and Karnik, K. *Planning for Satellite Broadcasting: The Indian Instructional Television Experiment* (New York: UNESCO, 1977).

Communications in the Space Age: The Use of Satellites by the Mass Media (Paris: UNESCO Press, 1968).

Cooper, Bob, Jr. *Coop's Satellite Business Opportunity Journal* (Oklahoma: Satellites in Television Technology, 1980).

Cooper, Bob, Jr. *Coop's Satellite Operations Manual* (Oklahoma: Satellite Television Technology, 1980).

Cooper, Bob, Jr. *Home Satellite TV Reception Handbook* (Oklahoma: Satellite Television Technology, 1980).

Ethier, Nelson *The Nelson Parabolic TVRO Antenna Manual* (Oklahoma: Satellite Television Technology, 1980).

Gibson, Steve *The Gibson Satellite Navigator Manual* (Oklahoma: Satellite Television Technology, 1980).

Goddard Memorial Symposium (14th) *Satellite Communications in the Next Decade: Proceedings* (American Astronaut, 1977).

Gould, R. G., and Lum, L. F. *Communications Satellite Systems: An Overview of the Technology* (Inst. Electrical, 1976).

Howard, M. Taylor *The New Howard Terminal Manual* (Oklahoma: Satellite Television Technology, 1981).

Kinsley, Michael E. *Outer Space and Inner Sanctums: Government, Business, and Satellite Communications* (New York: Wiley, 1976).

Lukashok, Alvin *Communication Satellites: How They Work* (New York: Putnam, 1967).

Mueller, George Edwin, and Spangler, E. R. *Communication Satellites* (New York: Wiley, 1964).

Martin, James *Communication Satellite Systems* (New Jersey: Prentice-Hall, 1978).

Newman, Joseph (editor) *Wiring the World: The Explosion in Communications* (Washington: books by *U.S. News and World Report,* 1971).

Paul, Gunter *The Satellite Spin-Off: The Achievements of Space Flight* (New York: Robert B. Luce, 1975).

Policies for Regulation of Direct Broadcast Satellites (Washington, D.C.: FCC Staff Reports, 1980).

Queeney, K. M. *Direct Broadcast Satellites and the United Nations* (London: Sijthoff and Noordbaff, 1978).

Reed, Stephen *Satellite Television Handbook and Buyers Guide* (Maitland, Fla.: Reed Publications, 1980).

Salvati, M. J. *TV Antennas and Signal Distribution Systems* (Indianapolis: Howard W. Sams and Co., 1979).

Satellite Teleconferencing: An Annotated Bibliography (Wisconsin: University of Wisconsin Press, 1972).

Signitzer, Benno *Regulation of Direct Broadcasting from Satellites: The U.N. Involvement* (New York: Praeger, 1976).

Snow, Marcellus *International Commercial Satellite Communications: Economic and Political Issues of the First Decade of Intelsat* (New York: Praeger, 1976).

Stockholm International Peace Research Institute *Communication Satellites* (Stockholm: Almqvist and Wikwell, 1969).

Van Trees, Harry *Communications* (New York: Wiley-Interscience, 1980).

Washburn, Clyde *Washburn High Performance Receiver Manual* (Oklahoma: Satellite Television Technology, 1980).

Articles from Periodicals

Hobbyist and Technical Periodicals

"A Satellite-Dish Odyssey," *Video Review,* Oct. 1981, p. 68.

"All Feedhorns Are Not Alike," *Videoplay Magazine,* Oct./Nov. 1981, p. 26.

"Backyard Satellite TV Receiver," *Radio-Electronics,* March 1980, p. 42, and April 1980, p. 47.

"Backyard Satellite TV Reception: Fact or Fantasy," *Radio-Electronics,* June 1980, p. 68.

"Building a Home Satellite Receiver That Really Works," *Video Review,* Aug. 1980, p. 40.

"Buyer Beware," *Videoplay Magazine,* April/May 1981, p. 64.

"Compact Backyard Antennas Deliver Programs to Wider Audience," *Parade,* Sept. 28, 1980.

"Directory of Home Satellite Equipment," *Videoplay Magazine,* April/May 1981, p. 72.

"DXing Those TV Satellites," *Popular Electronics,* Oct. 1981, p. 49.

"Earth Station Can Sit in the Window," *Electronics,* July 31, 1980, p. 71.

"Earth Station Technology Developments Spark Interest in Satellite Terminals," *EDN,* Sept. 5, 1980, p. 70.

"Europeans Building TV Satellite System," *Aviation Week,* April 21, 1980, p. 167.

"French, Germans to Begin TV Satellite Effort," *Aviation Week,* Dec. 10, 1979, p. 69.

"Germans, French Plan TV Satellite," *Electronics,* Sept. 27, 1979, p. 46.

"Get the Signal: Shopping for an Earth Station," *Video,* Aug. 1981, p. 50.

"Guide to Satellite Dishes," *Video Review,* Jan. 1982, p. 58.

"Home Earth Station Tunes Into Satellites," *Electronics,* Sept. 27, 1979, p. 46.

"Home Earth Stations for Satellite Transmissions," *Popular Electronics,* July 1981, p. 24

"Home Reception Using Backyard Satellite TV Receivers," *Radio-Electronics,* Jan. 1980, p. 55.

"Home Reception Via Satellite," *Radio-Electronics,* Aug. 1979, p. 47; Sept. 1979, p. 47; Oct. 1979, p. 81.

"Home Viewing Via the Satellite," *Engineering,* July 17, 1980, p. 35.

"Living with Satellite TV," *Video,* Jan. 1982, p. 64.

"Low Cost Backyard Satellite TV Earth Station," *Radio-Electronics,* Feb. 1980, p. 47.

"Nordic TV Satellite System Expected," *Aviation World,* June 11, 1979, p. 195.

"North America Plays King-Pin in Satellite TV," *New Scientist,* Nov. 8, 1979, p. 442.

"Precise Real-Time Dissemination Over the TV Broadcasting Satellite," *Radio Science,* July 1979, p. 685.

"Satellite Dish Live: Ready for Prime Time," *Video Review,* March 1981, p. 50.

"Satellite Earth Station Now in Operation," *Engineer,* April 19, 1979, p. 11.

"Satellite Receivers," *Videoplay Magazine,* Oct. 1980, p. 77.

"Satellite TV," *Videoplay Magazine,* April 1980, p. 32.

"Satellite TV Antenna: Part I," *Radio-Electronics,* August 1981, p. 45.

"Satellite TV Antenna: Part II," *Radio-Electronics,* September 1981, p. 59.

"Satellite TV Antenna: Part III," *Radio-Electronics,* October 1981, p. 48.

"Satellite TV News," *Radio-Electronics,* March 1980, p. 12; April 1980, p. 12; May 1980, p. 16; June 1980, p. 26; July 1980, p. 16.

"Something for Everyone—Programming Service Available Via Satellite," *Videoplay Magazine,* April/May 1981, p. 58.

"So You Want to Build a Home Satellite Earth Station," *Videoplay,* Aug./Sept. 1981, p. 62.

"Space Telecommunications," *IEEE Communications,* Sept. 1980, p. 5.

"TV Culture Invasion from Space," *New Scientist,* Feb. 28, 1980, p. 645.

"TeleSat Canada Plans for New Satellite Systems," *Journal of Spacecraft and Rockets,* March 1980, p. 75.

"The 'Eight-Ball' Satellite Antenna Kit," *Videoplay Magazine,* Oct. 1980, p. 52.

"The Home Satellite Receiver Scene," *Videoplay Magazine,* Oct. 1980, p. 43.

"Tracking the Birds," *Videoplay Magazine,* June/July 1981, p. 61.

Popular Periodicals

"Are You a Signal-Napper?" *HiFi,* Feb. 1981, p. 2.

"Continent's Turning On to Terrible Ted," *Macleans,* Feb. 26, 1979, p. 46.

"Direct Satellite TV: Canada," *Science Quest,* Jan. 1980, p. 5.

"Going Super with Ted: Feeding Independent Station Programs to Cable Systems Via Satellite," *Newsweek,* Jan. 1, 1979, p. 61.

"How to Dish Up a Skyful of TV," *Newsweek,* Oct. 27, 1980, p. 101.

"Install a Backyard Antenna to Tune-In Satellite TV," *Popular Science,* March 1980, p. 122.

"Next, Television Goes Into Orbit (Communications Satellite Corporation's Direct-to-Home Service)," *U.S. News & World Report,* April 14, 1980, p. 76.

"Now: Receive Home TV Directly from Satellites," *Popular Mechanics,* Sept. 1980, p. 112.

"Open Up the Satellites," *Newsweek,* Nov. 3, 1980, p. 13.

"Plowboy Interview: 'Television . . . a Medium for and by the People." *Mother Earth News,* May/June, 1980, p. 16.

"Revolution in TV Watching: Programming From Satellites," *Science Digest,* Dec. 1979, p. 25.

"Satellite Fever," *Washington Journalism Review,* March 1980, p. 18.

"Satellite-Sent Pictures for TV," *Mechanics Illustrated,* May 1980, p. 141.

"TV's Future: The World on a Dish," *Macleans,* Oct. 8, 1979, p. 45.

"Ted Turner's True Talent: WTCG, Atlanta," *Esquire,* Oct. 10, 1978, p. 34.

"The Collector Brings Chotsie the Satellite," *TV Guide,* June 20, 1981, p. 34.

Newspaper Articles

"American Express Company to Buy 50% of Warner Cable TV Subsidiary," *Washington Post,* Sept. 15, 1979.

"Black Entertainment Television Cable Network Head Featured," *Washington Post,* Nov. 30, 1979.

"Comsat Requests FCC Permission to Start Direct Satellite-to-Home Subscription Service," *Washington Post,* Dec. 17, 1980.

"Comsat Unveils Plan for 3-Channel Satellite-to-Home Broadcasting Service," *Washington Post,* Dec. 18, 1980.

"Comsat's Proposed Direct Broadcast Satellite System Examined," *Washington Post,* Oct. 1, 1980.

"Corporation for Public Broadcasting Investigating Satellite-to-Home Technology," *Washington Post,* April 18, 1980.

"FCC Proposed Deregulation of Cable TV Industry," *Washington Post,* April 29, 1979.

"FCC Seeks Public Comments on Direct Satellite Service," *Washington Post,* Oct. 3, 1980.

"FCC Staff Studies Possible Effects of Satellite Broadcasting on TV Industry," *Washington Post,* April 3, 1980.

"Staff of FCC Urges Agency to Adopt 'Hands Off' Policy Toward Development of Newest Frontier of TV Broadcast Technology—Direct Broadcasts Home Via Satellites," *New York Times,* Oct. 3, 1980.

"Ted Turner III Established Cable News Network," *New York Times,* Feb. 17, 1980.

Business Periodicals

"AT&T Plans Big Satellite Role," *Advertising Age,* Aug. 20, 1979, p. 53.

"Affiliates Told NBC Will Study Usage of Satellite Technology," *Advertising Age,* May 21, 1979, p. 3.

"American Satellite to Invest in Westar," *Aviation Week,* Nov. 26, 1979, p. 48.

"Arranging Satellite Broadcasting Practically," *Electronics and Power,* Nov.–Dec. 1979, p. 733.

"BBC Wants DBS," *Broadcasting,* Dec. 8, 1980, p. 78.

"Basic Overview of Equipment for Video Earth Station Applications," *Communication News,* April 1980, p. 74.

"Battle of the Satellites," *Infosystems,* May 1981, p. 54.

"Black Entertainment TV to Debut in January," *Advertising Age,* Nov. 26, 1979, p. 22.

"Bob Wold: Satellite Star," *Marketing and Media Decisions,* June 1979, p. 66.

"CBS Aims for the Stars on Satellite TV," *Broadcasting,* Oct. 13, 1980, p. 23.

"C-SPAN: Carving Out a New Programming Niche," *Broadcasting,* Nov. 3, 1980, p. 48.

"Cable Systems Find Some of the Best Things Can Be Free Via Satellite," *Communications News,* Feb. 1980, p. 65.

"Cable TV Firms in Canada Seek $2.5 Million in Satcom Gear," *Electronic News,* April 30, 1979, p. 50.

"Cable TV: The Great Leap Forward," *Dun's Review,* Oct. 1979, p. 70.

"California Microwave to Build Earth Stations for AP Net," *Electronic News,* April 9, 1979, p. 60.

"Channel for Women Wants Package Goods," *Advertising Age,* Oct. 22, 1979, p. 78.

"Chinese Formulate Requirements for Broadcast Satellite System," *Aviation Week,* March 31, 1980, p. 63.

"Communications Satellites: The Birds Are in Full Flight," *Broadcasting,* Nov. 19, 1979, p. 36.

"Comsat Gets FCC Nod for International TV Service," *Telephony,* Feb. 11, 1980, p. 63.

"Comsat Promise of Pay TV Systems Stirs Up Industry," *Advertising Age,* Aug. 6, 1979, p. 1.

"Comsat, Sears Getting Cozy Over Satellite-to-Home Service," *Broadcasting,* Jan. 7, 1980, p. 28.

"Comsat, Sears Near Agreement on Direct-to-Home TV Service," *Aviation Week,* Jan. 28, 1980, p. 66.

"Comsat to Ask Direct-Broadcast Rights," *Aviation Week,* Dec. 8, 1980, p. 61.

"Curtain's Going Up on DBS: Television's Next Frontier," *Broadcasting,* Sept. 15, 1980, p. 36.

"DBS/Direct Broadcast Satellites: The New Frontier," *Television International,* Sept./Oct. 1980, p. 27.

"DBS Study Missed the Point, Say NAB and NBC," *Broadcasting,* July 14, 1980, p. 57.

"Direct Satellite Broadcasting: An Up and Coming Medium, Says FCC," *Broadcasting,* April 7, 1980, p. 38.

"Dishy Offer for Satellite Television," *Economist,* July 7, 1979, p. 103.

"Downconversion Techniques in Earth Station Video Receivers," *Communications News,* April 1980, p. 66.

"Earth Station Gear to Total $1.1 Billion Over Next Decade," *Microwave,* July 1979, p. 17.

"Earth Station Use Rules Sought by UPI, AP, Papers," *Electronic News,* April 14, 1980, p. 47.

"Earth-Terminal Hardware Market Growing," *Telecommunications,* April 1979, p. 31.

"Europe Pioneers Direct-Broadcast TV Satellites," *Interavia,* July 1980, p. 645.

"Everybody Wants a Piece of the Sky," *Broadcasting,* May 5, 1980, p. 38.

"Expect West Germans to OK Plan for TV Satellites in '82," *Electronic News,* Aug. 20, 1979, Supp. 6S.

"FCC Extracts RCA from Tangle of SATCOM Claims," *Broadcasting,* June 30, 1980, p. 70.

"FCC Gets the Ball Rolling on DBS," *Broadcasting,* Oct. 6, 1980, p. 24.

"FCC Weighs Regulations on Direct Satellite TV," *Electronic News,* Oct. 6, 1980, p. 12.

"From Out-of-the-Blue: COMSAT Designs Direct-to-Home Subscription TV," *Broadcasting,* Aug. 6, 1979, p. 27.

"Getty Cable Plan Draws Fire," *Advertising Age,* April 28, 1980, p. 2.

"Getty Oil Prospects for Pay Cable's Gold," *Advertising Age,* April 21, 1980, p. 3.

"Ground-Control System for Satcom Satellites," *Journal of Spacecraft and Rockets,* July 1979, p. 273.

"Ground Station Sales Swelling," *Advertising Age,* Aug. 13, 1979, p. 20.

"Group-W Buys Into Satellites," *Advertising Age,* July 30, 1979, p. 2.

"Group-W Hitches Program Wagon to a Westar," *Broadcasting,* July 30, 1979, p. 34.

"Homesat: A New Market for Dealers," *Two Way Radio Dealer,* Feb. 1981, p. 41.

"Homesat on the Range," *Advertising Age,* Sept. 3, 1979, p. 26.

"In Search of the Bird," *Advertising Age,* Dec. 17, 1979, p. 63.

"Industry Watches Superstation Fray," *Advertising Age,* Aug. 27, 1979, p. 32.

"Influx of Cash Set for Westar," *Broadcasting,* Nov. 5, 1979, p. 63.

"Intelsat Earth Station TV Performance," *Electronic Engineering,* May 1979, p. 63.

"Joe Charyk and the Gleam in Comsat's Eye," *Broadcasting,* Sept. 15, 1979, p. 50.

"Local TV Not Threatened: FCC Urges More Cable Decontrol," *Advertising Age,* April 31, 1979, p. 3.

"Lot of Ifs (Wall Street Analysts on Comsat's Plan for Direct Satellite-to-Home Service)," *Broadcasting,* Aug. 13, 1979, p. 38.

"NBC-TV Looks Up to the Sky," *Broadcasting,* May 21, 1979, p. 84.

"Networks Ho-Hum as Possibility of AT&T Satellite Is Opened to Them," *Broadcasting,* July 30, 1979, p. 71.

"No Big Deal: Technology Is Least of Comsat's Worries About Satellite-to-Home," *Broadcasting,* Aug. 13, 1979, p. 55.

"No Consensus in DBS's Line of Sight," *Broadcasting,* Oct. 13, 1980, p. 25.

"Nuclear Powered Communications Satellite for the 1980's," *Journal of Spacecraft and Rockets,* July 1979, p. 268.

"Owning Your Own Earth Station," *Advertising Age,* Sept. 17, 1979, p. 72.

"PBS, WU Deal Looks Good at the FCC," *Broadcasting,* Feb. 4, 1980, p. 87.

"P.O.'s Largest Satellite Earth Terminal Operation: Medley 1," *Wireless World,* June 1979, p. 56.

"Program Service Aims for Blacks By Satellite," *Broadcasting,* Sept. 3, 1979, p. 60.

"RCA and UPI Plan Satcom Network," *Electronic News,* March 5, 1979, p. 67.

"RCA Offer Signals Vast TV Industry Change," *Advertising Age,* March 19, 1979, p. 1.

"RKO Radio Ready to Leap on AP's Satellite System," *Advertising Age,* Aug. 27, 1979, p. 71.

"Satellite and Pay Cable Spur CATV," *Communications News,* April 1979, p. 61.

"Satellite Called Symbol of '80s," *Advertising Age,* June 18, 1979, p. 3.

"Satellite Communications and the Growth of Earth Stations," *Telecommunications,* April 25, 1979, p. 25.

"Satellite Links Get Down to Business," *High Technology,* June, 1980, p. 49.

"Satellite Transmission Takes Off," *Advertising Age,* Oct. 8, 1979, Sec. 2, S2.

"Satellite TV Causes Static at the Networks," *Business Week,* Oct. 27, 1980.

"Satellites Provide New Horizons for Cable TV," *Communications News,* April 1979, p. 65.

"Satellites Strong in Broadcast Services," *Communications News,* March 1979, p. 70.

"Sears, Comsat Plan '83 Air Date," *Advertising Age,* Jan. 7, 1980, p. 2.

"Should the Satellite Pie Be Cut Into More Slices?" *Broadcasting,* Sept. 11, 1978, p. 64.

"Sky's the Limit for Expanding Earth Station Market," *Advertising Age,* June 25, 1979, p. 51.

"Sky's the Limit for Growing Usage of Satellite Technology," *Communications News,* Sept. 1979, p. 22.

"Spanish TV Network Traces Growth to Satellite Usage," *Communications News,* Feb. 1980, p. 68.

"Storer Stations Look to the Sky," *Broadcasting,* Aug. 6, 1979, p. 96.

"Superstations Draw Fans and Pans," *Advertising Age,* May 14, 1979, p. 12.

"Superstations Find Progress Steady," *Advertising Age,* July 30, 1979, p. 101.

"Ted Turner's Latest," *Broadcasting,* Nov. 20, 1978, p. 70.

"Ted Turner's Satellite Squeeze-Play," *Media Decisions,* Aug. 1978, p. 68.

"Television: Look Out for Footprints," *Economist,* Dec. 29, 1979, p. 37.

"Threat to Networks from Super Cable-TV," *Business Week,* Nov. 27, 1978, p. 36.

"U.S. Abandons Its Evolutionary Approach to Space Broadcasting," *Aviation Week,* May 5, 1980, p. 74.

Appendix 5
Federal Communications Commission and Copyright Requirements

The purpose of this appendix is to review some of the specific legal and technical requirements that are placed upon the mini-CATV operator by the Federal Communications Commission and the Copyright Law. The first section reviews the FCC requirements and the second section reviews the copyright requirements.

I. The FCC Cable Television Requirements

In most cases the mini-CATV operator, apartment building owner, hotel or trailer court operator, and condo association will be completely exempt from all of the FCC rules and regulations, and will not have to file any forms or follow any of the FCC's requirements. In some other instances a complicated set of FCC procedures must be followed. This section should be reviewed carefully before any reader of this book operates a TVRO earth station for a business purpose. Of course, home users of TVRO earth stations are not, by definition, commercial users and need not be concerned with the FCC's "Cable Television Rules."

1. Definition of Cable Television System

The first point that should be addressed is whether or not a particular mini-CATV system is included in the Commission definition of a cable television system. If it does not fall within the Commission definition, it is not subject to any of the FCC regulations.

For regulatory purposes, the rules distinguish between the terms "cable television system" and "system community unit." A cable television system is described as a "nonbroadcast facility consisting of a set of transmission paths and associated signal generation, reception, and control equipment, under common ownership and control, that distributes or is designed to distribute to subscribers the signals of one or more *television broadcast stations. . . .*"

Often called the "headend" or "physical system" definition, it is used basically in provisions related to network program nonduplication protection, exemptions, record maintenance, certain FCC reporting forms, cross-ownership, and technical standards. In terms of this definition, each earth station would be analogous to a headend. Therefore, each earth station with its associated cables and paths would be considered a separate cable television system.

The term "system community unit" applies to registration, signal carriage, and network program nonduplication protection. "System community unit" means a cable TV system, or portion of a cable system operating within a separate and distinct community or municipal entity, including unincorporated areas.

- *This definition excludes any facilities that serve fewer than 50 subscribers.* Therefore, a mini-CATV company that has 49 subscribers need not file any forms with the FCC and need not comply with Commission regulations. It is simply not considered a cable television system for FCC purposes. If your mini-CATV system meets this test, then most of the information in this appendix will not apply.
- *More important, this definition excludes any such facility that serves only subscribers in one or more multiple-unit dwellings under common ownership, control, or management.* It is expected that most mini-CATV systems will fall within this exclusion. Each earth station and its associated transmission paths (cables) will generally be considered a separate mini-CATV system. If the mini-CATV system serves only one set of multiunit buildings (such as a condominium, apartment house, or hotel) owned or controlled by the same entity, then that mini-CATV system is not covered by the Commission's "Cable Television Rules" and it need not comply with any of the provisions of these rules.

In those instances where one mini-CATV system (earth station) will serve two or more multiunit dwellings that are owned or controlled by different entities, the "Cable Television Rules" must be followed. The origin and present requirements of those rules are set forth below.

2. Development of FCC Cable Television Rules

In March 1966 the Commission adopted regulations for all cable systems [*Second Report and Order in Docket 14895, 2 FCC 2d 725 (1966)*]. The initial rules imposed a myriad of signal carriage regulations on all cable television systems. These rules were challenged in the courts. In June 1968 the U.S. Supreme Court affirmed the Commission's jurisdiction over cable [*United States v. Southwestern Cable Co.,* 392 U.S. 157 (1968)]. The court found that the FCC required jurisdiction over cable systems to assure the preservation of local broadcast services and to effect an equitable distribution of broadcast services among the various regions of the country.

On December 13, 1968, the Commission invited comments on a major revision and expansion of its cable rules (Docket 18397) and adopted "interim procedures" for use during the rule-making proceeding. These procedures included suspension of distant signal hearings, deferral of processing of all petitions or applications seeking authorization of service inconsistent with the proposed signal carriage rules, and grant of carriage requests consistent with the proposed rules.

A general outline of the proposed resolution of this proceeding was submitted to Congress on August 5, 1971, in a "Letter of Intent." The rules adopted six months later generally adhered to those proposals and contained modifications to reflect the communications industry consensus agreement on exclusivity and distant signal importation.

The rules the FCC adopted, which became effective March 31, 1972, were the most comprehensive compilation of regulations ever issued on cable television. [See *Cable Television Report and Order,* 36 FCC 2d 143 (1972).]

Current Rules. Since 1972 the FCC has modified or eliminated many of the rules. Among the more significant actions, the Commission deleted most of the franchise standards; substituted a signal registration process for the certificate of compliance application process; and eliminated the distant signal carriage restrictions and syndicated program exclusivity rules. In addition, court actions led to the deletion of the pay cable programming rules in 1979. Rules regarding required carriage of local broadcast signals, network program nonduplication protection, sports program blackouts, cross-ownership, equal employment opportunity, origination cablecasting, technical standards, cable television relay service (CARS), and record-keeping and reporting comprise the current cable television rules. Each of these topics is described more fully in this appendix.

3. Registration for All Systems Serving 50 or More Subscribers

As a precondition to commencing operation or adding any television broadcast signals to existing operations, a cable system operator must separately register each system community unit with the Commission. If a cable television facility serves more than 50 subscribers total, but fewer than 50 subscribers in any given community, it must register each community with the FCC. To register, a cable television operator must send the following information to the Secretary of the Commission:

1. The legal name of the operator, Entity Identification or Social Security number, and whether the operator is an individual, private association, partnership, or corporation; if the operator is a partnership, enter the name of the partner responsible for communications with the Commission.
2. The assumed name (if any) used for doing business in the community.
3. The mailing address, including zip code, and the telephone number to which all communications are to be directed.
4. The date the system provided service to 50 subscribers.
5. The name of the community or area served and the county in which it is located.
6. The television broadcast signals to be carried that previously have not been certified or registered.
7. For a cable system (or an employment unit) with five or more full-time employees, a statement of the proposed community unit's equal employment opportunity program, unless such program has previously been filed for the community unit.

The cable television operator is not required to serve the registration statement on any party and may begin operation or add new signals immediately upon filing the registration statement. However, commencement of operation is entirely at the risk of the system operator; if violations of the rules are subsequently discovered, appropriate regulatory sanctions may be employed.

No further filings, other than required annual reports, are necessary unless new signals are added or the system changes ownership.

4. Requirements for Small Systems

Cable television systems serving fewer than 1000 subscribers are exempt from most of the Commission's rules. Generally, only the following requirements apply to smaller cable systems:

1. Comply with the registration requirements described above.
2. Comply with requests from local television stations for carriage on the cable system.

3. Comply with the Commission's technical standards for cable television systems (except that annual proof of performance tests are not required).
4. Correct and/or furnish information in response to the following forms sent to the cable operator annually by the Commission:
   Form 325: "Annual Report of Cable Television System" (Schedules 1 and 2 only)
   Form 326: "Cable Television Annual Financial Report"
   Form 395A: "Annual Employment Report"

5. Franchising and Local Regulation

The Commission adopted a regulatory plan allowing local or state authorities to select a franchisee and to regulate in any areas that the FCC does not preempt. Typically, local and state governments have adopted laws and/or regulations on franchising, basic subscriber rates, theft of service, taxation, and pole attachments. These vary from state to state. The Commission preempts local regulation of signal carriage, pay cable, and technical quality.

6. Rates for Service

The FCC does not regulate rates for cable television service. Local government may decide whether or not to regulate most rates; however, the FCC preempts local regulation of pay cable rates.

Most systems have a rate for basic subscriber service, one or more charges for pay cable services, and charges for installation and additional hook-up. An increasing number of systems have several tiers of programming, each with separate charges. The use of computerized billing methods is facilitating more complex rate structures.

7. Signal Carriage

The 1972 carriage rules set up very complicated requirements for the carriage of broadcast signals by cable television systems. However, these requirements have been eliminated (pending an appeal to the Supreme Court). Currently, a cable television system may carry whichever broadcast signals it desires subject only to the FCC's Mandatory Carriage Rules and the Network Nonduplication Rules.

8. Mandatory Carriage

Certain television broadcast signals are required to be carried by cable systems, if the broadcast station requests carriage. Systems serving communities in major or smaller television markets must carry, on request, signals from the following broadcast sources:

- All television stations licensed to communities within 35 miles of the cable system community.

- Noncommercial educational television stations within whose "Grade B contour" the system community is located.
- Commercial TV translator stations with 100 watts or higher power serving the community of the system and noncommercial translators with 5 watts or higher power serving the community.
- Stations whose signals are significantly viewed in the community. A list of significantly viewed signals on a countywide basis is contained in Appendix B of the *Memorandum Opinion and Order on Reconsideration of the Cable Television Report and Order,* FCC 72-530, 36 FCC 2d 326 (1972), and in the FCC's Cable Television Rules. The method for establishing significantly viewed status for signals not listed in Appendix B is contained in Section 76.54 of the Cable Television Rules. Stations that were not on the air at the time of the survey are permitted to submit countywide data; other stations must submit community surveys.
- All Grade B signals from other small markets (applies to systems serving communities in smaller markets only).

In communities located outside of all markets, cable systems must carry signals from the following sources:

- All stations within whose Grade B contour the system community is located.
- Commercial translator stations with 100 watts or higher power and noncommercial translators with 5 watts or higher power serving the community.
- Stations whose signals are significantly viewed in the community.
- All educational stations licensed to communities within 35 miles of the system community.
- All educational stations within 35 miles of a cable system community located outside of all television markets and all educational stations that place a Grade B contour over the community of a system in a major or smaller market must be carried by the system. Also, cable systems serving communities in states where noncommercial educational stations are run by a state agency may carry any such stations. Any additional educational stations may be carried if there are no convincing objections.

9. Network Program Nonduplication

Nonduplication protection is afforded a local television network station program when a cable system does not duplicate the same program carried simultaneously on a distant station of the same network. This protection is a requirement of FCC rules. The rationale for the nonduplication rules is that duplication of the local station's programming through carriage of distant signals could economically injure local broadcasters and the service they provide to the public.

10. Radio Programming

The Commission places no restrictions on either the carriage of radio stations or cable system origination of radio programming.

11. Sports Programming

Cable television systems serving 1000 or more subscribers may not carry local sports events broadcast by distant television stations when the events are "blocked out" on local television stations. The holder of broadcast rights is responsible for notification of programming deletions at least no later than Monday preceding the calendar week during which deletions are desired. Programming from any other television station may be substituted for the deleted program.

12. STV Stations

Cable operators are not required to carry the subscription, or scrambled, programming portion of a subscription television station (STV). However, there is nothing in the rules that prohibits an STV station from making an independent contractual agreement with a CATV system for the CATV system to carry the STV station's signal for a fee.

13. Equal Employment Opportunity

All operators of cable systems are required to afford equal opportunity in employment to all qualified persons and are prohibited from discrimination in employment because of race, color, religion, national origin, or sex.

The rules require all cable operators to establish and maintain programs designed to assure equal opportunity for females, blacks, Hispanics, American Indians or Alaskan natives, and Asians or Pacific Islanders in recruitment, selection, training, placement, promotion, pay, working conditions, demotion, layoff, and termination.

In addition, if systems have five or more full-time employees, they must keep on file with the Commission an up-to-date statement of their equal opportunity program and file any changes to existing programs on or before May 31 of each year. They must also submit an annual employment report (FCC Form 395A).

14. Reports and Forms

The Commission sends computer printed forms on the following subjects and asks for verification of their accuracy: (1) Schedule 1 of Form 325 (Annual Report of Cable Systems); (2) Form 326 (Annual Financial Report), Schedule 1 of which is preprinted; (3) Form 395A (Annual Employment Report); and (4) for systems with 1000 or more subscribers, Schedules 3 and 4 of Form 325 (Ownership Information).

Whenever a change occurs in a system's mailing address, operator's legal name, or operational status—for example, it obtains 50 or 1000 subscribers—the operator must notify the Commission of the change within 30 days. The operator must furnish the following information: (1) operator's legal name and structure (individual, private association, partnership, corporation), including the EI (Employment Identity) number or the Social Security number (if the company is an individual or partnership); (2) assumed name, if any; (3) mailing address; (4) nature of operational status change; and (5) the names of the system communities affected and their FCC identifiers (e.g., CA0001), which are assigned routinely when the system registers with the Commission.

Various information reports are available to the public based on data collected from cable companies. Photocopies of these reports, such as general statistical information, TV station distribution, mailing addresses, and many others, may be obtained through the Downtown Copy Center (telephone 202/452-1422) for a nominal charge per page. Some reports are available on microfiche and magnetic tape from the National Technical Information Service (703/557-4780).

15. Cable System Ownership

The Commission prohibits cable system ownership by telephone companies within their local exchange areas, by television stations within the same local service area (Grade B contour), by national television networks anywhere in the country, and by television translator stations in the same community as the cable system.

In January 1980 the Commission instituted a new policy of reviewing waiver requests from telephone companies proposing to provide cable television service in areas with a density of fewer than 30 homes per square mile (CC Docket 78-219). The purpose of the policy is to encourage the development of cable service in rural areas.

16. Technical Performance Requirements

To assure the delivery of satisfactory television signals to cable subscribers, the Commission requires cable operators to meet certain technical performance requirements. In general, the rules require every cable operator to conduct annual performance tests and to use these results to determine whether the system is performing satisfactorily. Systems serving fewer than 1000 subscribers are subject to the technical standards but are not required to take annual measurements. The FCC preempts state or local regulation of technical performance requirements.

Major current technical standards include: (1) standards for signal leakage, (2) frequency channeling plans, and (3) standards for cable compatible receivers. Research in progress on the nature of signal leakage must precede frequency channeling decisions, which in turn must precede standardization of cable-compatible receivers.

17. State Regulations

The question of whether copyright or franchise fees may be added to subscriber rates is entirely within the purview of state or local regulatory authorities. A wide variety of laws and regulations for cable television exist at the state level. Eleven states regulate cable television on a comprehensive basis through a state agency. In addition, at least 30 other states have one or more laws specifically applicable to cable television, dealing most commonly with such subjects as franchising, theft of service, pole attachments, rate regulation, and taxation.

In most instances the regulations at both the state and city levels that define what a CATV system is (and thus subject to regulation) do not include apartment buildings, hotels, trailer courts, and other master distribution antenna systems, or mini-CATV systems *whose cables do not cross public property.* In a few states, however, the state's definition of a cable system could be construed to apply to an apartment house MATV system. Care should always be taken to consult local laws. In most cases a mini-CATV operator who has fewer than 50 customers or who serves a set of multiunit dwellings under common ownership and whose cable system is contained completely on private property and carries only pay TV signals or other nonbroadcast signals is probably exempt from most forms of regulation at the federal, state, and city levels.

This is the best of all worlds. Such an operator could construct and own literally dozens of such separate mini-CATV systems scattered throughout the city or state and still be exempt from governmental regulation.

18. How To Obtain the Rules

Every cable operator subject to the cable rules is required by the FCC to have an up-to-date copy of the "Cable Television Rules and Regulations" (47 CFR Part 76). These rules are available in looseleaf form on a subscription basis through the Superintendent of Documents, U.S. Government Printing Office, Washington, DC 20402. Changes in the rules are forwarded to subscribers within a few months after their adoption by the Commission, without additional charges. To order, request Volume XI of the "FCC Rules and Regulations," August 1976 edition (subscription price: $15). Detailed information about FCC regulation of cable television is available from the FCC in its Publication 18, available at no charge by writing to the Public Information Office, FCC, 1919 M Street, NW, Washington, DC 20054.

19. Further Information from the FCC

Members of the Bureau's staff welcome inquiries from the public. By contacting the appropriate division, assistance can be obtained.

TABLE A5–1.

Bureau Organization	Telephone
	(Area Code 202)
Office of Bureau Chief	632-6480
Policy Review and Development Division	632-6468
Research Division (including EEO unit)	632-9797
Records and Systems Management Division (including the Public Reference Room)	632-7076
Compliance Division	632-7480
Complaints and Information Branch	632-9703
Enforcement Branch	254-3407
Special Relief and Microwave Division	632-8882
Special Relief Branch	632-8882
Microwave Branch	254-3420

A detailed package of all forms, regulations, and other documents pertaining to the establishment of a mini-CATV system is available from the CATV/Satellite Information Center, P.O. Box 7104, San Francisco, CA 94120, for a fee of $20. Ask for the "Mini-CATV Report." The report gives specific detailed "how-to" information complete with all necessary forms, rules and regulations, and examples.

II. Copyright Compliance

Introduction

On October 19, 1976, a new Copyright Act was signed into law. This Act completely revised the old 1909 Copyright Law. Cable television operators paid nothing for the carriage of broadcast signals under the old law. Under the new Act, however, CATV systems are required to pay copyright fees for the carriage of broadcast signals. This liability began on January 1, 1978. Regulations implementing the law may be obtained from the Copyright Office, which administers the law. The Office is a part of the Library of Congress, Washington, DC 20559 (telephone 703/557-8731).

Though cable systems were never specifically required to pay copyright fees for the use of broadcast programming prior to the new copyright law, the new law does have advantages for cable operators. It provides for a compulsory licensing system whereby a cable system is entitled to a license to carry any of the broadcast signals permitted under FCC regulations. This means that the cable operator does not have to bargain with each broadcaster or program supplier for the right to carry (retransmit) that broadcaster's programming on the cable system. Therefore the broadcaster cannot deny a cable operator access to broadcast programming and the fee paid by the cable operator for use of the programming is fixed in the Copyright Law itself.

1. Utilization of the Compulsory License

The compulsory license entitles the cable operator to use only broadcast programming (local and distant television and radio programming), including broadcast programming that is also distributed via satellite such as Ted Turner's WTBS-TV, Channel 17, in Atlanta; WOR-TV in New York; and WGN in Chicago. (You should note, however, that the microwave carriers that transmit these signals to the satellite have a separate fee for that service that they charge everyone receiving it commercially.) Also included as broadcast signals are the "hybrid" signals such as the PTL religious network, which obtains programming from a regular broadcasting station. The compulsory license does not entitle the cable operator to use privately distributed nonbroadcast programming such as HBO, Showtime, or Home Theatre Network. Such nonbroadcast programming can only be utilized pursuant to negotiation of a viewing rights contract with the program copyright holder.

The Copyright Law contains a very broad definition of what constitutes a cable television system: earth station reception systems distributing satellite signals to anywhere from one to one million individual television sets can be considered cable television systems under the Copyright Law. As such it is recommended that all earth stations register with the Copyright Office as mini-CATV systems and follow the procedures to obtain a compulsory license. This will legally entitle the earth station to receive any broadcast programming from any of the domestic satellites.

Mini-CATV systems may register as cable television systems with the Copyright Office even though they are excluded from the FCC definition of a cable television system. Why register at all with the Copyright Office as a cable system if a mini-CATV system is excluded from FCC regulations? The answer is that broadcast programming is copyrighted and it is protected. It cannot be used by a mini-CATV system without authorization. A compulsory license is the simplest form of authorization a mini-CATV system can obtain. If a compulsory license is not obtained, authority from each program copyright owner must be obtained before that program can be used by the mini-CATV system.

The Copyright Law defines individual cable television systems by headend. The headend is the piece of equipment that collects and processes all signals to be distributed by the cable system. In the case of a mini-CATV system the headend would be analogous to the earth station. Therefore each earth station should be registered as a separate mini-CATV system.

2. Obtaining the Compulsory License

In order to obtain a compulsory license, the cable operator must file an *Initial Notice of Signal Registration* and a semiannual *Statement of Accounts* with the Copyright Office as well as maintain certain records. All filings should be made with the Licensing Division, Copyright Office, Library of Congress, Washington, DC 20557.

A. Initial Notice of Signal Registration A suggested format for the *Initial Notice* appears in Figure A5-1. The *Initial Notice* must contain the following information:

1. *Owner.* The full legal name of the owner, any assumed names used for doing business, and the address of the owner.
2. *System.* The full tradename of the system and the mailing address of the system.
3. *Area served.* The name of the community served by the system.
4. *Regularly carried signal carriage complement.* List the call sign, type (AM-FM-TV), and city of license of all broadcast signals regularly carried by the cable system. If a translator is carried, enter the call sign of the translator, not the call sign of the original broadcast station.

   If the cable system provides all-band FM service, indicate this on the *Initial Notice* and list all of the FM stations generally received. The best way to determine generally receivable signals is to scan the FM radio dial and list any signals received.

   Television signals are "regularly carried" if the cable systems provides the signal at least one hour per week for 15 or more consecutive weeks. Frequently this includes signals that are substituted for other signals that are deleted according to FCC requirements.
5. *Signature.* The *Initial Notice* must be signed by the owner or a duly authorized representative. It is not required that the signature on the *Initial Notice* be notarized.

   The *Initial Notice* must be filed at least 30 days prior to the commencement of operation of the mini-CATV system.

B. Notice of Change Any time there is a change in ownership or signal carriage complement in the system, a *Notice of Change* must be filed within 30 days after such an event in order to keep the compulsory license current. For instance, if you switch independent signals when you turn on your TVRO earth station, you must notify the Copyright Office.

APPENDIX 5

FIGURE A5–1.
Cable Television
Initial Notice of Identity and Signal Carriage Complement

To Copyright Office/Washington, DC 20559

---

1. OWNER
   (A) Full Legal Name: (B) Assumed (dba) Name: (if any)

   (C) Mailing Address:

2. SYSTEM (if same as given above just check box) ☐
   (A) Full Trade Name: (B) Mailing Address:

3. AREA SERVED (List all communities or areas served by the system)

4. REGULARLY CARRIED SIGNAL CARRIAGE COMPLEMENT
   [List indicating call sign, type (AM-FM-TV), and City of License]

   Note: If system carries all-band FM, check here ☐—individual FM stations "generally receivable" must also be listed separately. Attach a separate sheet if necessary.

   CALL SIGN & TYPE CITY CALL SIGN & TYPE CITY

5. Signed:____________________________
   (Owner or duly authorized representative)

   ____
   Type or print name and title

   Date:____________________________

NOTE: If you wish a Certified Receipt, enclose $3.00 with Notice. ☐

If the ownership changes, you must list the name of the person or entity that formerly owned the cable system, and the same information about the "new owner" and the "system" given above in the *Initial Notice.* The effective date of the change of ownership should also be entered.

If there is a change of signal carriage complement, you should enter the information about the "owner," the "system," and the names and locations of the primary transmitters whose signals have been added, as well as any that may have been deleted, and the approximate date of each addition or deletion. The *Notice of Change* should be signed and dated as set out above. If the "owner" identified in *Notice of Change* is different from the "operator" in the *Initial Notice,* put that person's name down next to the "owner" signature.

A *Notice of Change* is not required if the business name of the owner or cable system is changed, the address of the owner or cable system changes, the call letters of carried stations change, or changes occur in the names of the communities served by the system. To repeat, the *Notice of Change* is filed only if there is an actual change of ownership or signal carriage complement. See Figure A5-2 for an example of a typical *Notice of Change.*

C. Permissive Amendment of Notices If there is an error in an *Initial Notice* or a *Notice of Change,* the Copyright Office will permit correction by amendment. Clearly label the submission as an amendment to whichever notice is being amended. Specify the amendment and sign and date the submission. A fee of $10 must accompany such an amendment. This is the only filing fee required of a CATV system anywhere in the rules. See Figure A5-3 for an example of a typical *Notice of Amendment.*

D. Required Amendments of Notices Some CATV systems are required to file amendments in certain circumstances. An amendment is required if the *Initial Notice* indicated carriage of FM stations but failed to identify the specific stations "generally receivable" by the system. Any cable system that is unable to list the FM signals carried in an all-band situation in an *Initial Notice* must, within 50 days after filing of the *Initial Notice,* submit an amendment listing the signals.

All of the *notices* and *amendments* described above will be placed in the public files of the Copyright Office. Upon payment of $3, the Copyright Office will issue a certified receipt for any *notice* or *amendment* submitted. Although this is optional, it is a wise thing to do. To repeat, no filing fee must be paid for these *Required Amendments,* or for the *Initial Notice* or a *Notice of Change;* there is only a filing fee ($10) for a permissive amendment to correct an error. See Figure A5-4 for an example of a *Notice of Amendment.*

E. Statements of Account Every CATV system must file a Statement of Account twice a year. A completed Statement of Account form and a check for the copyright royalty payment must be received by the Copyright Office 60 days after the close of each of the payment periods, which run from January 1 to June 30 and from July 1 to December 31.

FIGURE A5–2.
Cable Television
Notice of Change

To: Copyright Office/Washington, DC 20559

TYPE OF CHANGE: ☐ Ownership ☐ Signal Carriage Complement

OWNER–FORMER OWNER (Fill in for either type of Notice)
(A) Full Legal Name: (B) Assumed (dba) Name: (if any)

(B) Mailing Address:

Note: If pre-1978 Notice did not list same name as above, list that name here.

NEW OWNER (Fill in for ownership changes only)
(A) Full Legal Name: (B) Assumed (dba) Name: (if any)

SYSTEM (If same as given above just check box) ☐
(A) Full Trade Name: (B) Mailing Address:

SIGNAL CARRIAGE COMPLEMENT CHANGES [List call sign, type (AM-FM-TV), and City of License]

ADDED		DELETED		
SIGN & TYPE	CITY	SIGN & TYPE	CITY	DATE

Signed:______________________________
(Owner or duly authorized representative)

____
Type or print name and title

Date:______________________________

NOTE: If you wish a Certified Receipt, enclose $3.00 with Notice. ☐

Therefore, the due dates for copyright payments are August 29 and March 1.

1. Forms There are three Copyright Office-supplied forms:

"Short Form" (CS/SA-1) for cable systems with semiannual "gross receipts" of $41,500 or less.

FIGURE A5–3.
Cable Television
Amendment to: (Check One)
☐ *Initial Notice of Identity and Signal Carriage Complement*
☐ *Notice of Change of Identity or Signal Carriage Complement*

To: Copyright Office/Washington, DC 20559

NOTICE TO BE AMENDED:
Name Notice Was Filed Under: File Number (if known):

Date Notice Was Filed:

NATURE OF AMENDMENT:

Signed:______________________________
(Owner or duly authorized representative)

____
Type or Print Name or Title

Date:______________________________

Note to Operators: This Amendment form is for PERMISSIVE amendments only. It should only be used to correct errors or omissions in the information given in an earlier document. Do not use this form for information changes (such as changes in signal carriage or ownership). A $10.00 fee is required for any permissive amendment. Make checks payable to the Register of Copyrights.

Note: If you wish a Certified Receipt, enclose $3.00 with Notice. ☐

- "Intermediate Form" (CS/SA-2) for cable systems with semiannual "gross receipts" of over $41,500 but less than $160,000.
- "Long Form" (CS/SA-3) for cable systems with semiannual "gross receipts" of $160,000 or more.

Copies may be obtained by writing the Copyright Office.

2. Determining Which Form to Use Which of the three copyright Statement of Account forms is to be used by a cable television system depends on the system's "gross receipts."

The majority of the readers of this book (who decide to set up their own for-profit CATV systems) will likely qualify as very small or mini-CATV companies, and these will be exempt from filing most of the forms described in this appendix. Mini-CATV companies with

FIGURE A5–4.

Cable Television—Special Required Amendment to Initial Notice of Identity and Signal Carriage Complement

---

To: Copyright Office/Washington, DC 20559

NOTICE TO BE AMENDED:

Name Notice Was Filed Under: File Number (if known):

Name of OWNER (if different from above):

Date Notice Was Filed:

SPECIFIC IDENTIFICATION OF FM SIGNALS GENERALLY RECEIVABLE:
(Attach separate page if needed)

Call Sign	Location	Call Sign	Location

Signed:________________________
(Owner or duly authorized representative)

____
Type or Print Name and Title

Date:________________________

Note to Operators: This Amendment is required of (1) any system that filed an Initial Notice prior to February 10, 1978, and did not list the specific FM signals generally carried on an "all-band" basis. (A list of all FM stations in the country is NOT accepted by the Copyright Office for this purpose.) All such systems must file this Amendment by June 30, 1978. (2) Systems filing Initial Notices after February 10, 1978, but not able to include a list of "generally receivable" FM signals must file this Amendment within 120 days of the Initial filing. No fee is required for either filing. The Amendment should be signed by the same person who signed the Initial Notice.

Note: If you wish a Certified Receipt, enclose $3.00 with Notice. ☐

semiannual gross receipts *under* $41,500 will, however, have to file form CS/SA-1 as described below, and pay the $15 semiannual fee. The other sections of this appendix apply to those enterprising individuals who plan to operate cable TV companies earning revenues above these amounts. (Good luck to you.)

3. Calculating Gross Receipts New rules introduced after 1978 include some clarifications and changes in defining the "gross receipts" for secondary transmissions of a cable television system.

First, if the system makes a separate charge for convertors, revenues from that charge are included in gross receipts if any standard television signals are involved. If the convertor is used solely for pay channels or other nonbroadcast service, the fees are not included as gross receipts. Second, the entire basic subscriber fee must be included as gross receipts. Portions that cover nonbroadcast services cannot be excluded unless billed and paid separately. Third, there are some clarifications about accounting methods. Systems may report either on a cash or an accrual basis, depending on how the revenue accounts are maintained.

4. Completion of the Statement of Accounts Form The three Statement of Accounts forms request similar information, including:

- Owner, operator, system, and area served. All of this information should be completed in the same manner as for the filing of the *Initial Notice.*
- Under the "channels" heading enter the actual number of channels on which broadcast signals have been carried during the six-month period.
- In the "subscriber information" category complete the number of subscribers to the basic service in each category listed (e.g., private home, motel, apartment, second set) and the charge per subscriber for each category.
- List the "rates" charged to a subscriber for any service other than the basic signal retransmission service (e.g., per movie channel = $8 per month).
- Under the designation "primary transmitters" list all broadcast signals carried. For each television station (including translators) enter the call sign, name of community of license, channel number on which the station transmits, whether the station is network, independent, or educational, and whether the station is a "distant" signal (remember that any signal that is not a "must carry" or "local" signal under FCC rules is "distant"). Systems with semiannual revenues of less than $41,500 need not identify which signals are distant or provide the information on part-time carriage.
- List each radio station carried by call sign, AM or FM, and community of license. If all-band FM is carried, indicate this and then list the "generally receivable" stations in the manner indicated in the instructions for the *Initial Notice.* Describe the monitoring activity, including frequency, method, and equipment used.
- If for some reason there is program substitution on the mini-CATV system, a log must be maintained and submitted with a special statement. This applies only to the use of distant nonnetwork television programs. Logs are suggested if any signals are carried on a part-time basis, but only the program substitution log is mandatory. Refer to the suggested logging procedure in Figure A5-5.

5. Calculating Payment For mini-CATV systems payment is made twice a year and is figured on a per-system basis as shown in the following table. Note that all CATV systems pay something.

The semiannual copyright royalty fee must accompany the filing. The fees must be paid by certified check, cashier's check, or money

FIGURE A5–5.
Instructions to Recommended Log

1. *Key to Symbols*

IC Insufficient Channels—Signals carried on a part-time basis where full-time carriage was not possible because of insufficient activated channel capacity.

SP Specialty Programming—Signals carried pursuant to the part-time specialty programming rules.

LN Late Night—Signals carried pursuant to the late-night programming rules.

M Mandatory—Signals carried because programming on the regularly carried signal was required to be deleted under the syndicated program exclusivity rule, i.e., programming substituted because of a mandatory deletion.

D Discretionary—Signals carried when programming on the regularly carried distant independent signal was permitted to be deleted because such programming was primarily of interest to the distant station's own community, i.e., a discretionary deletion.

2. *General Instructions*

No entries need to be made on the log for full-time signals regularly carried, since it is assumed that you will know from other sources what signals are being carried on this basis.

You do not need a separate log for each cable channel. Just enter substituted programs and part-time signals chronologically throughout the day, in each instance indicating the *cable channel* involved.

3. *Instructions for All Cable Systems with Respect to Substituted Programming*

If you check M in column 8 or D in column 9 for substituted programming, you must complete columns 1–4 and, in addition, you must indicate the name of the substituted program in column 10, and also indicate in column 11 or 12 whether the program substituted was live (L) or nonlive (N).

4. *Instructions with Respect to Part-Time Signals*

Only for systems which have gross receipts exceeding $41,500 for the six-month reporting period, limited to receipts attributable to the basic service of delivering broadcast signals.

If you check IC, SP, or LN in column 5, 6, or 7, then the only other columns you need to complete are columns 1–4.

(This form is provided by courtesy of NCTA Associate Member George Shapiro.)

order made payable to the Register of Copyrights. If any other form of payment is used, the Copyright Office will not accept it.

6. Computing Distant Signal Equivalents (DSEs) If a mini-CATV system grosses over $160,000, payment of copyright fees is based on the number of distant signals carried. Only those using the long form SA-3 (gross receipts over $160,000) actually compute these DSEs.

1. In assigning values to a translator, treat it the same way as the station it

TABLE A5–2.

Semiannual Subscriber Revenue (basic monthly and second TV set fees)	Semiannual Payment Rate	Computation
Under $41,500	$15	$15
$41,500–$80,000	0.5% of "adjusted revenue." Method for calculating adjusted revenue:	That is, revenue = $50,000
	(1) Subtract actual revenue from $80,000, and then	$80,000 −50,000 = $30,000
	(2) Subtract result of step 1 from revenue to determine "adjusted revenue"	$50,000 −30,000 "Adjusted revenue" = $20,000
		$20,000 × 0.005 Payment + $100
$80,000–$160,000	0.5% of gross subscriber revenue up to $80,000. 1% of gross subscriber revenue between $80,000 and $160,000	That is, revenue = $100,000 $80,000 × 0.005 = $400 $20,000 × 0.010 = $200 Payment = $600
Over $160,000	Payment based on number of distant signal equivalents 0.675% of revenue of 0–1 DSE 0.425% of revenue for each DSE over 1 to 4 0.2% of revenue for each DSE over 4 (calculation of DSEs is contained in the next section)	That is, revenue = $500,000 DSE = 2.5 $500,000 × 0.00675 × 1 + $3,375 $500,000 × 0.00425 × 1.5 = $3,188 Payment = $6,563

retransmits. A translator is given full station status for copyright purposes.

2. The formula for computing the DSE for substituted programs is determined by dividing the number of qualifying substitute programs carried from each station by the number of days in the year, then multiplying the result by the "type value" of the station. For example, if your system carried five such programs for station A (an independent with a type value of 1), its DSE would be 0.014 (five programs divided by 365 days). Basing the formula on the number of programs substituted instead of the number of days that substitution occurred results in a DSE count that we believe is equitable.
3. In calculating the DSE of a station carried part-time (category A, B, or C stations) include the total number of hours the station was on the air during the accounting period. The most accurate method to determine this is by checking the station's program logs. Some television stations are willing to figure it out; but they are under no obligation to do so. Other ways to find the information include checking the station's public file or using a *TV Guide*.

7. Changing Signals A frustrating problem arises when a cable system adds a distant signal, drops one, or decides to change signals. *The Copyright Office has interpreted the law to require payment of copyright fees for the full six-month period if a cable system carries a distant signal for any part of the six-month accounting period.* This is an obvious inequity but it is currently being enforced.

3. How to Lose the Compulsory License

The compulsory license may be lost in a number of ways: (1) where the CATV system has not complied with the filing requirements; (2) where a CATV system engages in commercial substitution (there is a market research exception to this provision); (3) where the CATV system engages in "willful or repeated" carriage of a broadcast signal that has not been authorized by the FCC; (4) where a CATV system does not carry a broadcast signal on a simultaneous basis; (5) where a system violates the signal carriage rules for Canadian and Mexican signals. Any of these situations could result in the loss of a compulsory license for one or more of the broadcast signals being carried by a CATV system. In such a case the CATV system would have to enter into a marketplace agreement or cease carrying the affected signal or signals. It is unclear at this point as to when and how a compulsory license, once lost, can be regained. Presumably the loss of a compulsory license would happen only after a court order. The court can also impose fines—up to $50,000 for statutory damages, plus actual damages and court costs—so compliance is worth the trouble.

4. Additional Information

This appendix is not intended as a legal document, of course, and the interested mini-CATV entrepreneur should investigate the possibility of joining either the CATA or NCTA (or both) to benefit from the support services provided by these organizations. (See Appendix 4 for their addresses.) Portions of this appendix have been provided by courtesy of the NCTA and CATA as a copyright primer for their members.

Starting your own mini-CATV system can be very profitable, as well as fun. If your own system has less than $41,500 in semiannual gross revenues, you need only complete one form and send a semiannual payment of $15 along with a second form to the Copyright Tribunal in Washington, D.C. If your system earns more than $41,500 every six months, a somewhat more complicated procedure (described above) will apply.

Of course, if you carry *no* broadcast signals (including any local TV stations) but only nonbroadcast signals such as HBO, Showtime, CNN,

C-SPAN, or ESPN, you are not classified by the Copyright Office as a CATV system at all—and don't have to file any forms with them.

A detailed step-by-step package of all forms, regulations, and other documents pertaining to the establishment of your own mini-CATV system is available from Satellite Center, P.O. Box 7104, San Francisco, CA 94120, for a fee of $20, and is recommended for anyone interested in starting a mini-CATV system of his or her own. Ask for the "Mini-CATV" package.

Appendix 6
How to Find the Birds

This appendix includes an Apple II Plus Applesoft BASIC program, which will automatically calculate the necessary antenna azimuth and elevation rotation and tilt angles for any given latitude and longitude on the earth. A pocket calculator program written for the Texas Instruments' Programmable Scientific Calculator has also been included, along with several examples showing how to use it. Finally, an AZ/EL Approximation Map will give you a quick estimate (to the nearest degree) of your necessary antenna pointing angles.

Finding the bird in space is one problem. Aligning the antenna to true north or south is another. Whether you put together your own satellite antenna from its modular components or hire a qualified TVRO installing dealer on a "turnkey" basis, this information should be of interest (you never know when you'll need to find true north someday, when lost in the wilderness.).

Basic Geometry

Sailors have always needed to know where they were in relation to the land. Sailing any great distance out of sight of shore posed great problems in antiquity. Thus an accurate map of the high seas was vital. But how could one map the waves? There were no fixed points on water. So cartographers turned to the skys, and mapped the heavens. During the Middle Ages it had been discovered that these sky maps varied from location to location on the earth's surface. By the time of Napoleon, the process of map-making and navigation had progressed to the point that the surface of the earth, by then recognized as a sphere, could be divided accurately into units of angular measurement. The British proceeded to establish two systems of angular measurement, called latitude and longitude. Beginning at the Greenwich Observatory in England, the earth was divided into 360 degrees of longitude. As one traveled westward from the Greenwich Meridian (zero degrees longitude), the longitude increased, and was called west longitude. East longitude progressed the opposite direction around the globe, heading eastward from Greenwich.

The earth was further divided into north and south angular measurements, beginning at the equator, known as zero degrees latitude. As one traveled further north, the latitude increased until the north pole was reached at +90 degrees north latitude. The south pole had a latitude of −90 degrees south latitude. This system of angular measurements is retained to this day. (See Figure A6-1.) To convert degrees of latitude or longitude to statute miles for a sphere the size of the earth, simply multiply by 69.057. That is, each degree on a globe or map represents approximately 69 miles. Therefore the circumference of the earth is approximately 24,860 miles, and its radius is 24,860 divided by 2 pi, or 3957 miles.

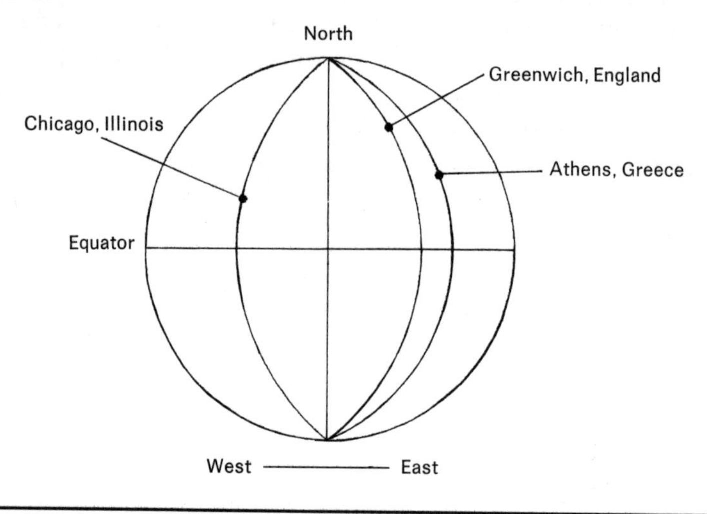

Figure A6-1. Latitude and longitude of the earth. (Courtesy of James Lentz)

A communications satellite circling the earth is in a geosynchronous orbit 22,300 miles above the equator. Because of the law of gravity any object, large or small, placed at this altitude above the earth's surface will revolve around the earth precisely once every 24 hours. If the orbit is exactly over the equator, then the earth, also turning on its axis once per day, will appear to remain fixed to a viewer sitting on the satellite.

Conversely, the satellite will appear to remain motionless in the heavens above to a viewer stationed on the earth.

While points on earth are measured in terms of degrees of latitude and longitude, satellite antennas are rotated and tilted in terms of azimuth and elevation angles. Several different earth station antenna mounts have been developed to allow the antenna's pointing direction to be easily changed. Figure A6-2 illustrates the polar and AZ/EL mounts, which are the most commonly used. The main advantage of the AZ/EL mount is that it provides the most stable type of mount from a mechanical point of view. Two axes of rotation must be varied in order to change the pointing direction from one satellite to another. The polar mount provides a single axis of motion (hour angle axis) to repoint the antenna. The main disadvantages of the polar mount are that the reflector must be rotated around the polar axis, so that for satellites that are more than 20 degrees from due south for a particular location, the structural load exerted on the polar axis by the reflector is somewhat unbalanced. Also, the polar axis is not truly polar, so that a slight adjustment has to be made in the other axis (declination).

Figure A6-2. Popular satellite TV antenna mounts. (Courtesy of James Lentz)

The elevation angle of the antenna in an AZ/EL mount is simply its angle of upward or downward tilt. Pointed toward the horizon, the elevation of an antenna is zero degrees. Pointed straight up, its elevation is 90 degrees. The antenna's azimuth is the angle of left–right rotation, measured clockwise starting from north. That is, when the antenna is pointed north, its azimuth angle is zero. East is 90 degrees, south is 180 degrees, and west is 270 degrees.

To relate these angles correctly to the earth's surface, the antenna must be aligned with true north. Once this is done, and the location of the antenna in latitude and longitude is found, then a simple set of geometric equations can be employed to determine the correct azimuth and elevation angles necessary to point the antenna at a satellite located at any specific west longitude in the geosynchronous Clarke orbit.

To accomplish this task, let's take the polar mount antenna as an example. The direction of true north can be found in several ways. One technique consists of using a simple magnetic compass. Unfortunately, true north is not quite the same as magnetic north, and the error can be as large as 23 degrees or more. To correct for variations from place to place, a magnetic compass correction factor is necessary. Figure A6-3 presents a table of corrections for each of the states in the United States and the major provinces of Canada.

Another way to locate true north is to find the North Pole Star, or Polaris. Figure A6-4 is a map of the North Sky, which can be used as

FIGURE A6–3.
Magnetic Compass Corrections for North America

Alabama	2E	Kentucky	1E	North Dakota	11E
Alaska	26E	Louisiana	6E	Ohio	3W
Arizona	14E	Maine	20W	Oklahoma	9E
Arkansas	6E	Maryland	8W	Oregon	20E
California	17E	Massachusetts	15W	Pennsylvania	8W
Colorado	14E	Michigan	3W	Rhode Island	15W
Connecticut	13W	Minnesota	6E	South Carolina	2W
Delaware	10W	Mississippi	5E	South Dakota	11E
Washington, D.C.	8W	Missouri	6E	Tennessee	1E
Florida	2E	Montana	18E	Texas	10E
Georgia	0	Nebraska	11E	Utah	15E
Hawaii	11E	Nevada	17E	Vermont	15W
Idaho	19E	New Hampshire	16W	Virginia	6W
Illinois	2E	New Jersey	11W	Washington	22E
Indiana	0	New Mexico	13E	West Virginia	5W
Iowa	6E	New York	10W	Wisconsin	2E
Kansas	9E	North Carolina	5W	Wyoming	13E
Alberta	22E	Manitoba	10E	Saskatchewan	17E
British Columbia	23E	Ontario	8W	Quebec	17W

Directions: For a specific city, adjust the outer ring of your compass so that its needle is pointing to the corresponding degree (east or west) printed on the ring. The compass "north" pointer or line will then be pointing at the approximate true north for that city.

an aid to finding the Big and Little Dipper star patterns. The North Star is the brightest star at the very tip of the Little Dipper's handle.

An effective way of finding true north or south is to use the sun. This

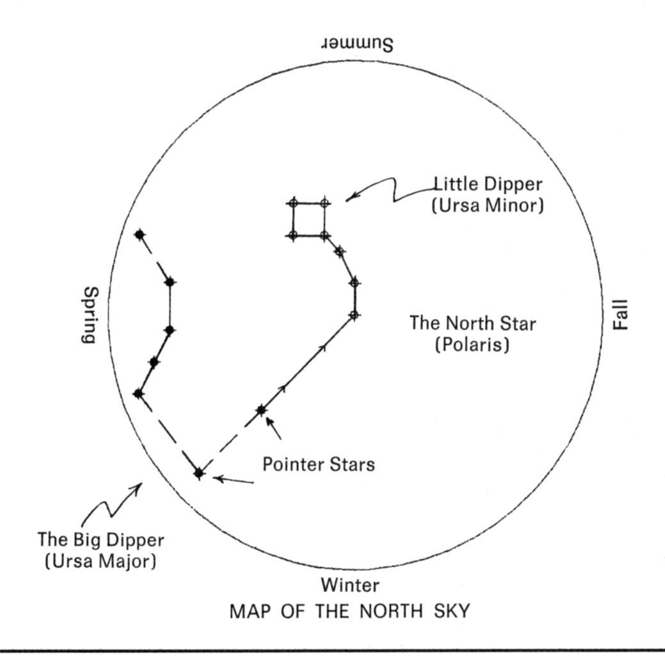

Figure A6-4. To find the North Star (True North): Rotate the map so that the particular season is at the bottom of the page. Map is valid for latitudes of 25–50 degrees north. (Courtesy of James Lentz)

method simply uses the sun's shadow at the time of local noon to fix the position of due south. Local noon is computed by taking the times of local sunrise and sunset from newspaper data (usually printed for one week at time). For example, if local sunrise is 5:58 a.m. and local sunset is 7:50 p.m., then local noon is halfway between those times, which is computed as

$$\frac{12{:}00 - 5{:}58 + 7{:}50}{2} + 5{:}58$$

or 12:54 p.m. The shadow cast by a vertical rod will give the true north/south line for that location. The only problem with this method is that it does require a sun unobstructed by clouds. Once the direction "true north" is located, the satellite antenna polar axis needs to be aligned with it. Since the antenna itself is not sitting on top of the north pole of the earth, an elevation-type "declination adjustment" will have to be made when installing the antenna so that the antenna will point at the satellite orbital arc. A true polar axis antenna would have its polar axis parallel with the earth's polar axis and would have an adjustment declination angle to point it at the orbital arc. Virtually all "polar" axis antennas on the market are actually "approximate" polar axis antennas, simply because it is very difficult to build a true polar axis mount that will support antennas 10 feet in diameter or larger.

Computation of declination angle for a near polar mount antenna for a given location is computed from the latitude position according to the formula:

$$\text{Declination angle} = 90^\circ - \tan^{-1} \frac{3964 \sin L}{22{,}300 + 3964(1 - \cos L)}$$

where L is the latitude in degrees and 3964 is the radius of the earth in miles. (See Figure A6-5.)

Once the antenna is aligned with north, an inclinometer can be easily bolted to the side of the antenna directly to read off the "declination"

Figure A6-5. Polar mount showing declination angle offsetting north. (Courtesy of James Lentz)

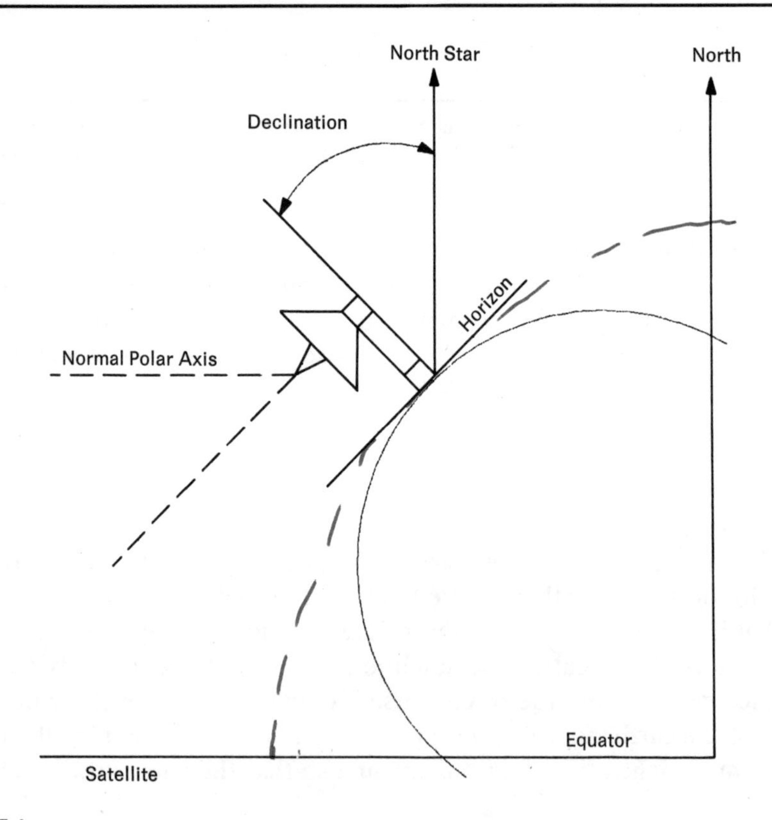

in degrees. Figure A6-6 presents one such inclinometer, which can be cut out and pasted to a piece of stiff cardboard or wood. Sears, Roebuck and Montgomery Ward sell professional inclinometers used in surveying work, which are readily usable for satellite antenna positioning.

The polar antenna mount, once aligned to north, and with its proper declination angle, will cut out an arc in space almost exactly (but not quite) the arc of the satellite belt. (See Figure A6-7.) Thus, once the proper tilt angle is established, the antenna can be swung from left to right, varying its "hour angle" by rotation about the polar axis. If the antenna was precisely on the equator, then pointing it directly up and

Figure A6-6. Inclinometer. (Courtesy of James Lentz)

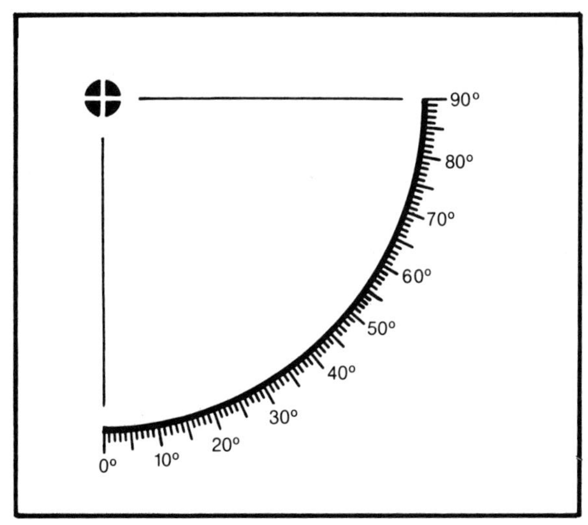

Attach at ⊕ to Earth Station Antenna (as Below) to Adjust Antenna Elevation.

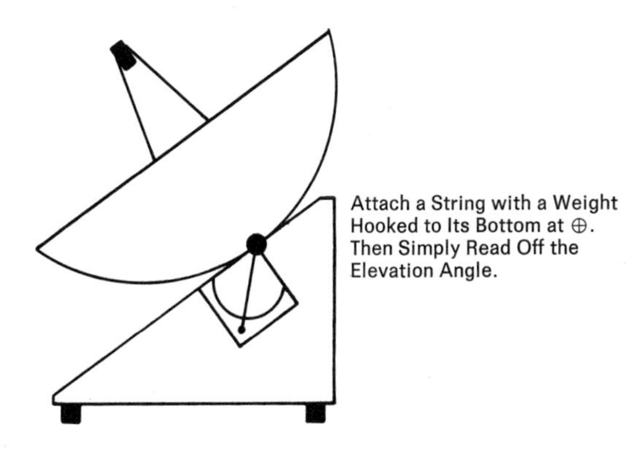

rotating it from east to west would allow the user to pick up each of the satellites whose west longitude would exactly match the antenna's azimuth setting. Finding a geosynchronous satellite when the TVRO antenna is sitting on the equator is a snap. Hot and muggy, but easy. As the antenna's latitude is increased above the equator, the error between the arc cut out in space by the antenna and the Clarke orbit satellite belt becomes greater and greater. So even with a polar antenna mount, when changing from satellite to satellite some slight adjustment will be required when the hour angle swing is large enough. Of course, if you are fortunate enough to have a motorized antenna mount, changing the antenna's pointing direction is a piece of cake. A microcomputer can even be programmed to steer the antenna to the proper position simply by pushing a button. Several of the installing dealers listed in Appendix 4 provide these deluxe motorized antenna mounts.

Figure A6-7. The satellite arc versus the polar mount arc. (Courtesy of James Lentz)

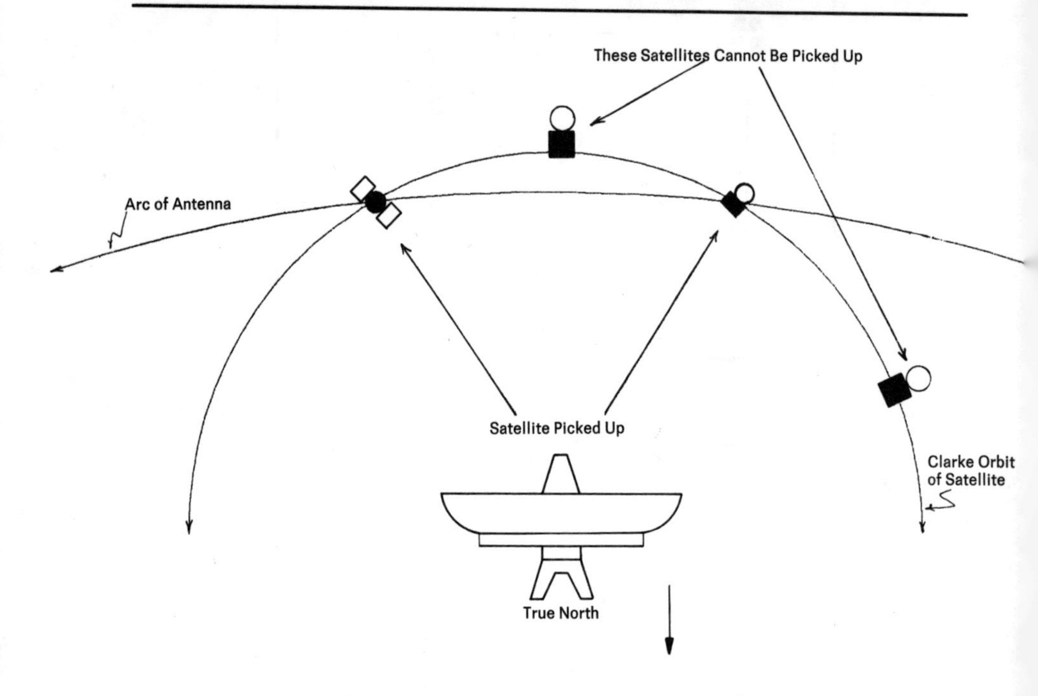

Locating the Satellites

There are a number of different techniques available to determine where to point the satellite antenna to pick up a given satellite. One can use a microcomputer or a programmable scientific calculator, perform the calculations by hand, look up the angles for an AZ/EL mount on an AZ/EL map plotted for a specific satellite, or use a computer satellite tracking bureau that can ship you a satellite printout tailored to your specific location. This section gives you the specific information on all of these options.

The formulas to find the proper antenna azimuth and elevation angles given the known west longitude of the satellite and the earth station's latitude and longitude are as follows:

$$\text{Azimuth } (AZ) = 180^\circ + \arctan(\tan B/\sin A)$$

for the northern hemisphere, or

$$AZ = \arctan(\tan B/\sin A)$$

for the southern hemisphere, where A = latitude of earth station (north is +, south is −), and B = longitude of earth station (east is +, west is −) minus the longitude of the satellite.

The range (RA) between the ground station and the satellite is found by the law of cosines:

$$RA = \sqrt{R^2 + (R + H)^2 - 2R\,(R + H)\cos C}$$

where

R = radius of the earth (3957 miles/6367 km).

H = height of the satellite above the earth (22,245 miles/35,800 km)

C = central angle between the earth station and the satellite given by Equation 3.

$$C = \arccos(\cos A \times \cos B)$$

The elevation angle (EL) is derived from the range (RA):

$$EL = \arccos\,[(RA^2 + R^2 - (R - H)^2)/2 \times RA \times R] - 90^\circ$$

Examples of Hand Calculations

A ground station at San Francisco, California, has a latitude of 37°45′. Its longitude is 122°26′. From the equations, to locate SATCOM F-1 at 135° west longitude,

$$A = 37°45'$$

$$B = -122°26' - (-135°) = 12°34'$$

Therefore,

$$\text{Azimuth } (AZ) = \tan^{-1} [\tan (12°34')/\sin (37°45')] = 199.8°$$

Assuming that the radius of the earth is 3957 miles, and the height of the satellite above the earth is 22,245 miles, then

Range (RA) =

$$\sqrt{(3957)^2 + (3957 + 22{,}245)^2 - 2 \times 3957 \times (3957 + 22{,}245) \times \cos C}$$

$$= 23{,}283 \text{ miles}$$

and

$$C = \cos^{-1} [\cos(37°45') \times \cos (12°34')] = \cos^{-1} [0.7906 \times 0.9706]$$

$$= 39°29'$$

Finally,

elevation angle (EL) =

$$\cos^{-1}\left[\frac{(23{,}283)^2 + (3957)^2 - (3957 - 22{,}245)^2}{2 \times (23{,}283) \times (3957)}\right] - 90° = 44.3°$$

In computing C, if its value is ever less than 81.3 degrees, then the satellite is actually below the horizon of the earth station and can't be seen. Also, if B ever turns out to be less than 81.3 degrees, the satellite would also not be visible because that difference of longitude would put the satellite below the horizon even if the earth station physically were sitting on the equator!

To help in determining whether your answers derived from the formulas are right or not, simply scan over the SATCOM F-1 AZ/EL map in Figure A6-8. This map shows the correct azimuth and elevation angles for SATCOM 1 in 5-degree increments as plotted on a map of the United States and Canada. The map can also be used to obtain the approximate AZ/EL information. By carefully interpolating the distance between the lines, one should be able to get within 1 degree or so, certainly close enough to find the satellite when pointing the antenna in its direction with a little bit of rocking the antenna back and forth a degree or so.

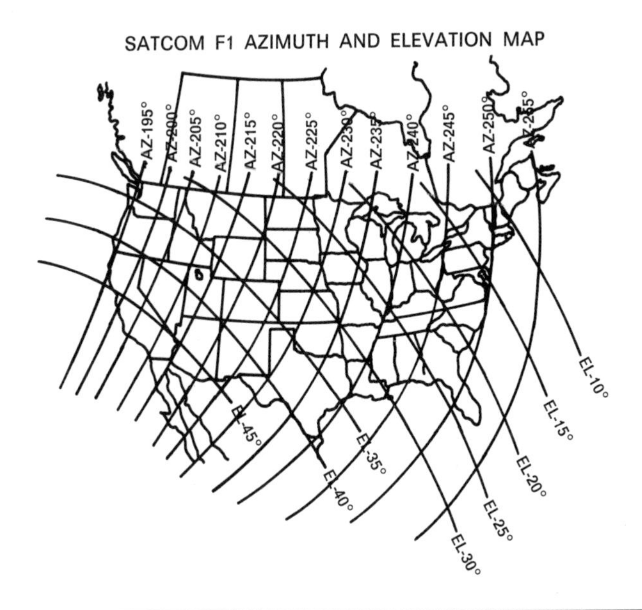

Figure A6-8. Approximate azimuth and elevation antenna angles for the United States and Canada. To use, simply read off the approximate azimuth and elevation angles for any specific location. Accuracies to 1 degree should be possible—close enough to aim the satellite antenna to get a usable picture. Rocking the antenna back and forth through its AZ/EL adjustments will then center it for maximum signal. (Courtesy of James Lentz)

Programmable Calculator Program

Unless you have a book of trig tables handy, calculating AZ/EL angles is most easily done on a portable scientific calculator. Companies such as Radio Shack sell these calculators for under $50, and one can often find a favorite brand on sale for less than $25. One of the nicer scientific calculators is the Texas Instruments Model 58C, which has a built-in memory and the ability to store several hundred programming steps to eliminate the manual button pushing. Another TI calculator, the Model 59, can even store these steps on a small magnetic memory strip.

The program in Figure A6-9 is written for these TI calculators. To use it, simply load the program steps into the calculator following the instructions that come with your specific calculator. This program can also be readily adapted for use by another programmable scientific calculator, such as the Hewlett-Packard series. It will work for any satellite longitude and any earth station longitude and latitude, north or south. Thanks to James Kennedy of Gaithersburg, Maryland, an old friend and owner of a satellite television mini-CATV apartment building installation dealership, for use of this program. Jim originally wrote it for use in his own business!

Figure A6-9. TI-58/59 calculator satellite locator programs (listings #1 and #2). (Copyright © 1981 by J. P. Kennedy. Used by permission. All rights reserved.)

LISTING #1: Printing-Calculator Program Version

```
000  43  RCL
001  01   01
002  88  DMS
003  42  STO
004  00   00
005  43  RCL
006  03   03
007  88  DMS
008  75   -
009  43  RCL
010  06   06
011  95   =
012  42  STO
013  02   02
014  00   0
015  32  X:T
016  43  RCL
017  02   02
018  67   EQ
019  00   00
020  41   41
021  53   (
022  53   (
023  43  RCL
024  00   00
025  39  COS
026  54   )
027  65   ×
028  53   (
029  43  RCL
030  02   02
031  39  COS
032  54   )
033  54   )
034  22  INV
035  39  COS
036  42  STO
037  04   04
038  61  GTO
039  00   00
040  53   53
041  00   0
042  32  X:T
043  43  RCL
044  01   01
045  22  INV
046  77   GE
047  00   00
048  70   70
049  43  RCL
050  00   00
051  42  STO
052  04   04
053  43  RCL
054  00   00
055  30  TAN
056  55   ÷
057  43  RCL
058  04   04
059  30  TAN
060  95   =
061  24  CE
062  94  +/-
063  22  INV
064  39  COS
065  42  STO
066  05   05
067  61  GTO
068  00   00
069  73   73
070  00   0
071  42  STO
072  05   05
073  53   (
074  03   3
075  06   6
076  00   0
077  75   -
078  43  RCL
079  05   05
080  54   )
081  42  STO
082  07   07
083  43  RCL
084  04   04
085  39  COS
086  42  STO
087  09   09
088  43  RCL
089  09   09
090  33  X²
091  94  +/-
092  85   +
093  01   1
094  95   =
095  34  √X
096  35  1/X
097  65   ×
098  53   (
099  43  RCL
100  09   09
101  75   -
102  93   .
103  01   1
104  05   5
105  01   1
106  03   3
107  54   )
108  95   =
109  24  CE
110  22  INV
111  30  TAN
112  42  STO
113  08   08
114  00   0
115  32  X:T
116  43  RCL
117  02   02
118  77   GE
119  01   01
120  29   29
121  43  RCL
122  07   07
123  58  FIX
124  00   00
125  99  PRT
126  61  GTO
127  01   01
128  34   34
129  43  RCL
130  05   05
131  58  FIX
132  00   00
133  99  PRT
134  43  RCL
135  08   08
136  99  PRT
137  22  INV
138  58  FIX
139  98  ADV
140  91  R/S
141  81  RST
142  00   0
143  00   0
144  00   0
145  00   0
146  00   0
```

LISTING #2: Non-Printing Calculator Program Version

```
000  43 RCL
001  01  01
002  88 DMS
003  42 STO
004  00  00
005  43 RCL
006  03  03
007  88 DMS
008  75  -
009  43 RCL
010  06  06
011  95  =
012  42 STO
013  02  02
014  00  0
015  32 X:T
016  43 RCL
017  02  02
018  67  EQ
019  00  00
020  41  41
021  53  (
022  53  (
023  43 RCL
024  00  00
025  39 COS
026  54  )
027  65  ×
028  53  (
029  43 RCL
030  02  02
031  39 COS
032  54  )
033  54  )
034  22 INV
035  39 COS
036  42 STO
037  04  04
038  61 GTO
039  00  00
040  53  53
041  00  0
042  32 X:T
043  43 RCL
044  01  01
045  22 INV
046  77  GE
047  00  00
048  70  70
049  43 RCL
050  00  00
051  42 STO
052  04  04
053  43 RCL
054  00  00
055  30 TAN
056  55  ÷
057  43 RCL
058  04  04
059  30 TAN
060  95  =
061  24 CE
062  94 +/-
063  22 INV
064  39 COS
065  42 STO
066  05  05
067  61 GTO
068  00  00
069  73  73
070  00  0
071  42 STO
072  05  05
073  53  (
074  03  3
075  06  6
076  00  0
077  75  -
078  43 RCL
079  05  05
080  54  )
081  42 STO
082  07  07
083  43 RCL
084  04  04
085  39 COS
086  42 STO
087  09  09
088  43 RCL
089  09  09
090  33 X²
091  94 +/-
092  85  +
093  01  1
094  95  =
095  34 √X
096  35 1/X
097  65  ×
098  53  (
099  43 RCL
100  09  09
101  75  -
102  93  .
103  01  1
104  05  5
105  01  1
106  03  3
107  54  )
108  95  =
109  24 CE
110  22 INV
111  30 TAN
112  42 STO
113  08  08
114  00  0
115  32 X:T
116  43 RCL
117  02  02
118  77  GE
119  01  01
120  29  29
121  43 RCL
122  07  07
123  58 FIX
124  00  00
125  91 R/S
126  61 GTO
127  01  01
128  34  34
129  43 RCL
130  05  05
131  58 FIX
132  00  00
133  91 R/S
134  43 RCL
135  08  08
136  99 PRT
137  22 INV
138  58 FIX
139  98 ADV
140  91 R/S
141  81 RST
142  00  0
143  00  0
144  00  0
145  00  0
146  00  0
```


Apple II Computer Program

For those readers who own or have access to a microcomputer, the author has developed a generalized BASIC program that runs on an Apple II Plus computer under Applesoft. With minor modification the program will also run on a Radio Shack Level II TRS-80 computer, and on a CPM/Z-80 microcomputer system.

To use it, simply load the program into the computer's memory and *run* it. The program automatically prompts the user for information, and the REMARK statements imbedded within the program should help to clarify what's going on.

Basically, the Satellite Tracking Program will point out the required azimuth and elevation, range, and distance from the earth station to the satellite subpoint (that point on the equator directly underneath the satellite) for a TVRO earth station located anywhere in the world.

Over 67 satellites are presently in the database of the program, including the existing U.S. and Canadian TV satellites, global satellites, and future (planned for launch by 1985) satellites. The program provides a number of output options, including obtaining the AZ/EL angles for U.S. and Canadian satellites only, for all satellites, or for a single satellite, whose west longtitude can be entered from the keyboard if the satellite cannot be found in the computer's memory. An additional option allows the user to obtain a list of all the satellites in the database. The program will work with both printers and CRTs, and compute distances in either miles or kilometers.

Figure A6-10 is a listing of the Satellite Tracking Program. Figure A6-11 provides a sample printout for San Francisco, California.

The list of satellites in the database is given in Figure A6-12.

FIGURE A6–10.
Texas Instruments Calculator Program (for Model 58C or 59)

There are two versions of this program. Listing 1 allows the program to be used with the TI printing cradle option, which permits the calculations to be output to a small printer that plugs into the calculator. Listing 2 allows the program to be used with a stand-alone TI calculator. It will display the AZ/EL answers on the built-in calculator display.

Example 1

In operation, both programs are identical. The latitude of the earth station TVRO antenna is entered in degrees, minutes, and seconds to location 01 of the calculator memory. North latitude is input as a positive number, and south latitude as a negative number. For example, a TVRO antenna at 38°57′21″ would be entered as 38.5721. Next, the longitude of the earth station TVRO antenna is entered in degrees, minutes, and seconds to location 03 of the calculator memory. West longitude is positive, east longitude is negative. For example, TVRO antenna at 77°84′47″ is entered as 77.8447.

Finally, the location of the desired satellite's west longitude in *decimal format* is entered into memory location 06. For example, SATCOM F-1 at 134.9 degrees would be entered as 134.9.

To execute the program, the user simply presses the RUN/START key. The calculator will automatically calculate the azimuth and elevation angles, rounded to the nearest degree, and display them in the sequence AZIMUTH followed by ELEVATION.

To change satellites, the user need only reenter the new west longitude of the new satellite into location 06, and depress the RUN/START key again.

(Note: The *AZ* and *EL* answers are stored in memory locations 07 and 08, respectively, for locations computed in North America, and may also be recalled by depressing the RECALL button followed by the memory location desired.)

Example 2

Compute *AZ/EL* for TVRO antenna at San Francisco, California, 37°45′25″N; 22°26′75″ W, for COMSTAR D-2 at 95 degrees west longitude.

STEP 1 LOAD 37.4525 into LOCATION 01
STEP 2 LOAD 22.2675 into LOCATION 03
STEP 3 LOAD 95.00 into LOCATION 06
STEP 4 RUN/START
STEP 5 Read answers: *AZ* = 140 (degrees)
EL = 38 (degrees)

Apple II Plus Applesoft BASIC Satellite Tracking Program.

```
]LIST

10  REM : SATELLITE TRACKING PROGRAM
20  REM : FOR GEOSTATIONARY SATELLITES
21  REM -------------------------
25  REM : PRINTS OUT AZIMUTH & ELEVATION
26  REM : ANTENNA POINTING ANGLES FOR OVER 50 SATELLITES
27  REM : IN GEO-STATIONARY ORBITS AROUND THE EARTH, OR
28  REM : CALCULATES AZ/EL FOR A SPECIFIC SATELLITE
29  REM : WHOSE W. LONGITUDE IS ENTERED BY THE USER.
30  REM
35  REM : COPYRIGHT 1981, A.T. EASTON
40  REM : ALL RIGHTS RESERVED
45  REM
50  REM : THIS PROGRAM MAY BE USED ONLY
51  REM : BY THE PURCHASER OF THIS BOOK
52  REM : FOR HIS OR HER OWN PERSONAL
53  REM : USE, AND IS NOT TO BE SOLD OR
54  REM : USED COMMERCIALLY IN ANY WAY.
55  REM
56  REM : FOR A COMPLETE SATELLITE PRINT-
57  REM : OUT OF ALL SATELLITES VISIBLE
58  REM : FROM YOUR SPECIFIC LOCATION,
59  REM : SEND $7 TO:
60  REM
61  REM : THE SATELLITE CENTER
62  REM : SATELLITE AZ/EL PRINTOUT
63  REM : P.O. BOX 7104
64  REM : SAN FRANCISCO, CA, 94120
70  REM
80  REM : A COMPLETE PACKAGE OF USEFUL
81  REM : COMPUTER PROGRAMS INCLUDING THIS ONE AND
82  REM : OTHERS RELATED TO HOME AND 'MINI-CATV'
83  REM : SATELLITE TV USE (SUCH AS A PROGRAM SCHEDULE
84  REM : ROUTINE, ETC) IS AVAILABLE FROM THE SATELLITE CENTER
85  REM : FOR APPLE II OR TRS-80 COMPUTERS ON 5" FLOPPY
86  REM : DISKS, COMPLETE WITH DETAILED INSTRUCTIONS.
87  REM : TO ORDER, SEND $20 TO:
88  REM
89  REM : THE SATELLITE CENTER
90  REM : COMPUTER DISK PACKAGE I
91  REM : P.O. BOX 7104
92  REM : SAN FRANCISCO, CA, 94120
93  REM
94  REM : SPECIFY WHICH COMPUTER YOU HAVE.
95  REM : PLEASE ALLOW 4-6 WEEKS FOR SHIPPING.
96  REM
300  LET K = 0
310  LET K1 = 0
320  LET K2 = 0
399  REM -------------------------
400  REM  INPUT USER OPTIONS & LOCATION
401  REM
410  REM : 'ALL' PRINTS AZ/EL FOR ALL SATELLITES
411  REM : IN THE DATA BASE
420  REM : 'US' PRINTS US & CANADIAN SATELLITES ONLY
430  REM : 'KYBD' ASKS USER FOR A SPECIFIC SATELLITE NAME.
431  REM : IF NAME NOT FOUND, PROGRAM THEN REQUESTS THE
432  REM : USER THE W.LONGITUDE OF THE SATELLITE.
440  REM : 'LIST' PRINTS A LIST OF ALL SATELLITES IN THE
441  REM : DATA BASE, THEIR NAMES & W. LONGITUDES
445  REM
446  PRINT
```

```
447  PRINT "-----------------------------------------"
448  PRINT
45Ø  PRINT "SATELLITE TRACKING PROGRAM"
455  PRINT "COMPUTES ANTENNA AZ/EL"
46Ø  PRINT
465  PRINT "COPYRIGHT 1981, A.T. EASTON"
5ØØ  PRINT "-----------------------------------------"
51Ø  PRINT "OUTPUT OPTION (ALL,US,KYBD,LIST)";
52Ø  INPUT T$
521  PRINT
524  IF T$ = "LIST" THEN 4ØØØ
525  PRINT "ENTER NAME OF CITY, STATE";
526  INPUT D$,E$
527  PRINT
53Ø  IF T$ <  > "KYBD" THEN 6ØØ
54Ø  PRINT "NAME OF DESIRED SATELLITE";
55Ø  INPUT V$
6ØØ  PRINT
61Ø  PRINT "PRINTER OR CRT (P OR C)";
62Ø  INPUT U$
63Ø  PRINT
64Ø  PRINT "MILES OR KILOMETERS (M OR K)";
65Ø  INPUT W$
655  PRINT
66Ø  IF W$ = "M" THEN 69Ø
67Ø  REM : RADIUS OF EARTH & DISTANCE TO SATELLITE
671  REM : BELT FROM EQUATOR IN KILOMETERS
68Ø  LET Z1 = 6367
685  LET Z2 = 358ØØ
687  LET Z3 = 111.136
688  PRINT "DISTANCES ARE IN KILOMETERS"
689  GOTO 7ØØ
69Ø  REM : RADIUS OF EARTH & DISTANCE TO SATELLITE
691  REM : BELT FROM EQUATOR IN MILES
695  LET Z1 = 3957
697  LET Z2 = 22245
698  LET Z3 = 69.Ø57
699  PRINT "DISTANCES ARE IN MILES"
7ØØ  PRINT "--------------------------"
994  REM
995  REM : GET ANTENNA COORDINATES IN
996  REM : DEGREES (NOT RADIANS)
997  REM : NOTE - TRIG FNS COMPUTE IN RADIANS ONLY
998  REM : THAT'S WHY .Ø174533 & 57.29587 CONVERSION
999  REM : MULTIPLIER CONSTANTS ARE IN THE FORMULAS
1ØØØ  PRINT "ENTER LATITUDE IN DEGREES, MINUTES, "
1ØØ1  PRINT "SECONDS, N OR S ";
1Ø1Ø  INPUT M1,M2,M3,M$
1Ø2Ø  PRINT
1Ø3Ø  PRINT "ENTER LONGITUDE IN DEGREES, MINUTES, "
1Ø31  PRINT "SECONDS, E OR W ";
1Ø4Ø  INPUT L1,L2,L3,L$
1Ø5Ø  LET L = L1 + (L2 + (L3 / 6Ø)) / 6Ø
1Ø6Ø  LET M = M1 + (M2 + (M3 / 6Ø)) / 6Ø
1Ø7Ø  IF L$ = "E" THEN 1Ø9Ø
1Ø8Ø  LET L =  - L
1Ø9Ø  IF M$ = "N" THEN 12ØØ
11ØØ  LET M =  - M
12ØØ  PRINT
12Ø1  PRINT "-----------------------------------------"
12Ø2  PRINT
12Ø5  PRINT "LOCATION: ";D$;", ";E$
12Ø7  PRINT
12Ø8  PRINT
121Ø  IF U$ = "C" THEN 2ØØØ
122Ø  PRINT "SATELLITE"; TAB( 8);"DEGREE"; TAB( 9);"AZIMUTH";
1221  PRINT  TAB( 4);"ELEVATION"; TAB( 4);"DISTANCE";
1222  PRINT  TAB( 4);"DIST-TO-SUBPOINT"
```

```
1230  PRINT "---------"; TAB( 8);"------"; TAB( 9);"-------";
1231  PRINT  TAB( 4);"---------"; TAB( 4);"--------";
1232  PRINT  TAB( 4);"----------------"
1240  PRINT
1999  REM ------------------------
2000  REM : READ LIST OF SATELLITES
2001  REM : J1=# OF US/CAN COMSATS, J2=TOTAL #
2002  REM : N$=SAT NAME, N=SAT DEGREE W. LONGITUDE
2003  REM
2005  RESTORE
2010  READ J1,J2
2011  IF T$ <  > "US" THEN 2014
2012  LET J = J1
2013  GOTO 2015
2014  LET J = J2
2015  FOR I = 1 TO J
2020  READ N$,N
2025  LET C1 = 0
2026  LET C2 = 0
2027  LET C3 = 0
2028  LET C4 = 0
2035  IF T$ <  > "KYBD" THEN 2040
2038  IF V$ <  > N$ THEN 3000
2039  REM
2040  REM : CHECK IF SAT IS BELOW HORIZON
2041  REM
2044  LET K1 = 1
2045  LET L4 = L + N
2050  IF L4 <  = 180 THEN 2070
2060  LET L4 = L4 - 360
2070  IF L4 >  =  - 180 THEN 2090
2080  LET L4 = L4 + 360
2090  IF L4 < 81.3 THEN 2110
2095  IF U$ = "P" THEN 3000
2100  LET X = L4 - 81.3
2105  PRINT N$;" AT ";N;" W. LONGITUDE"
2106  PRINT "IS BELOW THE LOCAL HORIZON BY"
2107  PRINT  INT (X + .5);" DEGREES. NOT VISIBLE."
2108  GOTO 2452
2110  LET Y = M
2120  LET X = L4
2199  REM ------------------------
2200  REM : CALCULATION ROUTINE FOR
2201  REM : GREAT CIRCLE MEASUREMENTS
2210  LET C1 =  COS (0.0174533 * X) *  COS (0.0174533 * Y)
2219  REM
2220  REM : COMPUTE GREAT CIRCLE ANGLE BETWEEN SITE
2221  REM : AND POINT ON THE EQUATOR BELOW SATELLITE
2225  REM : INVERSE COSINE FORMULA FOR C2
2226  REM
2230  LET C2 = 57.29578 * ( -  ATN (C1 /  SQR ( - C1 * C1 + 1)) + 1.5708
2232  IF C2 < 81.3 THEN 2240
2234  IF U$ = "P" THEN 3000
2235  LET X = C2 - 81.3
2236  PRINT N$;" AT ";N;" W. LONGITUDE"
2237  PRINT "IS BELOW THE LOCAL HORIZON BY"
2238  PRINT  INT (X + .5);" DEGREES. "N$;" NOT VISIBLE."
2239  GOTO 2452
2240  REM
2241  REM : CALCULATE DISTANCE TO SUB-POINT
2242  REM
2250  LET C4 = Z3 * C2
2259  REM
2260  REM : CALCULATE AZIMUTH OF ANTENNA TO SATELLITE
2261  REM
2270  LET A = 180 + 57.29578 * ( ATN ( TAN (0.0174533 * X) /
          SIN (0.0174533 * Y)))
```

```
28Ø  IF A > Ø THEN 23ØØ
29Ø  LET A = A + 18Ø
299  REM
3ØØ  REM : CALCULATE RANGE FROM EARTH STATION
3Ø1  REM : TO SATELLITE IN ORBIT
3Ø2  REM
31Ø  LET R =  SQR (Z1 ^ 2 + (Z1 + Z2) ^ 2 - 2 * Z1 * (Z1 + Z2) *
          COS (Ø.Ø174533 * C2))
319  REM
32Ø  REM : CALCULATE ELEVATION FROM SITE TO SATELLITE
321  REM
325  LET C3 = ((R ^ 2 + Z1 ^ 2 - (Z1 + Z2) ^ 2) / (2 * Z1 * R))
33Ø  LET E =  - 9Ø + 57.29578 * ( -  ATN (C3 /  SQR ( - C3 * C3 + 1)) + 1.57Ø8)
41Ø  IF U$ = "C" THEN 245Ø
42Ø  PRINT N$,N,( INT (A * 1Ø)) / 1Ø; TAB( 7);( INT (E * 1Ø)) / 1Ø;
421  PRINT  TAB( 8); INT (R); TAB( 8); INT (C4)
43Ø  PRINT
44Ø  GOTO 3ØØØ
45Ø  PRINT N$;" ";N;"    AZ=";( INT (A * 1Ø)) / 1Ø;" E=";( INT (E * 1Ø)) / 1Ø
451  PRINT "RANGE ="; INT (R);" SUBPOINT ="; INT (C4)
452  PRINT
455  IF U$ = "P" THEN 3ØØØ
46Ø  LET K = K + 1
47Ø  IF K < 7 THEN 3ØØØ
48Ø  PRINT
49Ø  PRINT "MORE (Y OR N)";
5ØØ  INPUT Q$
51Ø  IF Q$ = "N" THEN 5ØØØ
52Ø  LET K = Ø
53Ø  PRINT
ØØØ  IF K2 = 1 THEN 5ØØØ
Ø1Ø  NEXT I
Ø15  IF K1 = 1 THEN 5ØØØ
Ø2Ø  PRINT
Ø3Ø  PRINT V$;" IS NOT IN THE COMPUTER'S MEMORY"
Ø4Ø  PRINT
Ø45  PRINT "WHAT IS ";V$;"'S WEST LATITUDE (TYPE Ø TO STOP)";
Ø5Ø  INPUT N
Ø55  IF N = Ø THEN 5ØØØ
Ø6Ø  LET N$ = V$
Ø7Ø  PRINT
Ø75  LET K2 = 1
Ø8Ø  GOTO 2Ø25
999  REM -------------------------
ØØØ  REM : SATELLITE DATA BASE LISTING ROUTINE
Ø1Ø  REM : READ NAMES & W. LONGITUDE OF SATELLITES
Ø15  RESTORE
Ø2Ø  READ J1,J2
Ø3Ø  PRINT
Ø4Ø  PRINT "US/CAN SAT";"  ";"LOCATION (W. LON)"
Ø5Ø  PRINT "----------";"  ";"-----------------"
Ø6Ø  PRINT
Ø7Ø  FOR I = 1 TO J1
Ø8Ø  READ N$,N
Ø9Ø  PRINT N$,N
1ØØ  PRINT
11Ø  NEXT I
12Ø  PRINT
13Ø  PRINT "OTHER WORLD SATS";"  ";"LOCATION"
14Ø  PRINT "----------------";"  ";"--------"
145  PRINT
15Ø  FOR I = 1 TO J2 - J1
16Ø  READ N$,N
17Ø  PRINT N$,N
18Ø  PRINT
19Ø  NEXT I
2ØØ  PRINT
```

```
5000  PRINT
5005  PRINT "----------------------------------------"
5010  PRINT "RUN AGAIN (Y OR N)";
5020  INPUT Q$
5030  IF Q$ = "Y" THEN 300
5040  PRINT
5099  GOTO 9999
5499  REM : #US/CAN SATS, TOTAL # SATS
5500  DATA  14,69
5596  REM -------------------------
5597  REM  DATA BASE OF SATELLITES
5598  REM
5599  REM : LIST OF OPERATIONAL US/CAN TV SATELLITES
5600  DATA  "SATCOM 4 (1982)",83
5610  DATA  "COMSTAR 3",86.9
5620  DATA  "WESTAR 3",91
5630  DATA  "COMSTAR 1/2",95
5640  DATA  "WESTAR 1",99
5650  DATA  "WESTAR 4 (1982)",99
5660  DATA  "ANIK 1",103.9
5670  DATA  "ANIK B",108.9
5680  DATA  "ANIK 2/3",113.9
5690  DATA  "SATCOM 2",118.9
5700  DATA  "WESTAR 2",123.4
5710  DATA  "COMSTAR 4",127
5720  DATA  "SATCOM 3 (1982)",132
5730  DATA  "SATCOM 1",134.9
5898  REM
5899  REM : LIST OF OTHER GLOBAL COM. SATELLITES
5900  DATA  "INTELSAT IV F5",1
5910  DATA  "INTELSAT IV F1",2.6
5920  DATA  "SYMPHONIE 1",11.6
5930  DATA  "STATSIONAR 4",14.2
5940  DATA  "SIRIC",15
5950  DATA  "MARISAT 1",15
5960  DATA  "INTELSAT I2",18.5
5965  DATA  "INTELSAT V F1",22
5970  DATA  "INTELSAT IVA F1",24.6
5980  DATA  "STATSIONAR 8",25
5990  DATA  "INTELSAT IVA F2",27.5
6000  DATA  "INTELSAT IVA F3",34.5
6005  DATA  "SMS-1 (12 GHZ)",66.7
6010  DATA  "ATS-5",70
6020  DATA  "LES-6  (12 GHZ)",74.4
6035  DATA  "RADUGA 4",87.1
6045  DATA  "SBS 2(12 GHZ)",97
6070  DATA  "FSC 1 (12 GHZ)",99.5
6080  DATA  "SMS-1",105
6085  DATA  "ATS-3 (12 GHZ)",105.2
6090  DATA  "SBS 1 (12 GHZ)",106
6120  DATA  "CTS",116
6170  DATA  "SMS-2",135
6180  DATA  "ATS-6",140
6190  DATA  "ATS-1",149.4
6195  DATA  "SYMPHONIE 2",161.3
6200  DATA  "STATSIONAR 10",170
6210  DATA  "INTELSAT IV F6",182.9
6220  DATA  "MARISAT 2",183
6230  DATA  "INTELSAT IV F8",186
6240  DATA  "STATSIONAR 7",220
6250  DATA  "CS",225
6260  DATA  "ETS",230
6270  DATA  "BSE (JAPAN)",250.9
6280  DATA  "STATSIONAR T",261
6290  DATA  "EKRAN 2",261
6300  DATA  "EKRAN 1",261
6310  DATA  "RADUGA 5",275.8
6320  DATA  "PALAPA 1",277
```

```
6330  DATA  "STATSIONAR 1",280
6340  DATA  "PALAPA 2",283
6350  DATA  "MARISAT 3",287
6400  DATA  "GHORIZONT 1",307
6420  DATA  "RADUGA 3",324.7
6496  REM
6497  REM : ADDITIONAL SATELLITES TO BE LAUNCHED
6498  REM : IN 1982 - 1985 TIME FRAME.
6499  REM
6500  DATA  "SPC 1 (1984)",70
6510  DATA  "HUGHES 1 (1984)",74
6520  DATA  "WU-TDRSS (1984)",79
6530  DATA  "ATT 1 (1982)",87
6540  DATA  "WU-TDRSS (1550  DATA  "COMSTAR 1 ('82)",95
6560  DATA  "SPC 2 (1984)",119
6570  DATA   "WESTAR 4 (1982)",123
6590  DATA  "HUGHES 2 (1984)",135
6600  DATA  "SATCOM 1 ('83)",139
6610  DATA  "SATCOM 2 ('83)",143
6699  REM
7000  REM : TO AELLITE TO THIS LIST PERIODICALLY,
7001  REM : SIMPLY PLACE IT AFTER THIS STATEMENT
7010  REM : IN THE FORMAT :7020 DATA 'SATNAME',DEGREE
7020  REM
7030  REM : BE SURE TO CHANGE J2 IN LINE 5500 TO INCREASE
7031  REM : THE TOTAL NUMBER OF SATELLITES IN THE LIST
9999  END
```

Figure A6-11. Satellite Tracking Program sample printout.

```
]RUN
----------------------------------------

SATELLITE TRACKING PROGRAM
COMPUTES ANTENNA AZ/EL

COPYRIGHT 1981, A.T. EASTON
----------------------------------------
OUTPUT OPTION (ALL,US,KYBD,LIST)?US

ENTER NAME OF CITY, STATE?SAN FRANCISCO, CALIFORNIA


PRINTER OR CRT (P OR C)?P

MILES OR KILOMETERS (M OR K)?M

DISTANCES ARE IN MILES
---------------------------
ENTER LATITUDE IN DEGREES, MINUTES,
SECONDS, N OR S ?37,45,25,N

ENTER LONGITUDE IN DEGREES, MINUTES,
SECONDS, E OR W ?122,26,75,W

----------------------------------------

LOCATION: SAN FRANCISCO, CALIFORNIA
```

SATELLITE	DEGREE	AZIMUTH	ELEVATION	DISTANCE	DIST-TO-SUBPOI
SATCOM 4 (1982)	83	126.6	30.1	23992	3617
COMSTAR 3	86.9	130.5	32.7	23850	3450
WESTAR 3	91	135	35.3	23713	3286
COMSTAR 1/2	95	139.6	37.6	23594	3138
WESTAR 1	99	144.6	39.8	23490	3004
WESTAR 4 (1982)	99	144.6	39.8	23490	3004
ANIK 1	103.9	151.2	42.1	23383	2862
ANIK B	108.9	158.5	43.9	23298	2746
ANIK 2/3	113.9	166.1	45.3	23239	2663
SATCOM 2	118.9	174.2	46	23207	2617
WESTAR 2	123.4	181.5	46.2	23200	2608
COMSTAR 4	127	187.3	45.9	23211	2623
SATCOM 3 (1982)	132	195.3	45.1	23249	2677
SATCOM 1	134.9	199.8	44.3	23283	2725

For Further Information and Immediate Help

Several excellent articles have been written on this subject, including one by Bill Johnston, in the March 1978 issue of *QST* magazine, entitled "Locating Geosynchronous Satellites" (page 23). The May 1978 issue of *Ham Radio* includes an article on calculating antenna bearings using an HP-25 programmable calculator by H. Paul Shuch (page 67). Richard Cleis wrote an article in the March 1980 issue of *Ham Radio* discussing how the HP-29C programmable calculator can be used to project the orbits, not only of fixed-orbit geosynchronous satellites, but of moving satellites such as the Russians' domestic TV MOLYNIA series!

Stephen Gibson has written an excellent manual that tells everything you'd ever want to know about satellite navigation, and then some. Called the *Gibson Satellite Navigation Manual,* it's published by Satellite Television Technology (STT) in Arcadia, Oklahoma, for $30. (See Appendix 4 for address.) Finally, if you just don't want to fiddle with these calculations, or to type the BASIC program into your computer, help is at hand from The Satellite Center in San Francisco. For $7, the

Center will send you a detailed printout of all the satellites that you can see from your location, including their azimuth and elevation angles.

For $20, the Center will also ship you via airmail a complete 5¼-inch mini-floppy computer disk of over a dozen useful BASIC programs for the home satellite TV user and the mini-CATV operator. The Satellite Tracking Program is included on this disk, which also comes with a detailed user manual and full documentation. (See Appendix 4 for address.)

Note: The examples used in this appendix illustrate the calculations made for the old RCA "Cable Bird," the SATCOM F-1 satellite located at 135 degrees west longitude. To locate any other satellite, including the new RCA F-3 bird located at 131 degrees west longitude, simply substitute the desired satellite's longitude in the formulas where indicated.

Figure A6-12. Satellite database of geostationary communications satellites.

```
]RUN SATELLITE

----------------------------------------

SATELLITE TRACKING PROGRAM
COMPUTES ANTENNA AZ/EL

COPYRIGHT 1981, A.T. EASTON
----------------------------------------
OUTPUT OPTION (ALL,US,KYBD,LIST)?LIST


US/CAN SAT   LOCATION (W. LON)
----------   ----------------

SATCOM 4 (1982) 83

COMSTAR 3       86.9

WESTAR 3        91

COMSTAR 1/2     95

WESTAR 1        99

WESTAR 4 (1982) 99

ANIK 1          1Ø3.9

ANIK B          1Ø8.9

ANIK 2/3        113.9

SATCOM 2        118.9

WESTAR 2        123.4

COMSTAR 4       127

SATCOM 3 (1982) 132

SATCOM 1        134.9
```

OTHER WORLD SATS	LOCATION
INTELSAT IV F5	1
INTELSAT IV F1	2.6
SYMPHONIE 1	11.6
STATSIONAR 4	14.2
SIRIC	15
MARISAT 1	15
INTELSAT IV F2	18.5
INTELSAT V F1	22
INTELSAT IVA F1	24.6
STATSIONAR 8	25
INTELSAT IVA F2	27.5
INTELSAT IVA F3	34.5
SMS-1 (12 GHZ)	66.7
ATS-5	70
LES-6 (12 GHZ)	74.4
RADUGA 4	87.1
SBS 2(12 GHZ)	97
FSC 1 (12 GHZ)	99.5
SMS-1	105
ATS-3 (12 GHZ)	105.2
SBS 1 (12 GHZ)	106
CTS	116
SMS-2	135
ATS-6	140
ATS-1	149.4
SYMPHONIE 2	161.3
STATSIONAR 10	170
INTELSAT IV F6	182.9
MARISAT 2	183
INTELSAT IV F8	186
STATSIONAR 7	220
CS	225
ETS	230
BSE (JAPAN)	250.9
STATSIONAR T	261
EKRAN 2	261
EKRAN 1	261
RADUGA 5	275.8
PALAPA 1	277
STATSIONAR 1	280
PALAPA 2	283
MARISAT 3	287
GHORIZONT 1	307
RADUGA 3	324.7
SPC 1 (1984)	70
HUGHES 1 (1984)	74
WU-TDRSS (1984)	79
ATT 1 (1982)	87
WU-TDRSS (1984)	91
COMSTAR 1 ('82)	95
SPC 2 (1984)	119
WESTAR 4 (1982)	123
HUGHES 2 (1984)	135
SATCOM 1 ('83)	139
SATCOM 2 ('83)	143

Appendix 7

Footprint Maps and Antenna Sizes

This appendix will attempt to answer the question: How big an antenna do I need? The answer is: That depends on a number of factors, including your location, the satellite's signal strength, and noise figure (in degrees Kelvin) of the dish-mounted low-noise amplifier (LNA). It also assumes that the dish is correctly pointed at the satellite (an error of 1 degree or so can spell disaster).

How Much Gain Do I Need?

The typical satellite transponder puts out a signal of only 5 watts. That's pretty small to begin with, but by the time that this signal travels the 22,300 miles to earth, very little of it is left at all. Referring to the first "footprint map" for RCA SATCOM F1, this signal is about 34 dBw at New York City, well under a millionth of a watt of power available for detection by a satellite TVRO antenna.

At such low signal levels the minimization of unwanted "noise" becomes a major problem. The basic background noise of the earth itself must even be taken into consideration. The noise picked up from surrounding terrain is a very localized phenomenon, but, in general, the noise received by an antenna from the atmosphere and from surrounding earth sources is fairly insignificant, usually contributing less than 25 percent of the total receiving system noise temperature.

Also, at these low signal levels a typical satellite TV receiver would generate noise internally, due to the random thermally driven bumping together of electrons in its solid-state amplifiers. The noise of the receiver would totally mask the signal.

It is for this reason that the low-noise amplifier (or LNA) is used between the TVRO antenna and satellite receiver. The LNA is an ultra-low-noise solid-state device whose noise temperature is so low that it's measured in terms of degrees Kelvin absolute temperature! Thus LNAs are rated not in terms of their amplification (they all provide plenty of gain—about 50 dB), but in terms of their noise temperature. The standard noise temperature LNA in the industry is 120°K.

To receive a satellite television signal successfully, both the antenna gain (as measured in dBi) as well as the system's overall noise temperature need to be computed. The ratio of these two figures (in dB) gives the G/T that can be used to compute the carrier-to-noise ratio (CNR), which determines the ultimate TV signal-to-noise ratio.

Antenna Gain Formulas

To calculate the CNR at the receiver input, the following equation is used:

$$\text{CNR} = \text{Satellite EIRP (dBw)} - \text{Path loss in space (196.5 dB)} + \text{G/T} - 10 \log BW - K \text{ (Boltzmann's constant)} \quad \text{(eq. 1)}$$

Typical antenna gains of various-sized parabolic TVRO antennas together with G/T figures and resultant CNRs are found in Table A7-1.

TABLE A7–1.
Parabolic Antenna Performance

Antenna Size	Antenna Gain (dBi)	G/T (120°K LNA)	CNR (dB) (34 dBw EIRP)
6 foot/2 meter	35	13.1	(3.6)
8 foot/2½ meter	37.5	15.7	(6.2)
10 foot/3 meter	39.5	17.8	8.3
12 foot/4 meter	41	19.4	9.9
15 foot/5 meter	43	21.4	11.9
20 foot/6 meter	45.5	24.0	14.5

The satellite EIRPs can be found from the footprint map for that satellite and specific location. (See Figures A7-1 through A7-4 for footprint maps.)

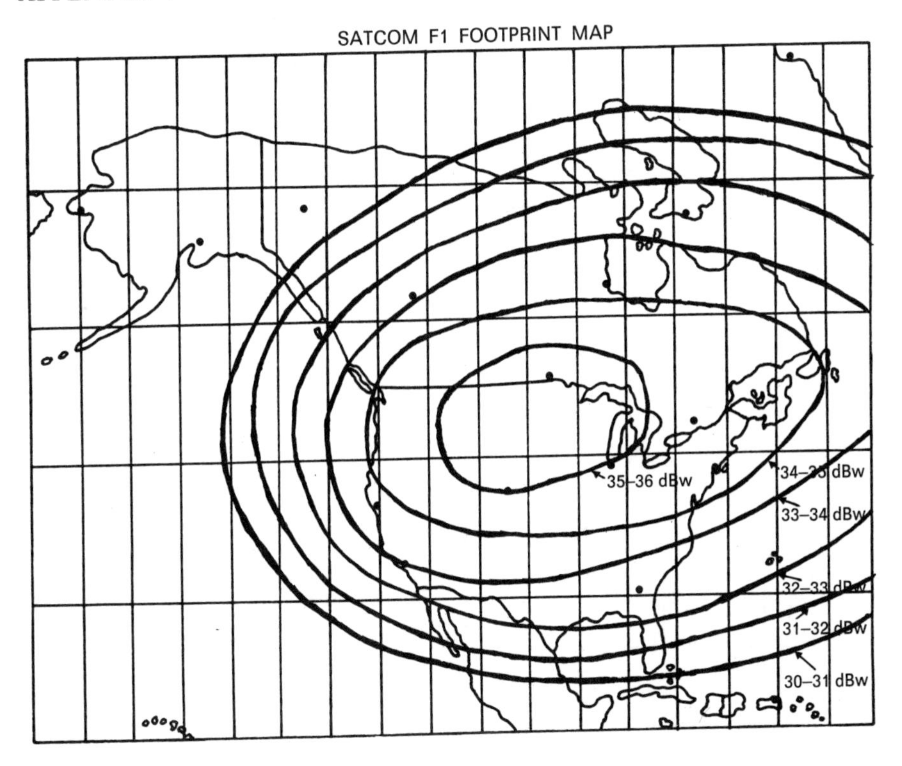

Figure A7-1. Satellite footprint map of EIRP power output as received on various points of the surface of the earth (measured in dBw). Radiation output patterns vary slightly from transponder to transponder. This is a typical footprint for RCA SATCOM F-1 satellite position 135 degrees west longitude. The new SATCOM F-1 "cable bird" positioned at 131 degrees west longitude has a somewhat better signal power level. (Courtesy of James Lentz)

For the 10-foot (3-meter) case the computation is as follows:

$$\text{CNR (dB)} = 34\text{ dBw} - 196.5\text{ dB} + 17.8 - 75.6 + 228.6 = 8.3\text{ dB}$$

Note that

$$10 \log \text{BW} = 10 \log 36\text{ MHz} = 75.6$$

$$\text{Boltzman's constant} = 228.6$$

The carrier-to-noise ratio (or CNR) relates this signal power level to the noise of the overall receiving system. To detect an acceptable video picture successfully, an optimum satellite receiver must have a CNR of at least 7.3 dB. Typically, a CNR figure of at least 8 dB must be obtained if the receiver is to operate without annoying noise "sparklies" appearing in the picture.

Figure A7-5 provides information concerning the INTELSAT satel-

COMSTAR D2 FOOTPRINT MAP

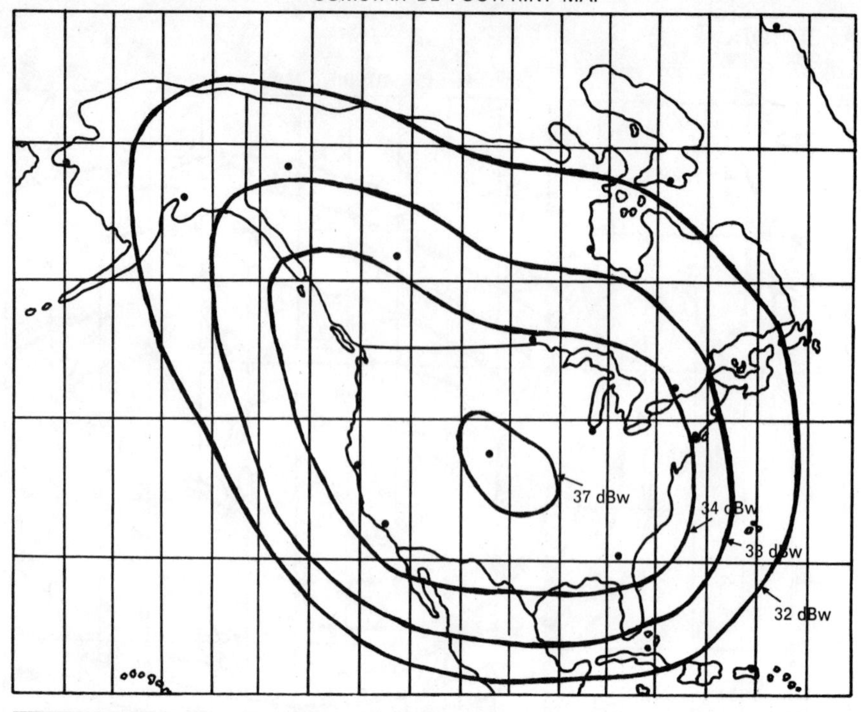

Figure A7-2. Satellite position: 95 degrees west latitude.
Satellite footprint map of EIRP power output as received on various points of the surface of the earth (measured in dBw). (Courtesy of James Lentz)

Figure A7-3. Satellite position: 91 degrees west latitude.
Satellite footprint map of EIRP power output as received on various points of the surface of the earth (measured in dBw). (Courtesy of James Lentz)

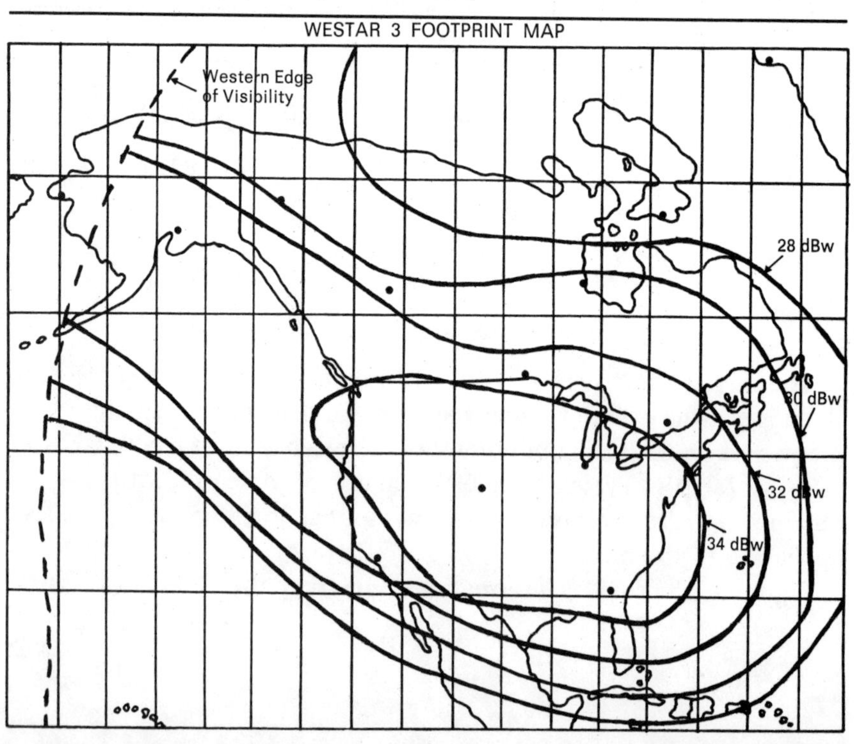

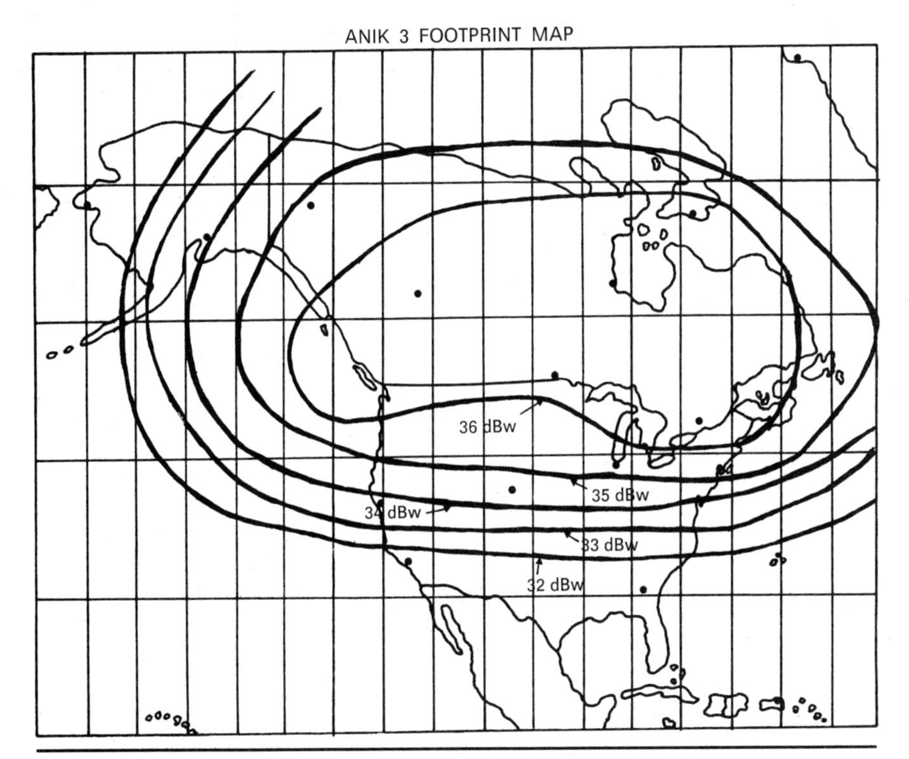

Figure A7-4. Satellite position: 114 degrees west latitude.
Satellite footprint map of EIRP power output as received on various points of the surface of the earth (measured in dBw). (Courtesy of James Lentz)

Figure A7-5. The INTELSAT V satellite utilizes both spot beams (at higher output power) and global/hemispheric beams. Rough effective coverage of various beams for the new INTELSAT V series of three-ocean satellites. (Courtesy of James Lentz)

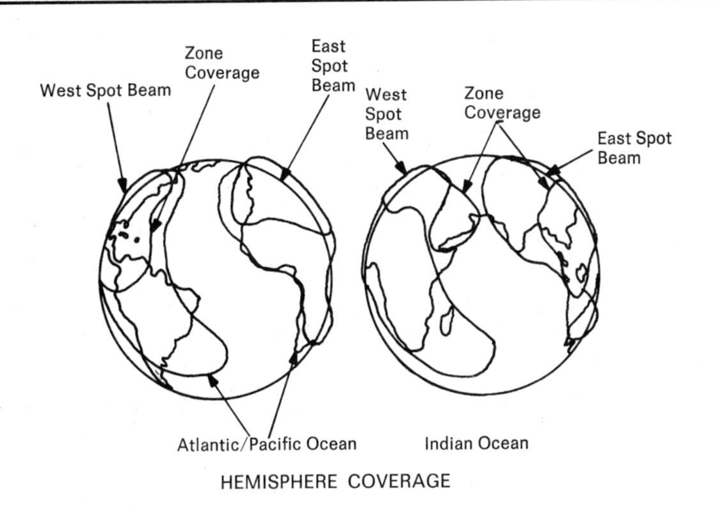

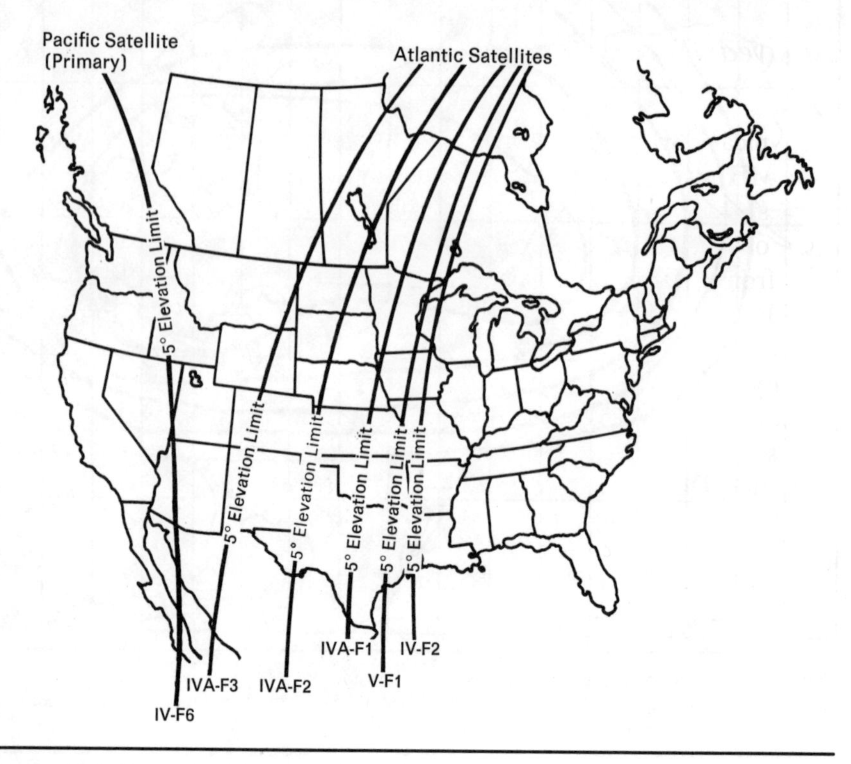

Fugire A7-6. The practical limits of visibility for the Atlantic and Pacific INTELSAT satellites from North America. Satellite locations (west longitude):
IV-F6 183 degrees
IVA-F3 34.5 degrees
IVA-F2 27.5 degrees
IVA-F1 24.6 degrees
V-F1 22 degrees
IV-F2 18.5 degrees
(Courtesy of James Lentz)

lite carriage for North America. The INTELSAT satellites put out much lower power (EIRP as received on the surface of the earth) and thus require a larger TVRO anntenna. A good size (minimum) for adequate reception of an INTELSAT satellite's signal throughout the United States and Canada is 18 feet, or 6 meters. Figure A7-6 indicates the practical limits of visibility of the various INTELSAT satellites throughout North America.

APPENDIX 7

Received Picture Quality

A typical satellite TV receiver accepts a signal that has a bandwidth of 36 MHz (the bandwidth of frequency that a satellite transponder radiates), and the standard NTSC color television signal has a bandwidth of 6 MHz. To successfully detect and separate the television signal from its background noise, the receiver must operate with at least a 37-dB signal-to-noise ratio (SNR). With a CNR of 10 dB, the SNR will be 47 dB, which will provide an excellent picture for the vast majority of viewers. Subjective tests relating SNR figures to viewer "picture quality ratings" have been conducted over the years by the Television Allocations Study Organization (TASO). The results of these studies are presented in Table A7-2.

TABLE A7–2.
Subjective Picture Quality vs. Received SNR Figures

CNR	SNR	Viewer Rating	Quality
8	45	70%—Excellent 99%—Fine	Home-TV quality
10	47	80%—Excellent	CATV quality at home distant from headend
12	49	90%—Excellent	CATV quality at home near headend
14	51	95%—Excellent	CATV quality at headend
16	53	99%—Excellent	Local broadcasting station quality
18	55		Network quality

Basically, the CNR of the TVRO system determines the overall picture quality. A marginal reception (but acceptable for most private home terminals) will be achieved with a CNR of 8 dB. The standard system reception will aim for 11 dB CNR, to allow for deterioration of system components (yours *and the satellite's!*) over time. Outstanding CATV-quality reception will occur when the CNR is 14 dB or greater.

The last part of the puzzle to be solved consists of determining what the CNR will be at any point on earth—quickly and graphically. This brings us back to a new EIRP footprint map calibrated not in dBw units of signal strength but in antenna size/LNA combinations.

Antenna Sizes for U.S. Cities

Given the preceding information, it is possible to determine accurately what combination of antenna size and LNA noise figure selection must be made to provide adequate reception at any location within the United States for any desired satellite.

Using the standard 120°K LNA, Table A7-3 presents the reception quality that can be expected for the RCA SATCOM 1 bird based on practical and "real world" installations rather than theoretical EIRP footprint maps.

TABLE A7–3.
SATCOM 1 Antenna Size vs. Geographic Location

Reception Quality	Geographic Location	Antenna Size
Acceptable (fair) (45 dB SNR)	Midwest	8 f
	East	12 f
	South	12 f
	West	10 f
Good (47 dB SNR)	Midwest	10 f
	East	15 f
	South	15 f
	West	12 f
Excellent CATV quality (50 dB SNR) Note: Using parabolic antenna with 120°K LNA	Midwest	12 f
	East	20 f
	South	20 f
	West	15 f

Figure A7-7 presents the footprint of SATCOM 1 in terms of antenna size, assuming that a standard 120°K LNA is used. Finally, Figure A7-8 provides a footprint map for SATCOM F-1 showing the minimum antenna size recommended for use in a mini-CATV system.

These footprint maps and tables are based on the SATCOM F-1 satellite, the original "Cable Net 1" bird in operation since the early 1970s. The new SATCOM F-3 satellite, which began operation in January of 1982, has since replaced the F-1 as the primary bird to watch. Similar, if somewhat better, viewing of the F-3 satellite is possible throughout the continental United States. Therefore, observations made using the F-1 satellite can apply as conservative guideposts in watching "sparkle-free" TV pictures from the F-3 bird.

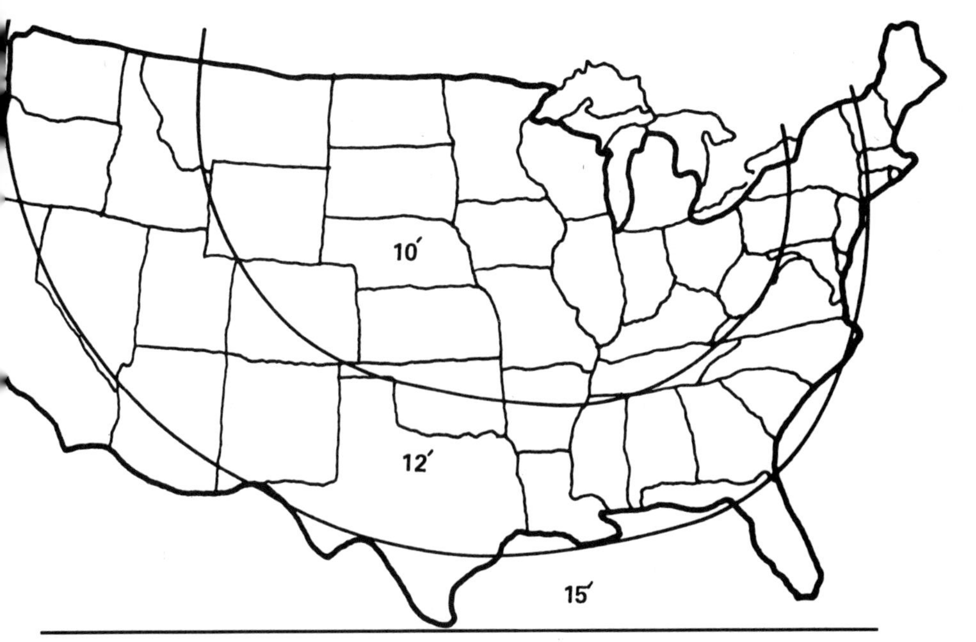

Figure A7-7. Minimum TVRO antenna size to receive SATCOM F-1, assuming that a good receiver with threshold extension and a 120-degree LNA combination are used in conjunction with an efficient parabolic antenna.
In certain central states, acceptable pictures can be received with antennas as small as 6 feet. The author has seen good pictures picked up by an 8-foot antenna located in the backyard of a suburban house in Sunnyvale, California.
(Courtesy of Robert Luly Associates)

Figure A7-8. Recommended minimum antenna size and LNA combination for a mini-CATV system will produce a broadcast-quality signal reception.
Note: Local conditions may require a larger antenna and/or lower-noise LNA.
The suggested minimum antenna and LNA are based upon the best available minimum EIRP data for all four beams from SATCOM I and the following criteria: 1. Receiver with 15-MHz noise bandwidth tracking IF filter (3 dB threshold extender).
2. C/N (carrier-to-noise ratio) is a minimum of 3.65 dB above receiver threshold.
(Courtesy of Robert Luly Associates)

4.6meter/120°K
4.6meter/100°K
6meter/120°K
6meter/100°K